COMPLEXITY AND SECURITY

NATO Science for Peace and Security Series

This Series presents the results of scientific meetings supported under the NATO Programme: Science for Peace and Security (SPS).

The NATO SPS Programme supports meetings in the following Key Priority areas: (1) Defence Against Terrorism; (2) Countering other Threats to Security and (3) NATO, Partner and Mediterranean Dialogue Country Priorities. The types of meeting supported are generally "Advanced Study Institutes" and "Advanced Research Workshops". The NATO SPS Series collects together the results of these meetings. The meetings are co-organized by scientists from NATO countries and scientists from NATO's "Partner" or "Mediterranean Dialogue" countries. The observations and recommendations made at the meetings, as well as the contents of the volumes in the Series, reflect those of participants and contributors only; they should not necessarily be regarded as reflecting NATO views or policy.

Advanced Study Institutes (ASI) are high-level tutorial courses to convey the latest developments in a subject to an advanced-level audience.

Advanced Research Workshops (ARW) are expert meetings where an intense but informal exchange of views at the frontiers of a subject aims at identifying directions for future action.

Following a transformation of the programme in 2006 the Series has been re-named and re-organised. Recent volumes on topics not related to security, which result from meetings supported under the programme earlier, may be found in the NATO Science Series.

The Series is published by IOS Press, Amsterdam, and Springer Science and Business Media, Dordrecht, in conjunction with the NATO Public Diplomacy Division.

Sub-Series

A.	Chemistry and Biology	Springer Science and Business Media
B.	Physics and Biophysics	Springer Science and Business Media
C.	Environmental Security	Springer Science and Business Media
D.	Information and Communication Security	IOS Press
E.	Human and Societal Dynamics	IOS Press

http://www.nato.int/science
http://www.springer.com
http://www.iospress.nl

Sub-Series E: Human and Societal Dynamics – Vol. 37

ISSN 1874-6276

Complexity and Security

Edited by

Jeremy J. Ramsden

Cranfield University, Bedfordshire, UK

and

Paata J. Kervalishvili

Georgian Technical University, Tbilisi

IOS Press

Amsterdam • Berlin • Oxford • Tokyo • Washington, DC

Published in cooperation with NATO Public Diplomacy Division

Proceedings of the NATO Advanced Research Workshop on Complexity and Security
Tbilisi, Georgia
26–30 March 2007

ISBN 978-1-58603-849-6
Library of Congress Control Number: 2008922758

Publisher
IOS Press
Nieuwe Hemweg 6B
1013 BG Amsterdam
Netherlands
fax: +31 20 687 0019
e-mail: order@iospress.nl

Distributor in the UK and Ireland
Gazelle Books Services Ltd.
White Cross Mills
Hightown
Lancaster LA1 4XS
United Kingdom
fax: +44 1524 63232
e-mail: sales@gazellebooks.co.uk

Distributor in the USA and Canada
IOS Press, Inc.
4502 Rachael Manor Drive
Fairfax, VA 22032
USA
fax: +1 703 323 3668
e-mail: iosbooks@iospress.com

PRINTED IN THE NETHERLANDS

A torz *az éppen, mit nem tűrhetek.*

—IMRE MADÁCH

Complexity and Security
J.J. Ramsden and P.J. Kervalishvili (Eds.)
IOS Press, 2008

Preface

IT CAN WELL BE ASSERTED that 'Complexity' and 'Security' are defining features of our world, by which we mean our global civilization at the beginning of the 21st century. The sheer number of actors and their technology-facilitated global interactions necessarily make it complex. As for security, *insecurity* is a major preoccupation both among the general population and among governments. Opinion polls merely confirm what can be shown anecdotally to be widely held truths. The proliferation of surveillance cameras, increasingly stringent airport security and electronic data protection are not only to combat what we shall call "acute" insecurity—the threat of terrorism—but also "chronic" insecurity—crime in the street—that is considered to be connected to delinquency, and somehow signifies a breakdown of Hobbes' social contract, with a concomitant return to egotistical fragmentation of human aims, everyman against everyman, standing in stark and curious contrast to the globalization of humanity that is such a feature of our present age. Yet that is far from being the only interpretation of insecurity. Man-made activities on an ever vaster scale, such as manufacturing industry and the agro-industrial complex, have created grave problems of pollution that already threaten security of access to those most basic of human needs, sunlight, clean breathable air, and clean potable water. The pooling of human resources required to create the concentration of capital without which industry could not exist at first created a certain security—the insurance industry can also be included here, and retirement pension funds—but increasing turmoil in the financial markets,[1] to a large extent the result of the growing complexity of financial instruments that can no longer be readily grasped and manipulated, has led to another kind of insecurity, financial insecurity, and even though we cannot eat, drink or breathe such things, they are so inextricably entangled within the daily fabric of our civilization that their instability also significantly contributes to insecurity; a perception of insecurity may be as harmful as concrete threats to security, "soft" social and psychological factors causing insecurity and no less "real" then the hard ones. And finally there are the vast forces of nature affecting weather, climate and the overall stability of our planet, that in the past seemed so well regulated, but now seems to be becoming upset, perhaps as a direct consequence of human actions, including the merciless exploitation of birds, beasts and fishes, many of which have been driven to extinction during the last few hundred years, the accelerated destruction of forests, and the relentless extraction and consumption of fossil fuels.

This is the state of affairs of the world in which this book appears: the world is complex and insecure. Implicit in the desire to deal with this topic

[1]See, for example, the 77th Annual Report, Chapter VIII. Basel: Bank for International Settlements (BIS) (2007).

is the notion that the insecurity at any rate is an undesirable state of affairs, antipathetic to human well-being, which perhaps means the progress of humanity to an ever-higher state of civilization. Some opinion holds that our present course threatens the very survival of humankind as a biological species. The reason for dealing with the (in)security issue is not just for the sake of fulfilling some academic exercise that will allow it to be understood, in the sense that the principle of gravity allows the structure and dynamics of the solar system to be understood, but also reflects a desire to alter that state of affairs in such a way as to make the situation lead to more desirable outcomes.

It is an intriguing and still open question is whether the two—complexity and security—are linked. In other words, were our world simple, would it be secure? A brief study of history does not encourage the idea of such a linkage. The civilization of the early Egyptians seems to have been incredibly complex, going by what we can glean about their religion. At the same time, there is nothing to suppose that life in early Egypt was particularly secure. There do seem to have been periods in the history of humanity when society was marked by a rude simplicity, for example in the Dark Ages—but was life less insecure then? If anything, such periods seem to show a relative absence of something that we could call a social contract, with concomitant individual insecurity. Hence simplifying our society, assuming it could be done, does not seem to offer a route to increased security.

We must also expect to have to deal with the question of how much security is desirable. Already groups of respectable and law-abiding citizens have questioned whether some of the more extreme security measures being introduced impose a heavier burden on civilization than is warranted by the benefit from the diminished insecurity.[2] Total security—although we have yet to define exactly what we mean by that—suggests an almost dead state in which no further human development can take place. On the other hand, general anarchy prevents the undertaking of the civilizing activities that we associate with the van of human development. The challenge therefore is to determine what is the optimum level of security. This is where mastery of the issue of complexity and security can be useful—to indicate what is that optimum level, and how it can be achieved, possibly by controlling the degree of complexity of our society.

We are constantly building representations and models of the world around us in an effort to grasp and understand it better. Great harm has been done in the past by the advocacy and adoption of models of an alluring simplicity that is, however, inadequate to describe reality.[3] One of the aims of this book is to equip the reader with the ability to create models appropriately complex for the situation they are trying to describe, but still useful (that is, simple enough to be handled) for practical purposes.

One candidate for rescuing humanity from the crisis of insecurity, including such aspects as desertification and species extinctions, is technology. The next technological revolution is supposed to be driven by nanotechnology, and will be even more far-reaching than its predecessors.[4] This assertion also requires crit-

[2] There now exists an approach to quantitatively assessing the benefits relative to the risks. See Thomas, P.J. et al., The extent of regulatory consensus on health and safety expenditure. Part 1: Development of the J-value technique and evaluation of regulators' recommendations. Trans. IChemE B 84 (2006) 329–336.

[3] E.g. equilibrium models in economics.

[4] See for example Kurzweil, R., The Singularity Is Near. New York: Viking Press (2005).

ical scrutiny. Technology gives us surveillance cameras but also chainsaws and dynamite; moreover there is a widespread perception that despite the incontrovertibly ever-higher levels of technology, security is progressively diminishing. One contributor to that diminution is the danger posed by those who see the sophisticated technological infrastructure of the country as a legitimate object of plunder for the sake of personal enrichment. For example, in Hungary roving bands of gypsies have been known to attack and destroy electrical transformers in substations in order to remove the metal in them for sale as scrap, or to dismantle lightly used railway lines for the same purpose. Our present society is well-nigh defenceless against such actions.

And what of motivations? The terrorist is driven by a "flag", which might be something simple and graspable such as the independence of Corsica, but also something more vague and intangible as in the new kind of global terror; the robber is presumably driven by simple greed to acquire something, and a similar motive presumably underlies the invention of a novel investment instrument. If these motivations can be better understood, and perhaps even subjected to some kind of natural selection process, society may be better able to defend itself against their consequences. Furthermore, motivations are not only individual, but can also be collective, and reflected in the institutions of human society. Certain aspects of our current state of affairs are particularly problematical, since they indicate an institutionalized disregard of personal security by government authorities. Numerous examples can be found within the European Union, which formerly prided itself on its stable, democratic institutions: in Hungary there have been several recent incidents in which the police attacked peaceful demonstrations against various aspects of government corruption, using water cannon and rubber bullets, in other words with a disproportionately exaggerated violence; in Great Britain, bungling incompetence occurs within the Home Office, seemingly at all levels from that of the most junior official up to that of the Secretary of State, including the employment of thousands of illegal immigrants in sensitive security jobs, even within government ministries, and the recent loss of millions of confidential personal data records—and this is apart from the fact that most actions of theirs are now mired in the appalling newspeak so well described in Orwell's *1984*; in Slovakia we have the astonishing case of Hedwig Malina, a student wantonly attacked and severely injured by Slovak youths, merely because she was speaking Hungarian in a public telephone cabin, and now accused by the Slovak government of having fabricated the incident and inflicted the injuries upon herself!

If one simply equates security with homeostasis, which might simply mean survival, then one can identify two fundamental kinds of responses to dangers impinging on the system attempting to survive (which might be an individual being, or a collective enterprise such as a firm, or a nation). The simplest strategy is simply to erect a barrier between the system and the rest of the world. This is a strategy of the tortoise, and the survival of the tortoise and other creatures that have a clear preference for this kind of strategy seems to be an indication of its success. The other kind of strategy is to respond to the danger with an appropriate countermeasure. Adaptation in fact means a response that is not only appropriate to the actual circumstance but also one that would have been appropriate had the initial circumstances been otherwise.[5]

[5] Sommerhoff, G., Analytical Biology. London: Oxford University Press (1950).

This is a strategy of the fencer, and more generally of human beings, which are above all distinguished from other creatures by the possession of the powerful brain needed to effectively implement such a strategy. Given that the dangers impinging on the system have great variety—and technology often increases that variety—it is essential when pursuing the second kind of strategy that the system has a sufficiently large repertoire in order to be able to adopt an appropriate countermeasure, i.e. to be said to be adapted.

These strategies have clear counterparts in society. The "barrier" type of strategy corresponds to the erection of a wall, or Marshal Graziani's 200 mile long barbed wire entanglement running along the Libyan-Egyptian frontier from the coast. Appreciation of the other kind of strategy is exemplified by the current desire of the Basel pharmaceutical company F. Hoffman-La Roche to build a new centre in which hundreds of formerly scattered researchers will be brought together under one roof, and hence presumably able to interact with each other more conveniently and effectively, "because of the complexity of the challenges facing the firm."

In this sense, complexity can be said to be directly linked to security, because variety of possible response is practically synonymous with complexity. What this actually means in practice is one of the issues that will be explored in this book. But we can already note that the biologist E.O. Wilson has identified loss of biodiversity (due to species extinctions) as one of the greatest challenges to the future survival of mankind, i.e. to security. In terms of the ideas developed in this book, the underlying reason for wishing to preserve biodiversity would be the enhanced repertoire of potential response that it confers upon the biosphere as a whole. Similarly, the success of nations such as Switzerland that are organized as confederations of small entities (communes grouped into cantons, in turn grouped into the country), keeping a large degree of autonomy regarding laws and customs, can be contrasted with the relative ineffectuality of large centralized countries that have sought to impose strict uniformity over a large population and territory; one can safely predict that even larger supranational conglomerates such as the European Union are likely to be even less effectual, and their predominant survival strategy (assuming that they do ultimately survive) is likely to be based on sheer bulk—which explains the eagerness with which new members are sought and accepted, with seemingly no regard for the consequent growing cumbrousness of the entire machinery of the organization.

However ingenious the models, however wise and farsighted the recommendations, the ultimate stumbling block is implementation. In the numerous past failures of implementing what have been so obviously the right things to do, one sees the limitations of the rational, enlightened self-interest that is supposed to underlie the philosophy of the social contract. Understanding these limitations and seeking alternatives are as yet largely unexplored research domains. The matter of implementation must therefore be largely left for the future: all we can say at present is that it is clearly not enough to set up expensive institutes producing so-called solutions and suppose that they will automatically lead to a better world.

This book is divided into a number of Parts, which represents a very imperfect attempt to impose some structure on what really is, in the spirit of a complex system, an indivisible whole. This Preface has been adapted from my opening speech. The first chapter, the General Survey, attempts to capture the overall result of the deliberations of the Workshop. The remaining chapters contain the papers contributed to the workshop (but the authors have been at liberty to revise them before submitting them in final form); they are grouped into Parts dealing with definitions of security and of complexity, including useful paradigms of complexity in well studied physical and biological systems; the problem of climate and energy; the contributions of technology; and the roles of economics, sociology and psychology in understanding security. The brief chapters at the beginning of each Part are intended as introductory material to place in context the work described later in more detail, and to capture some of the discussion that followed each paper.

It was a pleasure to have directed this Workshop together with my friend and colleague Paata Kervalishvili. The content of the entire book in its present form grew out of the preparations for the Workshop, including the advance circulation of discussion papers, the week of intensive discussion in Tbilisi, and a prolonged period of post-Workshop reflexion. All who participated in these processes, including those who finally were unable to be present in person, and those who were present but finally have been unable to include a paper in this written record, have, in that spirit of indivisibility alluded to above, contributed to the collective effort that you now see before you.

Acknowledgments

The editors would particularly like to record their thanks above all to the North Atlantic Treaty Organization (NATO) for having provided the basic finance for the workshop in Tbilisi; to the Georgian Technical University for having provided local facilities; to the following private companies and organizations in Georgia for having provided generous local support for the social events during the workshop, which provided indispensable occasions for debate and discussion in an informal setting: Aleksandreuli Wine Company, Centre Point Group Construction Company, Coca-Cola Bottlers Georgia, Georgian CODATA Committee, Geocell, Geoprogress Group, Instra Transport Company, Lilo City Trading Centre, Magticom, SakCementi Material Manufacturer, Tbilaviamsheni Air Company, and Telecom Georgia; and to the Collegium Basilea (Institute of Advanced Study) for having generously supported the preparation of this book.

Contents

Complexity and Security
J.J. Ramsden and P.J. Kervalishvili (Eds.)
IOS Press, 2008

Chapter 1

General survey

Jeremy J. Ramsden[1]

Cranfield University, Bedfordshire, MK43 0AL, UK

In this chapter, I attempt to effect a synthesis of the ideas developed in the entire Workshop, and provide a reflexion of the state of our knowledge at its conclusion.

One very important achievement was the formulation of intelligible definitions of complexity and security. Even if we can specify a procedure for measuring the complexity of the representation of something (descriptive or d-complexity), the definition remains elusive and verges on the privative. To acquiesce in the definition that a complex system is one that we cannot explain or understand is of course deeply dissatisfying to the scientist, but the working definition is only slightly less vague: something is complex if any assumptions made in producing a mechanistic description of it destroy the accuracy of that description, especially its ability to predict its future behaviour. These assumptions might be, in the order of successive restraint, a boundary, classification, the existence of average types, the existence of average events, and stationarity.[2] Practically it might be an insuperable challenge to create a mechanistic description without any of these assumptions, but at least this definition provides some inkling of what constitutes complexity.

Security can be clearly defined as a feeling of safety. Safety, albeit probabilistic and governed by the rules of risk management, is therefore closer to unconditional knowledge, whereas security lies in the psychological realm. Security is therefore usually founded on an actual state of safety (however imperfectly that state may be perceived). In this workshop, we could not be fully explore the fascinating link operating in the reverse sense between security and safety, namely that a particular level of security can have a concrete impact (beneficial

---

[1]The author would especially like to thank Workshop participants Peter Allen and Márta Szekeres for their inputs to this chapter.

[2]See Table 6.1 (p. 58) and Figure 6.3 (p. 56) in P.M. Allen, Complexity and identity: the evolution of collective self. In: J.J. Ramsden et al. (eds), Spiritual Motivation: New Thinking for Business and Management. Basingstoke: Palgrave (2007).

or deleterious) on the actual state of safety.

It is useful to define different types of safety and (in)security. Type A is acute, and comprises the effects of war and terrorism. Type C is chronic, and comprises individual threats arising from criminal behaviour, typically rooted in greed. Type D (for "dauer") refers to long-term threats such as climate change and unravelling the fabric of civilization.

Regrettably, it appears that governments are prime underminers of security. Clearly wars are declared at the level of the state. But through a thoughtless approach to the problem of terrorism, which leads them to impose policies of needlessly harsh repression, they actually increase the terrorist threat.[3] Chronic insecurity is amplified by penal systems apparently devoid of any robust underlying concept, thoughtlessly enacted legislation (Lord Hoffmann's remarks referred to above could apply to many countries), entrenched venality among the legislators (exemplified by members of parliaments who regularly vote to increase their emoluments), and bungling incompetence in government offices. When rational methods for optimizing the appropriateness of responses to risks are now available,[4] there is little excuse for the happy-go-lucky approach to regulation that has become an increasing burden to the working population, and indirectly, as well as in some cases directly, actually degrades safety.

High technology, a magnificent collective achievement of our society, is often seen as a solution to type D insecurity, but unfortunately all too often it leads to new problems. Human society does not seem to be regulated to guarantee its own survival, and there are plenty of examples from palaeontology of evolutionary dead ends. Many areas of current human activity seem to be heading in that direction.

Global warming is probably the most universally apprehended example of type D insecurity at the present time. Regardless of its origin, the surest guarantee of human survival in the face of such a threat is the maintenance of variety—of the repertoire of possible responses. The folly of monolithically developing organizations such as the European Union, which relentlessly crush variety (which is perceived as a threat to their own survival as an organization) is only too apparent in, for example, the Common Agricultural Policy. The enormous contraction of the variety of agricultural plants engendered by that policy means that the current agro-industrial complex that provides food to the bulk of the population is extraordinarily vulnerable to significant changes of temperature or rainfall. Although a low carbon economy is laudable in the sense of reducing pollution, in the sense of imposing a general reduction of human ac-

[3]These policies have wide ramifications. An example is provided by the laws permitting arbitrary arrest and detention of people suspected of activities or attitudes connected with terrorism, introduced in the UK a few years ago. As Lord Hoffmann pointed out during an appeal against such detention ([2004] UKHL 56 on appeal from [2002] EWCA Civ 1502), "The real threat to the life of the nation, in the sense of a people living in accordance with its traditional laws and political values, comes not from terrorism but from laws such as these." He further elaborated the meaning of "life of the nation": "The nation, its institutions and values, endure through generations. In many important respects, England is the same as it was at the time of the first Elizabeth ...the Armada threatened to destroy the life of the nation, not by loss of life in battle, but by subjecting English institutions to the rule of Spain and the Inquisition. The same was true of the threat posed to United Kingdom by Nazi Germany in the Second World War."

[4]Especially the J (judgment)-value, see P.J. Thomas et al., The extent of regulatory consensus on health and safety expenditure. Process Safety Environmental Protection 84 (2006) 329–336).

tivity on the developed world in particular it would exacerbate the difficulties of appropriately responding to climate change. Paraphrasing Ashby's law of requisite variety,[5] "to cope with a complex challenge, a complex response is needed".

Loss of variety is strongly driven by vested interests. The goal of the capitalist is of course monopoly (with a variety of one, or zero on a logarithmic scale). Although the so-called command economies developed in the Soviet Union and the member states of the Council for Mutual Economic Assistance (COMECON in English) were supposed to be built on rationality,[6] many of the resulting actions (such as the development of cotton monoculture in the Aral basin) were pure capitalism, which tends to be associated with greedy overexploitation. Variety within capitalism is supposed to come through the competition of monopolies, which leads to the constant disappearance of firms, even very large ones, and the emergence of new ones.[7] One notices a similar convergence in politics: although most of the COMECON countries were one-party states, nowadays the policies of the different parties in the so-called Western pluralist democracies are almost indistinguishable from one another. The degree of harmfulness of this loss of variety depends mainly on the influence that politics has on the life of the nation.

However clearly one may delineate safety, security and their links with complexity, this does not answer the more fundamental question, how desirable is safety? Danger is clearly necessary for maintaining a repertoire of effective responses, psychologically and sociologically as much as physically and biologically (one need only recall the importance of constant exposure to different antigens in order to maintain an effective immune system). In this regard, the traditional idea that the goal of human life is a state of repose as peaceful as that of cattle resting in a water meadow on a summer's day, as promulgated by organizations such as the Church of England,[8] seems to be quite contrary to the essence of human nature, which is to ever strive onwards to explore new territories. Too much security is stultifying, leading to boredom, with all *its* deleterious consequences both for society and for the individual. There can be little justification for the so-called precautionary (or "white queen") principle, of preventing or forbidding actions because they might be harmful. The appropriate degree of security and safety is that which maximizes human development potential—in its deepest and most expansive sense.

Directions for future research

The field of complexity, in particular as applied to questions of security, is very strongly multidisciplinary. Even a meeting such as this one could by no means encompass all the disciplines that should be involved. There is a definite need

[5] W.R. Ashby, Requisite variety and its implications for the control of complex systems. Cybernetica 1 (1958) 83–99.

[6] Always limited, of course, by the knowledge of the day, but the inertia of most economies means that a bad policy whose rationale vanishes due to new understanding might not be easy to divert. Perhaps this is one reason why the more intuitive approach taken by the capitalist system has sometimes been more successful.

[7] See also P.M. Allen, Complexity and identity (loc. cit.).

[8] The official Book of Common Prayer (1549) contains the homily "O God ... give unto thy servants that peace which the world cannot give; that our hearts may be set to obey thy commandments, and also that by thee we being defended from the fear of our enemies may pass our time in rest and quietness ...".

to continue these discussions between high-level specialists, for which purpose this NATO Advanced Research Workshop was indeed an excellent forum.

We identified a great and definite need for more research (at PhD level) into developing the ad hoc models that are necessary for capturing the real behaviour of complex systems. There is a very evident lack of expertise in this area. The most urgent need at present is simply to train more people capable of undertaking the kind of advanced modelling that is required. Because there is no fixed recipe for developing such models, such training is very difficult to fit in to the present structure of scientific research in the developed world, which is strongly organized around predefined research topics with precise deliverables.

Clearly there are some highly important specific technical areas needing urgent attention, where such modelling expertise is required. They are evident from some of the chapters in this book—soil (Chapter 8), climate change (all chapters in Part III), and energy (Chapter 11). To this list one might add environmental degradation caused by pollution and overexploitation of natural resources; and another indubitably complex current issue is demographic evolution, especially population growth and aging.

There is also a need for a deeper appraisal of technologies, in particular the implications of ever denser and more ramified networking. This more general research topic will require the close collaboration of different specialists from both "hard" and "soft" fields. As any physicist (for example) who has tried to collaborate with an economist (for example) will know, the difficulties of working together to produce genuinely new knowledge go far beyond those of language (using the same words with quite different meanings and connotations), but are affected by what seem to be fundamentally different approaches to phenomena.

Finally, there is the problem of implementation. The most perfect and incontrovertibly acceptable scientific edifices can crumble—in the sense of being ignored—before the onslaughts of ignorance and ill intentions. Here we are moving into the realm of ethics and motivations. Is scientific discourse still possible? In this Workshop, we have barely begun to touch this interface, but clearly it would be very important to explore it.

Advice to policymakers

It is an added bonus if a NATO Advanced Research Workshop can have some directly practical outcomes that could result in beneficial new policies being initiated. Here, in a few words, I give a kind of "executive summary" of some possible practical outcomes.

As Alan Greenspan has remarked, as wealth grows, the economy becomes less stable. *Pace* Gardner and Ashby,[9] as global networking grows, the economy becomes less stable. Traditionally, policymakers have relied on history as a guide to the future, with generally good results, and where monumental blunders were enacted (such as the succession of Franco-Prussian, First and Second World Wars), it was quite obvious that the catastrophes resulted from blind neglect of history. In economics too, one is accustomed to the "boom and bust" cycles, that are an elementary manifestation of Lotka-Volterra type of coupled differential equations.[10] The information revolution, and the following nano revolution,

[9] M.R. Gardner and W.R. Ashby, Connectance of large dynamic (cybernetic) systems: critical values for stability. Nature 228 (1970) 784.

[10] While the mathematical model may be elementary, the consequences may be appalling in

are changing the basis of our communications and hence the constraints of our civilization to such a degree that the change may have become qualitative, and hence the past is no longer a guide to the future. The formulation of wise policy—and ultimately human survival—and therefore requires an examination de novo of the problems with all the collective power of human thinking.

"Complex" means open-endedly evolving, indeterminate, unpredictable. The first lesson of complexity science is merely the appreciation of this open-ended indeterminacy, and the realization that models built on unverifiable constraining assumptions will almost certainly give results that are wrong both quantitatively and qualitatively. The second lesson is that there is (at least at present) no universal recipe for tackling complex problems: each one requires its own ad hoc solution, incorporating the best knowledge available.

A diagram such as Figure 1.1 can be useful for allowing self-reinforcing loops to be identified at a glance. Such loops can lead to an alarmingly rapid change of circumstances. Obviously, self-reinforcing aggravating loops are particularly dangerous. Even a brief inspection of Figure 1.1 reveals several such loops. At the very least, their identification shows what should be investigated in more detail as a first priority.

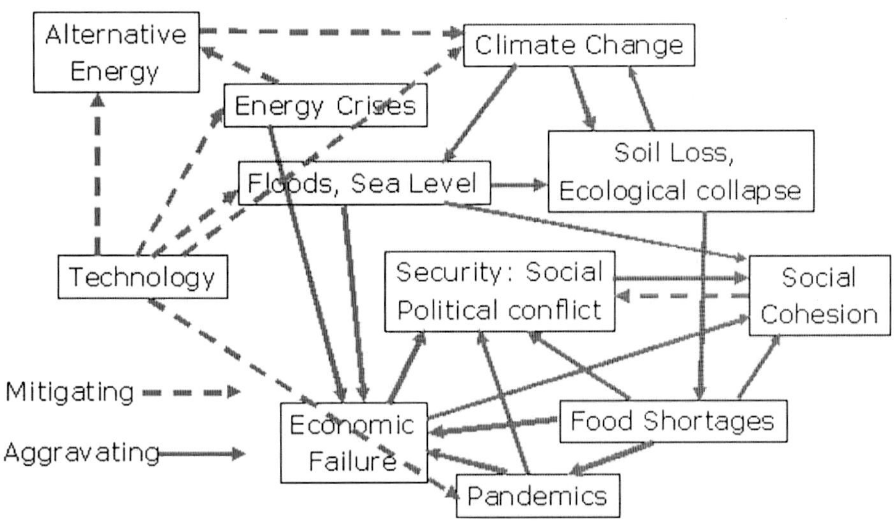

Figure 1.1: Global diagram of immediate effects.

Even though the ad hoc models may yield a reasonable and realistic answer, there remains an irreducible uncertainty of outcomes rooted in the *motivation* of the individual actors. One man is enthusiastic and energetic, another is indolent, yet another is imbued with a spirit of negation. Although our knowledge and understanding of human motivation can be deepened,[11] ultimately human

terms of human misery, and indeed one of the initial motivations of Marxist economics and the command system was to regulate the economy in order to eliminate these fluctuations. But the practical realization of the system involved the erection of a monolithic state apparatus that, as pointed out earlier in this Chapter, inevitably gravitates towards progressive loss of variety, and hence becomes unable to effect the desired regulation.

[11]See, for example, J.J. Ramsden, S. Aida and A. Kakabadse (eds), Spiritual Motivation, New Thinking for Business and Management. Basingstoke: Palgrave (2007).

motives are invisible.[12] One of the greatest enemies of security, safety and ultimate survival is greed, compounded by ignorance and distortion. And since our world system is not apparently globally regulated to ensure its own survival, *laissez-faire* is no guarantee that even a relatively benign status quo will persist.

Above all, variety must be preserved and enhanced. This needs to operate at many levels; the maintenance of biodiversity and variety of agricultural crops is only one facet, albeit an important one. Another important facets maintaining inclusiveness of participation in society. This was a particularly important feature in the technical and economic rise of Europe during the last 200 years, but as we again seem to be moving into an age in which real education is unaffordable for the poor, we should be aware that this may lead to a damaging loss of diversity of the reservoir of creativity possibly immanent in our society.

The individual chapters must be consulted regarding specific policy recommendations in specialized areas. For example, the work on terrorism (Chapter 3) suggests that less repressive policies than those typically pursued at present may diminish global terrorism, and an appraisal of the factors contributing to climate change (Chapter 10) suggests that top priority should be given to arresting and reversing deforestation.

We should also be very clear that these matters affect us all. Detachment is illusory. However distasteful a certain policy, however much we may personally disapprove, if it is enacted, we are to some degree responsible.[13]

This state of affairs underlines the importance, emphasized by Voltaire and others, of cultivating a habit of taking an independent and watchful interest in the transaction of national affairs, and a general consciousness of the duty of having some opinion on the business of the state, which is still the primary instrument for shaping our environment.

[12]Our insight into them is fundamentally limited by MacKay's Principle of Logical Indeterminacy (See J.J. Ramsden, Computational aspects of consciousness. Psyche: Problems, Perspectives 1 (2001) 93–100).

[13]Just as, in a democracy, all citizens are obliged to accept the results of a referendum, regardless of how they voted.

Part I

The Notion of Security

Complexity and Security
J.J. Ramsden and P.J. Kervalishvili (Eds.)
IOS Press, 2008

Chapter 2

Defining security

Jeremy J. Ramsden

Cranfield University, Bedfordshire, MK43 0AL, UK

The dictionary[1] divides the definition of security into groups: I. comprises (1) the condition of being secure; protected from or not exposed to danger; (2) freedom from doubt; well-founded confidence, certainty; and (3) freedom from care, anxiety or apprehension; a feeling of safety; and group II. includes (1) a means of being secure, i.e. something which makes safe; a protection, guard, defence; and (2) grounds for regarding something as safe or certain; an insurance, guarantee. The definition of secure, required for definition I.(1), is "without care, careless ... overconfident". A 1641 maxim is quoted: The way to be safe, is never to bee (*sic*) secure."

As in any language, of course the meaning of words evolves. The more recent (published in 2000) 10th edition of the Concise Oxford Dictionary still refers to security as the state of being or *feeling* secure, but also includes the safety of a state, company etc. against espionage, theft or other danger. Contemporary popular opinion seems to consider that insecurity is "anything that threatens what I want to do".

One notices the importance given to the psychological state of a person: feeling is as important as the actual physical means of ensuring absence of interference. One also notices the contrast with *safety*. This might be a useful distinction to emphasize: security is a psychological state of feeling safe, as well as the physical state of not being exposed to danger; whereas safety refers to the result of exposure to danger, a safe result meaning that one's essential variables are maintained within the boundaries of survival. (Note that one might define 'survival' rather broadly: as well as meaning literally staying alive, it could also mean maintaining a chosen way of life.) This is essentially homeostasis, or stability.

If it is accepted that the system under consideration is in some kind of equilibrium, it will be natural to assert that any perturbation potentially threatens stability, hence safety. *Resistance* refers to the ability of the system to withstand

[1] Shorter Oxford Dictionary. Clarendon Press (1933).

an imposed danger ("strain"), somewhat analogous to the moduli of materials science; *resilience* refers to the ability of the system to restore itself to its original state after the imposition of a danger (a "shock"); the higher the resilience, the more rapid the recovery. There is however ambiguity in the definition: sometimes the measure of resilience is the time taken to recover, but in some circumstances recovery may be only partial, in which case resilience might more usefully refer to the degree of restoration. Resilience also has its origin in materials science, but was adapted for use in ecology.[2] The Cranfield University Resilience Centre defines it as "the ability of an individual, institution, system or society to recover and develop after a shock, in order for it to keep going, or expanding, on its original or intended course."

The actual sources of danger are potentially very varied. They could be aggressive individuals, possibly armed; they could be pollution from traffic, industry or buildings (asbestos); they could be state policies leading to war. Crop failure, exhaustion of water resources, flooding or subsidence may perhaps be considered more indirect sources. Although some of them really are sudden shocks, others (such as climate change) are applied gradually, and useful mitigation of their potential effects could be achieved by simply slowing down their rate of application (this might be considered as a kind of friction at work).

The system to which we have referred is clearly a human system. This suggests that the system is essentially open-ended and indeterminate. Roger Penrose has constructed a powerful picture of the relationship between the physical, unconditional world; the world of our subjective experiences; and the world of timeless Platonic mathematical truths. Our subjective experiences clearly reside in our brain, which is a subset of the physical world, which is in turn a subset of the Platonic world, which is in turn a subset of our mental imaginings.[3] This truly Escherian perspective well illustrates the profound mystery in the interrelationship of those three worlds, and the difficulty that we would face in attempting to draw boundaries around the human part, to which we could then apply our concepts of resistance and resilience. Although they can be precisely expressed in the language of cybernetics for relatively simple systems such as a thermostat, to apply them to human systems would appear at first sight to be very difficult. It is however encouraging that resistance and resilience seem in fact to be special cases of the more general concept of adaptation, to which Sommerhoff has given a precise meaning suitable for applying to living systems,[4] and hence we should be able to describe human systems in these terms. The actual fulfilment of this research agenda nevertheless remains a challenge at present.

Any consideration of the governance of human behaviour must inevitably include some kind of ethical principle. It might be something seemingly very vague and tenuous, and hence perhaps difficult to incorporate in a formal scheme. For example, a very powerful restraint on violence against fellow human beings is that feeling of human solidarity that is so universal it might well be considered to be innate. Yet, as Konrad Lorenz has pointed out,[5] the training of soldiers

---

[2] C.S. Holling, Resilience and stability of ecological systems. A. Rev. Ecol. Systematics 4 (1973) 1–23.

[3] R. Penrose, Shadows of the Mind. Oxford: University Press (1994).

[4] G. Sommerhoff, Analytical Biology. Oxford: Clarendon Press (1950).

[5] K. Lorenz, Knowledge, beliefs and freedom. In: P.A. Weiss (ed.), Hierarchically Organized Systems, pp. 231–262. New York: Hafner (1971).

requires that feeling to be destroyed if the troops are to be effective. When seeking causes of diminished personal safety, it may not be irrelevant to note that the 20th century has seen unprecedentedly high fractions of the population systematically recruited into armies and subjected to the destruction of their feelings of human solidarity. Ethical matters will be explored in more detail in the final Part of this book.

Complexity and Security
J.J. Ramsden and P.J. Kervalishvili (Eds.)
IOS Press, 2008

Chapter 3

The sociophysics of terrorism: a passive supporter percolation effect

Serge Galam

Centre de Recherche en Epistémologie Appliquée (CREA)
Ecole Polytechnique, 1 Rue Descartes, 75005 Paris, France

3.1 Introduction

Terrorism has always existed at a given time in some parts of the world. But it has also been always confined within precise areas as with the traditional groups like the Basque, Irish and Corsican . A definition of terrorism could be the use of random violence against civilians (often to kill) as part of a global fight against some institutional political power (Sandler et al., 1983; Francart and Dufour, 2002).

Sometimes killing civilians is avoided and the destruction concentrates on property damage. After years of ongoing anti-terrorism, these traditional terrorism groups, which prosper in developed and democratic counties, are still not eradicated but reach some stable equilibrium. They became part of the political landscape of the corresponding countries.

At odds with this tradition was the terrorist attack on the USA that took place on 11 September 2001, which came as a sudden and dramatic blow to all experts on terrorism, intelligence services and military observers. It has marked the beginning of a new era in which terrorism is a daily reality of simultaneous horror and threat in many parts of the world, including rich and poor as well developed and underdeveloped countries. Today not a single country is immune from a possibly deadly terrorist attack. Terrorism can nowadays strike anywhere at any time. Its worldwide spread, as confirmed by the series of attacks on Bali (2002), Madrid (2004) and London (2005), have established the novel unbounded geographical status of a terrorist group.

Given such a permanent and intolerable threat, enormous efforts by military, police, financial institutions and information services are used and combined to try to curb or at least contain terrorist activities. Nevertheless it is unfortunately noticeable that in spite of a certain number of significant, but always limited, successes, current terrorism remains incredibly powerful and is omnipresent all over the world. It is an extremely complex, complicated and difficult phenomenon, and this, together with its usually fatal consequences makes any contribution that could shed some new light, even limited, on understanding it very valuable. One such contribution could be made within the realm of a new field of statistical physics denoted sociophysics that has emerged in the last few years and established itself as a very active field of research dealing with social and political behaviour (Galam et al., 1982; Galam, 2002a; Galam, 2002b). This has suggested for the first time the application of concepts and tools from the physics of disorder to tackle the problem of terrorism (Galam, 2002b; Galam and Mauger, 2003; Galam, 2003). We address some specific aspects of the terrorist phenomena without investigating either the terrorist net itself nor its internal functioning.

I suggest focusing on the social space in which the terrorists live, move and act. While much effort has been devoted to the study of terrorist networks, their structure, their means, their finances and their internal functioning (Sandler et al., 1983; Francart and Dufour, 2002), very little is known about the human and social environment in which terrorists evolve. This space includes the terrorists themselves, their potential targets and also each one of us. This work does not aim at an exact description of terrorism complexity (Sandler et al., 1983; Francart and Dufour, 2002; Ahmed et al., 2005), but studies the connexion between the terrorist range of destruction, i.e. its capacity to reach a given target, and the surrounding population's attitude towards it. In this chapter I show how some crude approximations allow the construction of a universal simple framework for terrorism, which in turn exhibits new and counterintuitive results. It emphasizes some of the invisible features behind terrorist activities by emphasizing their connexion to the existence of a distribution of passive supporters within the overall population.

3.2 The passive supporter attitude

The problem of terrorism is tackled here from a different viewpoint, which focuses on the *social space* in which terrorists can move freely, rather than on the terrorists themselves. We are not concerned with terrorist themselves, who exist with millions of nonterrorists around them.

We start from a given terrorist cause initiated in a local geographical area of some country. As soon as a terrorist group strikes for the first time, every person within that area has an opinion against on in favour of the terrorist act, but also against or in favour of the terrorist cause. The two opinions are not necessarily identical. We consider people who are in favour of the terrorist *cause*, independently of being either against or in favour of the use of violence to support that cause.

Social agents (i.e. people) are thus divided into supporters and opponents of terrorism. Usually, but not always, the division is unbalanced with most of the population being hostile to the terrorist cause. Simultaneously hoever there always exist some people who identify with its aspirations. They do not need to be active in their support; it is a passive unspoken support. We denote these agents *passive supporters*. They are normal people with no direct involvement in the terrorist activity. In general they are mainly concentrated within the terrorist home area.

Operationally, a passive supporter is an agent who, in a situation confronted with a terrorist move, will neither oppose nor disturb it. The chance of a passive supporter finding himself (or herself) in such a situation is extremely small. It is thus a dormant attitude, which results from a personal and often private opinion, and is almost never activated. Passive supporters do not need to explicitly express their position, neither in public nor privately. They constitute an invisible social group, which results in the natural creation of a friendly social space potentially open to terrorists. That space is essentially undetectable.

The passive supporters independently share an identical opinion but they do not need to communicate between themselves. They could be anybody anywhere. Therefore they must be regarded as randomly spread within the whole population. Only their density p can be roughly estimated using polls and election results. This nature of the problem makes the theory of percolation (Stauffer and Aharony, 1994), used in physics to study geometric connectivity of materials with randomly distributed active individual elements, a perfect tool to discover some of the properties of these open social spaces, in particular, their geographic extension.

3.3 Percolation theory: from physics to social properties

Assuming a random distribution of passive supporters, within the framework of percolation theory, we shall show how they create a series of finite-range connected social spaces that are each potentially open to terrorist activity. Within each one of these clusters, terrorists can move to hit any target located within the territory covered by the corresponding cluster. But they cannot cross from one cluster to another.

Percolation theory studies the macroscopic properties of connexion among a series of objects distributed randomly within a given space. This is accomplished according to the topology and the dimension of this space, as well as the density of the objects present. It provides in an exact quantitative way the dynamics of the spontaneous emergence of new collective macroscopic properties. These properties are of a long-range nature in contrast to the microscopic objects featuring only short-range properties.

The percolation-like description provides a universal explanation of the usual geographical confinement of past and traditional forms of terrorism. It is based on the current density p of the passive supporters within the global territory in which the fight is taking place. The number and geographical extension of these disconnected clusters are a function of that density.

However, percolation theory exhibits the existence of a critical threshold p_c which discriminates between two regions with very different qualitative properties. When $p < p_c$, the description of a series of isolated connected clusters holds as shown Figure 3.1. But above the threshold with $p > p_c$ one of these cluster extends from one side to the other of the total territory, as illustrated in Figure 3.2.

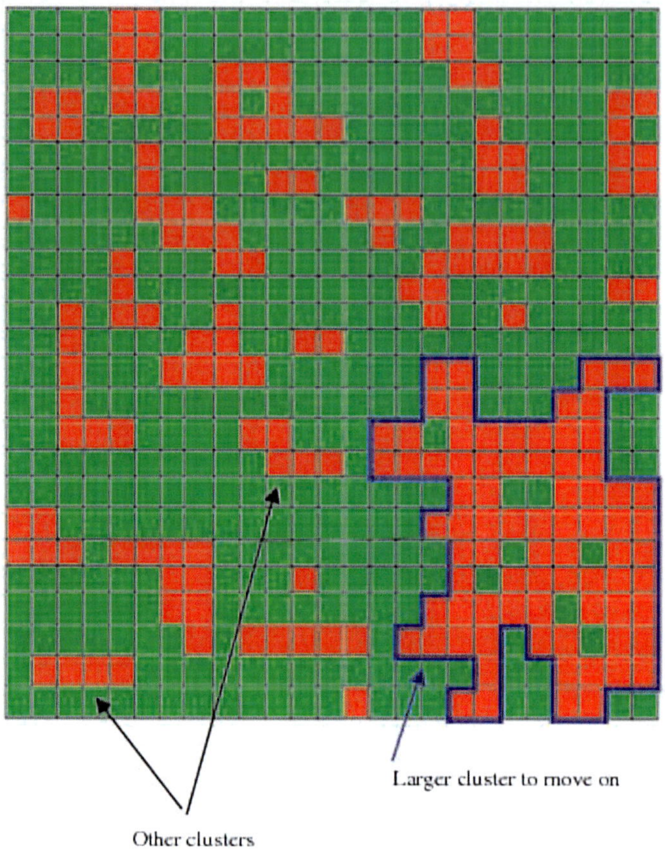

Figure 3.1: An illustration of a square lattice $p < p_c$. Only sparse little disconnected clusters of different sizes are present, shown in red. The cluster sizes are different with one big one(authud in blue).

Classical terrorism as in the Basque, Irish and Corsican cases corresponds to the situation with $p < p_c$. By contrast, the September 11 terrorist attack on the USA can be regarded as the signature of the first world percolation of a social space of passive supporters associated with the terrorist cause, i.e. with $p > p_c$. These people identify with at least some of the corresponding terrorism goals.

The transition from Figure (3.1) to Figure (3.2) is a sudden and sharp phenomenon. It is called a phase transition of second order. It produces a drastic qualitative change in the connectivity properties. It should be noted that the coverage p_c at which percolation occurs is a fixed value, much lower than one, i.e. 100% coverage. This value is a characteristic of the geometry of the problem and mainly the coordination number q, the maximum number of nearest

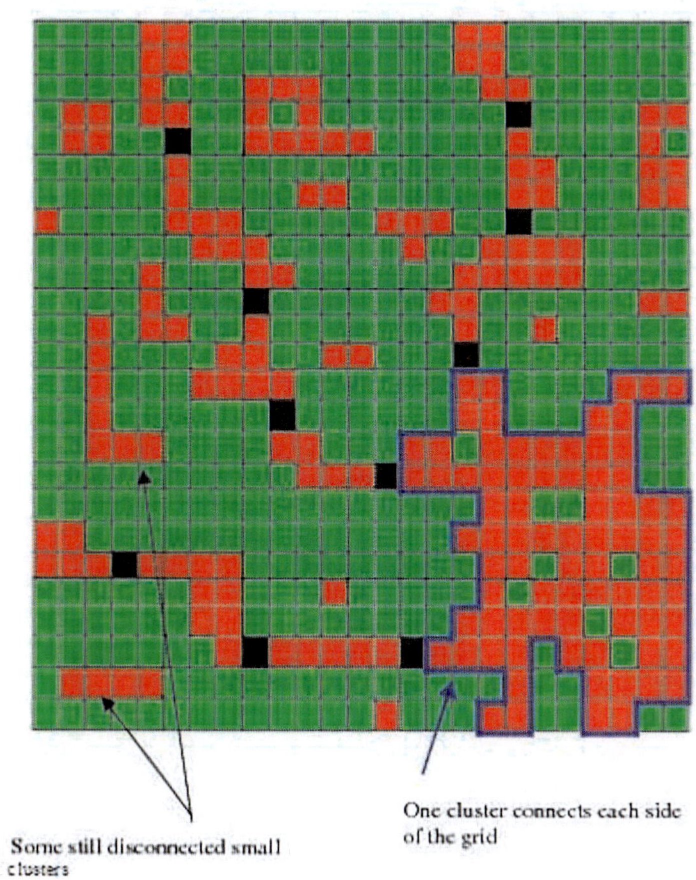

Figure 3.2: The same Figure as Figure 3.1 but now with $p > p_c$, obtained by adding the squares shown in black. The initial largest cluster now extends from one side of the grid to the other side. Some small isolated clusters are still present.

neighbours of a given site, and the space dimension d. In the case of Figures 3.1 and 3.2, $d = 2$ and $q = 4$. In physics cluster that connects from one side to the other is called the infinite cluster, because in an infinite space it would be itself infinite.

For the practical application of this idea, at stake is the determination of the percolation threshold value, which can either dramatically extend the terrorism range or shrink it to a limited geographic area. It is argued that it depends merely on various independent 'flags' (principles) around which terrorism articulates its cause. The relevant associated space in which percolation occurs is identified as a multidimensional social space. In addition to the physical space associated with the actual surface of the Earth, that space includes all the various independent flags displayed by the terrorist group.

In most cases only one flag, namely the independence of a territory, is activated. I shall show that the 'breakthrough' of current global terrorism is a result of the aggregation of a series of numerous independent flags. On this basis some hints can be given on how to shrink the geographical spreading of current terrorism. This could be achieved by setting a new strategic scheme to increase the terrorism percolation threshold, which in turn will suppress the percolation automatically.

In this paper we are dealing with passive supporters who have a positive benevolence toward the terrorist organization, but it could also be a passive support driven by fear, indifference or profit, as in in some known cases of underground activities or illicit practices. The model may be applied to a large spectrum of clandestine activities based on aggregated individual passivity such as guerilla warfare, black markets, corruption, illegal underground economics, tax evasion, illegal gambling, and illegal prostitution.

3.4 "Terrorists must be like fishes in water. But they must find that water"

This picturesque image was created by the Chinese leader Mao Zedong about revolutionaries who are like fish among the populous masses. It applies to terrorists with the major difference being that the latter need much less water. And above all the problem for them is to find the water. Indeed it could be said that we are now studying the various features of that metaphorical water.

In scientific terms we need to study the social permeability of a given society to the individual movements of terrorists. It results from a series of regions of passive support to the terrorist cause by some part of the population, which due to sympathy to the *cause* do not oppose a terrorist move (i.e. a sin of ommission not of commission). An open space emerges from the presence of local and individual passivities without connexions between them. It is a collective outcome from a series of individual attitudes from adjacent passive supporters. The geometrical aggregation, taking place by the simple adjacent juxtaposition of passive supporters, spontaneously creates a labyrinth of paths permeable to terrorist moves.

It is the degree of permeability of the social background that determines the limits of the social space that is open to terrorist action. Schematically, we can think of each individual looking at only his own window, and not beyond. Each person occupies and observes a certain portion of territory starting from his window. Each one of these people, independently from any other, can then decide to close or open his window curtains at the time of a suspicious observation on his portion of observable territory, see Figure 3.3.

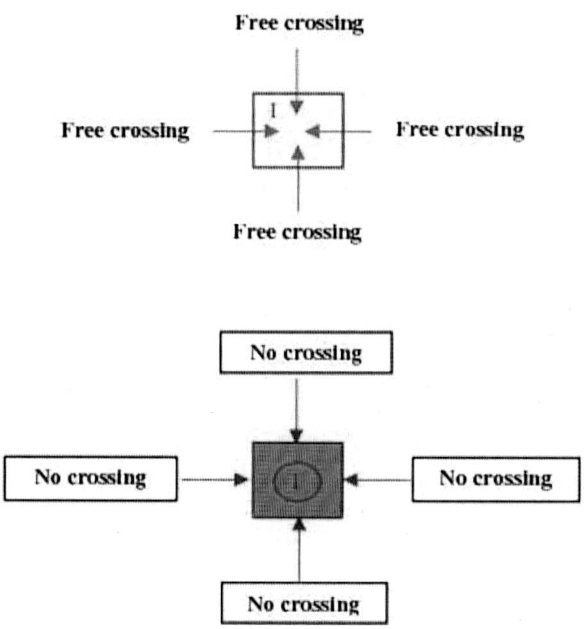

Figure 3.3: A schematic representation of an open local space in the upper part. The curtains are closed, a terrorist can cross the space from any of its four sides. In the lower part, the opposite individual attitude is shown. The curtains are open, forbidding a safe terrorist crossing from any of the four sides of the box.

Let me illustrate the notion of open space and path. First the world is mapped onto a two-dimensional grid where each living being is associated with a square. For simplicity we assume that each person has the same area, i.e. an identical square. We ignore empty spaces like the seas, deserts and high mountains. The grid is assumed to be totally occupied. It is worth noting that the space of one square is sufficient to contain several persons. Figure 3.4 represents a small area of the Earth viewed in this fashion.

To move on the grid, a terrorist must cross from one square to another, which is contiguous to it. From a given square denoted departure (D), an terrorist who wants to reach another further square denoted arrival (A) must follow some continuous path crossing a series of adjacent squares connecting these two extreme squares as shown in Figure 3.5. A large number of different paths are potentially available to join D and A. It is impossible to predict a priori which one is going to be followed by the terrorist in moving from D to A.

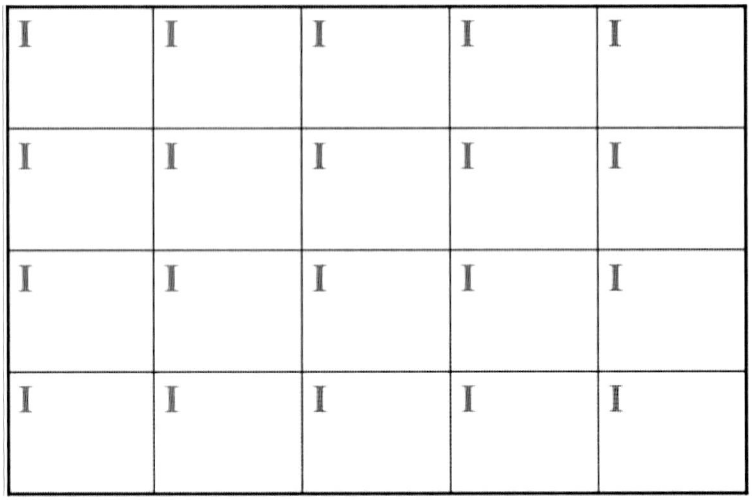

Figure 3.4: A tiny area of the Earth's surface. One individual I occupies each square. All spaces are open.

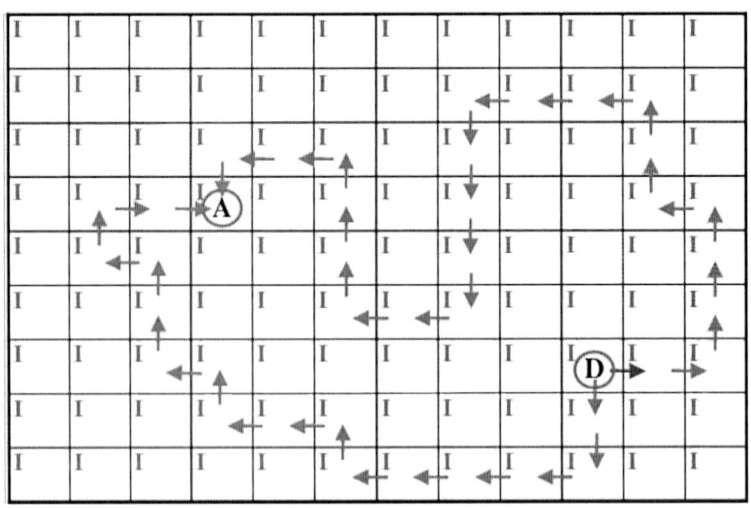

Figure 3.5: Several different paths can connect departure square D to arrival square A. Only two, out of a large number, are shown with the arrows indicating the direction of crossing in and out.

In Figure 3.5 each individual space is open to being crossed. Nevertheless an agent is aware of who is crossing through its square. However, that is not always the case. At any moment an agent can close its own square, for instance by reporting a suspicious activity taking place in it to the police. It makes a crossing by a terrorist a delicate move since he (or she) could be easily caught while crossing a non-friendly square. A terrorist will usually not force his way through. He needs to stay unnoticed to achieve his deadly goal.

It should be stressed that being passive while noticing a terrorist move has the same effect whether the motivation is support or indifference. It is necessary to act to close the square.

A given box is characterized by its four sides, each one being either open or closed according to the attitude, active or passive, of the corresponding occupant. For a passage between two squares to be open requires both occupants adjacent to the common side to be passive supporters. As soon as one opens his curtains, the passage is closed from the four sides, as seen in Figure 3.6.

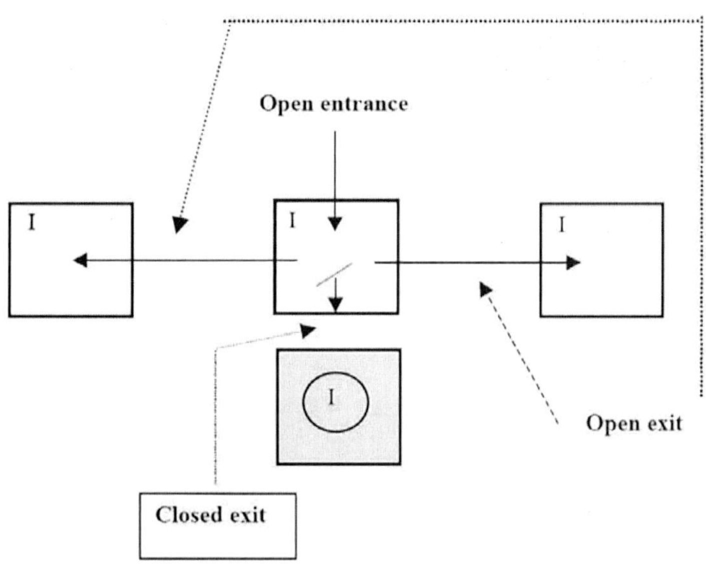

Figure 3.6: A space configuration with four adjacent squares, where three sides are open to a terrorist move and one is closed. Crossing the middle box from the top, a terrorist can proceed safely along either the right side or the left, but not downwards where the square is closed.

A hostile position is thus more efficient than a passive one, since in the first case the four adjacent squares are closed while in the second case an second open square is required to enable a crossing. The states of passive and hostile attitude are therefore asymmetric as indicated in Figure 3.7.

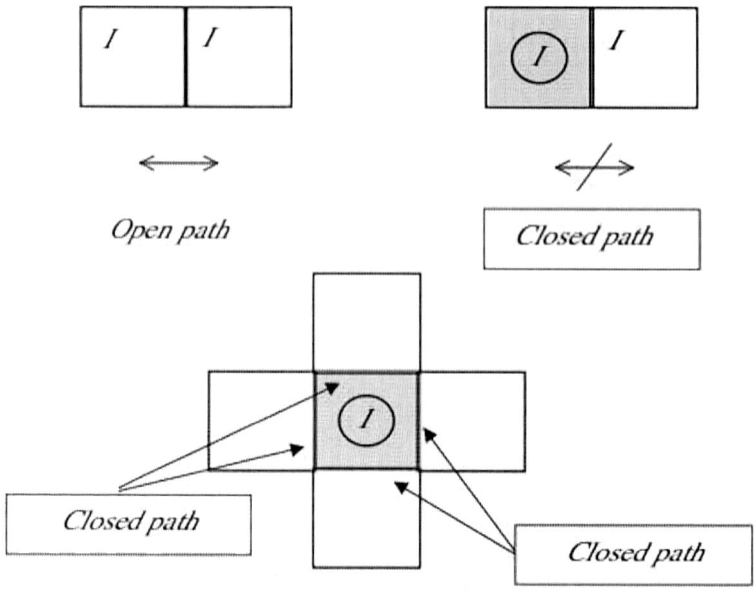

Figure 3.7: At the top left the path is open, while it is closed at the top right. Below, one closed box is shown to block four paths from open contiguous boxes.

3.5 From individual shifts to global properties

The state of a given square is a priori unknown. It is dormant. Only the arrival of a terrorist 'activates' the dormant state. To reach a target starting from its initial square a terrorist looks for a way connecting contiguous open squares. A terrorist can also be obliged to back-track if no way to continue to the target is found.[1] He can also shift target for geometrical reasons. It is in fact all of the existing possible paths starting from the terrorist base (TB) that determine the social space open to terrorist action, the active open space (AOS). In general there exists simultaneously several open spaces (OS), which not being connected to the terrorist base, they are not accessible to terrorist action, see Figure 3.8.

The localization of these open spaces is impossible to predict by the nature of their existence. They nevertheless determine the whole spectrum of targets potentially accessible to terrorist action. No attempt to circumvent the threat can be successful without an evaluation of the associated degree of social permeability. It is both a strategic and a conceptual challenge. It is here that the physics of disorder, in particular the theory of percolation, proves extremely useful since it deals precisely with such problems of connective geometry.

Passive supporters constitute the main part of the "terrorist water" in Mao Zedong's metaphorical picture. They directly determine the capacity of destruction of the corresponding terrorist organization. Some terrorists (including the so-called 'sleepers', who are in fact terrorists, but merely await the call to action before springing into action) are localized, by chance often (since the general

[1] An amusing account of such activity can be found in R.L. Stevenson, The Dynamiter. London: Longmans, Green & Co. (1885) *(Editor's note)*.

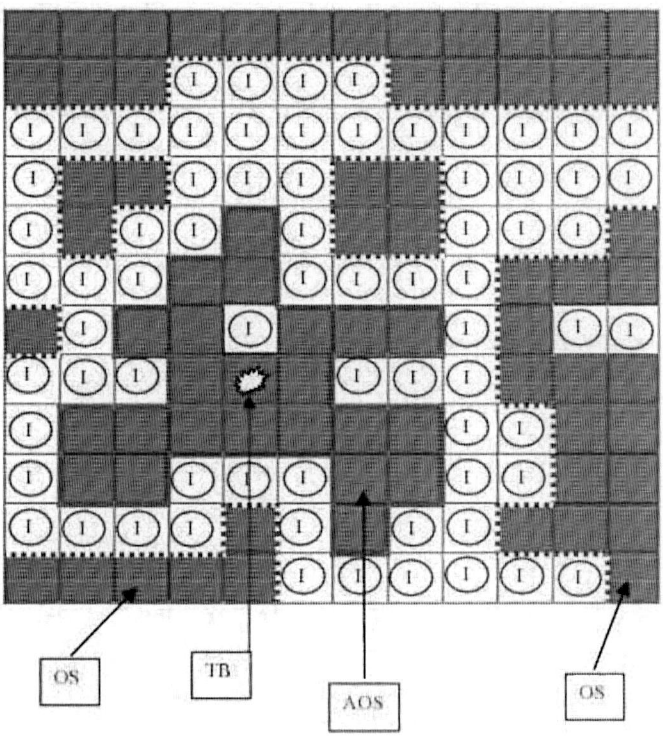

Figure 3.8: A portion of territory fragmented within a grid with 12×12 boxes. White squares have open curtains closing any terrorist move. Dark squares are occupied by passive supporters. They are open to terrorist crossing. Open spaces (OS) with respect to terrorist action are seen to result from the nearest neighbour juxtaposition of dark squares. Only the one including the terrorist base (TB) is active (AOS). There the terrorist threat is a reality. Other OS are inaccessible to terrorist action, making the terrorist threat there purely latent or virtual.

furtiveness that necessarily pervades these organizations often precludes knowledge of the whereabouts and even identity of fellow terrorists, who are known in totality only to the leader of the organization), side by side on adjacent squares, while others are isolated, surrounded by opponents to the terrorist cause. This random distribution of terrorists and passive supporters creates the map of the terrorist threat. A schematic representation of such an area is shown in Figure 3.9.

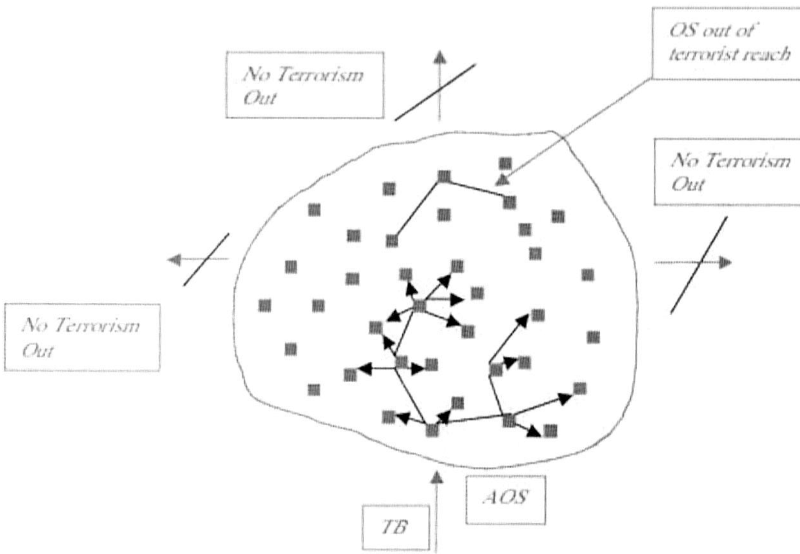

Figure 3.9: Example of a territory with two open spaces (OS), only one of which is active (AOS).

Away from the geographical area claimed by the terrorist, the number of people passively agreeing with the terrorist flag quickly decreases to zero. Far from it, people are not sensitive to the cause of terrorism, and may even be ignorant of it, save for the indefatigable efforts of radio and television journalists, who seemingly delight in bringing the most obscure and parochial events to the attention of an international audience. So the density of OS falls drastically to zero as soon as the frontiers of the terrorist area are crossed, see Figure 3.9. In addition, as noted earlier, the various OS are anyway inaccessible to terrorist action; potential targets within them are not in danger. Only those inside the space covered by the AOS are directly threatened.

But such a map of terrorist threat is not fixed over time. Its boundaries are dynamic and fragile. AOS extensability is volatile. It can grow or shrink depending on the evolution of the distribution of the passive supporters. A suicide bombing can cause the attitude of some passive supporters to become hostile. And an excessively violent military counteroperation can push agents initially hostile to the terrorist cause to join the passive supporters.

These individual attitude flips reshape the AOS and the OS, even if they are not numerous. Their effect depends on their respective locations. One individual shift can an OS to become connected to the AOS, thus making it extend over a much larger area, with many previously safe targets becoming immediately vulnerable. Such reshaping can occur even though little has changed either at the global level of support to terrorism, or with the terrorist infrastructure itself. One such extended AOS is shown in Figure (3.10). When such an extension does occur, every part of the territory is under threat. The AOS is said to have percolated within that area.

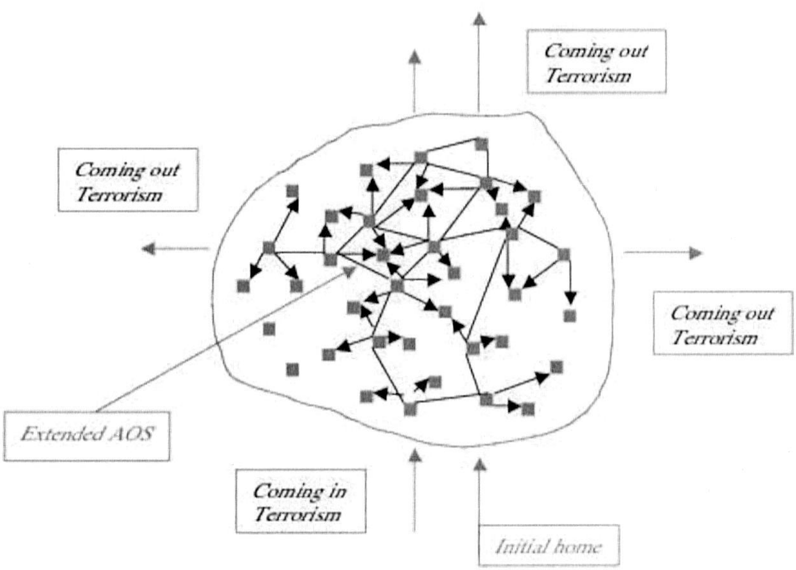

Figure 3.10: Schematic view of Figure 3.9 with an extended AOS, which now covers a large territory. All OS have been incorporated to the initial AOS. The global change has been driven by only a few individuals turning passive supporters.

To make this dynamical change clear, Figures 3.11, 3.12 and 3.13 exhibit the typical scenarios of global changes driven by individual attitude shifts. The first one shows an area with one active open space (AOS) and six open spaces (OS). The terrorist base (TB) is by definition located within the AOS. Figure 3.12 shows a one person shift. By becoming a passive supporter he extends the AOS to include a previous OS enlarging the current AOS with new potential targets. Now one larger AOS coexists with five open spaces (OS). An opposite shift is shown in Figure 3.13. Contrary to Figure 3.12, one passive supporter turns hostile to the terrorist flag. As a result, the extended AOS is shrunk by losing one part of its previous space. The broken part turns to OS status. All its associated targets are now out of reach of the terrorist group.

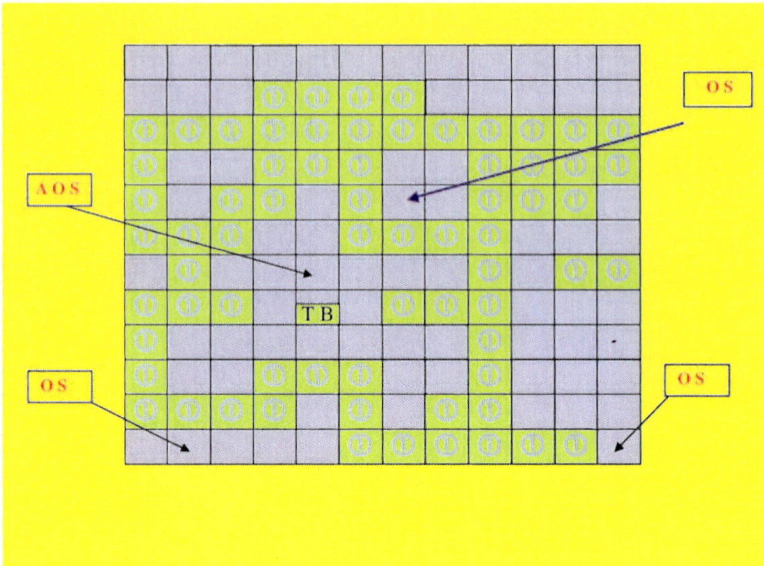

Figure 3.11: View of an area including one active open space (AOS), where the terrorist base (TB) is located, with six open spaces (OS).

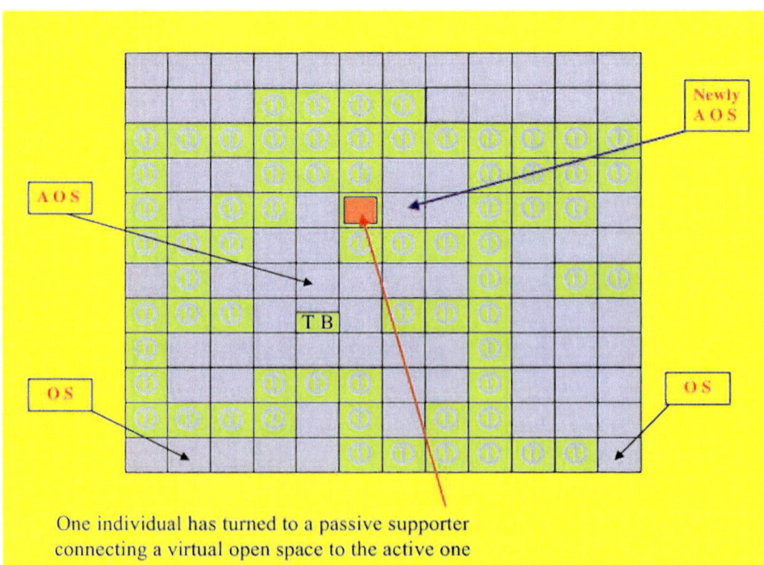

Figure 3.12: The same area as Figure 3.11 but with one person (marked in red) having shifted attitude. By becoming a passive supporter he extends the AOS to include a previous OS, thereby enlarging the current AOS with new potential targets. One AOS now coexists with five open spaces (OS).

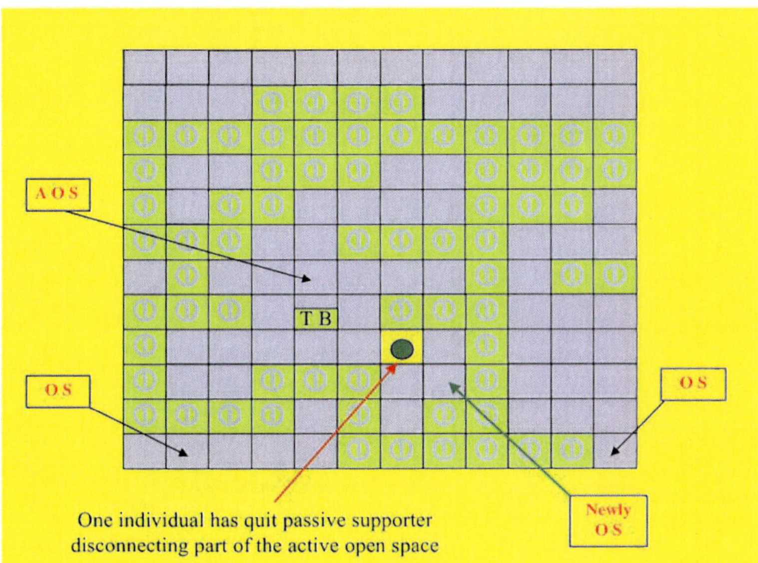

Figure 3.13: The same area as in Figure 3.12, with one formerly passive supporter turning hostile and opening his 'curtains'. He is indicated by a circle with a square. The AOS has been reduced by this individual shift. The active open space has shrunk and there are again six open spaces.

3.6 From the model to some universal features of terrorism

Traditionally, a given terrorist group is always attached to a geographical area with a territorial claim of independence. In democratic countries a substantial part of the corresponding population sympathizes with the terrorist cause but opposes the use of violence. However, this opposition is only formal, with nothing actively undertaken against the terrorist group and its members. In particular nobody denounces active nationalists involved in the terrorist network. This is the root of passive support.

Although extremely simple, our model provides an explanation for some basic features of the various forms of known terrorism. In particular, it explains two essential characteristics of terrorist activity.

The first one is the geographical anchoring of terrorism. The targets selected by the terrorist are in the immediate vicinity of his base, not primarily because of a reasoned choice, either ideological or strategic, but quite simply due to the physical constraint of possible access. This limitation seems clear with respect to classical terrorism, as found in Corsica and Ireland, see Figure (3.14).

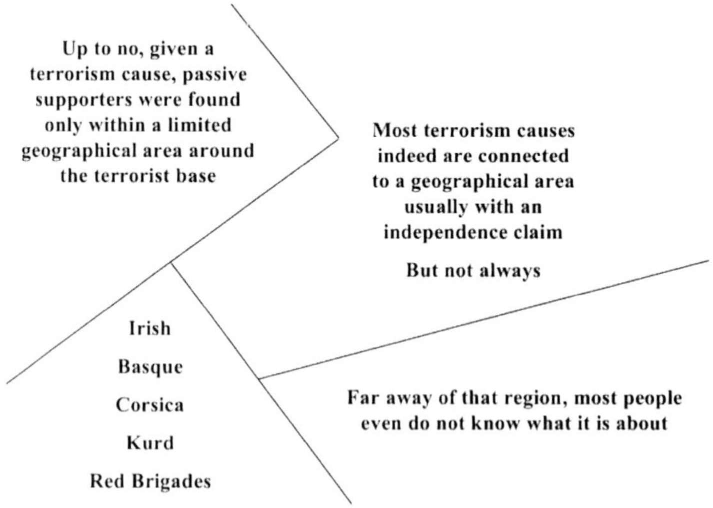

Figure 3.14: Local anchoring of traditional terrorism.

The second characteristic is the impossibility of determining precisely the space which is accessible to terrorist actions since this space is simultaneously multiple, invisible, dormant and volatile as shown explicitly in the series of Figures 3.11, 3.12 and 3.13. But from an overall assessment of the current support existing for a terrorist cause, percolation theory allows us to evaluate the number, the average size and the distribution of the OS within a well-defined geographical area.

We also see why the forms of terrorism that one could qualify as traditional or classical are confined geographically. They lack sufficient passive support outside their area of anchoring. For each terrorist group, there is one AOS concentrated in the heart of the disputed area, then some OS dispersed around the AOS in a more or less broad perimetaer zone. But beyond that claimed territory, only some tiny, fragmented open spaces exist.

It is then impossible for the terrorist capability to move beyond its natural territory. Figure 3.15 shows such an example of regional terrorism. The only possibilities of extension for the terrorist group are the connexions of the AOS to the other OS of the area. All the other parts of the world are completely out of reach of its operations. Three different, successively bigger, regions A, B, C can thus be defined, as depicted in Figure 3.16. The nonuniform distribution of passive supporters is the cause of this 'three region' setting: the region A where the terrorist threat is maximal; the region B where the terrorist threat can become real; and the region C where the terrorist threat does not exist at all.

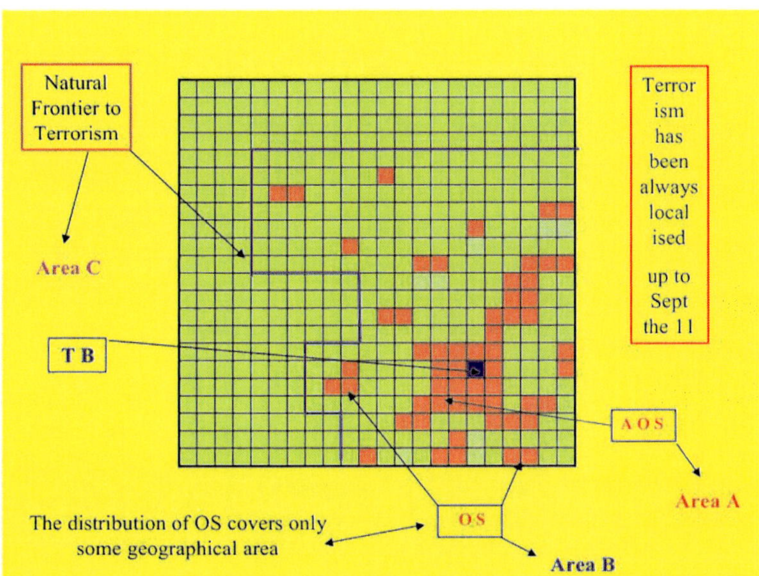

Figure 3.15: Division of a geosystem into three distinct areas A, B and C. Area A is under active threat since it is covered by an AOS. Area B is potentially under threat with a few OS. Area C is out of any danger.

3.7 What is novel in current global terrorism?

If we consider recent current global terrorism, which was able to strike with near-impunity in New York, Bali, Madrid and London, within the general approach described in this chapter, the implication is that its active open space covers the entire world. The inference would then be the existence of passive

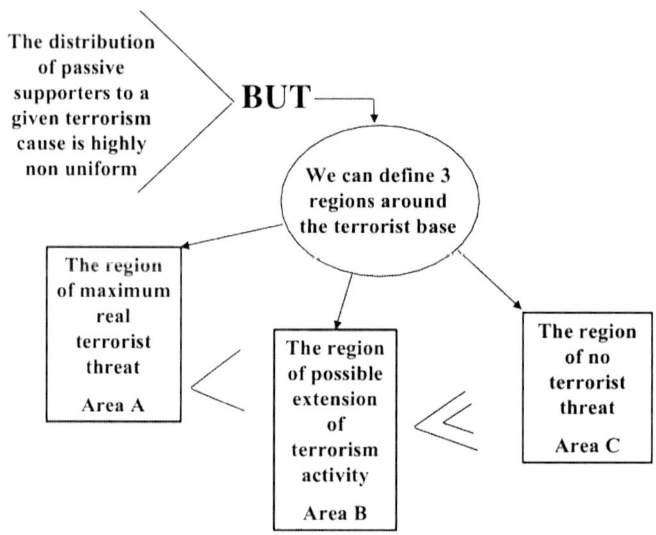

Figure 3.16: The nonuniform distribution of passive supporters engenders a three-region setting: the region A where the terrorist threat is maximal; the region B where the terrorist threat can become real; and the region C where the terrorist threat does not exist at all.

supporters spread all over the world, and they should be distributed and connected everywhere. Such a reality would require millions of passive supporters.

In many respects, current global terrorism has nothing novel with respect to traditional terrorism. The only significant difference is its capacity to have produced such a great number of passive supporters throughout the entire world. However this recruitment to its cause took time and was not obtained at once.

We could consider that before "September 11", many open spaces did exist all over the world but went unnoticed since they were disconnected from the AOS anchored well away from the the West. Therefore they stayed dormant, and the continuous dynamics of gaining more and more passive support, which may have lasted for years, caused neither concern nor worry of any kind. Numerous open spaces could have been coming into existence yet remained inaccessible to the terrorists. No change of the world level of safety would therefore have been noticed. But at a certain time, a sudden and brutal phenomenon occurred with respect to the connectivity of all these open spaces. It is called in physics a phase transition of the second order, and in a geometric context, a percolation.

With some small additional gain in passive supporters, at a precise threshold, the so called critical threshold p_c, we have the simultaneous fusion of several OS to the AOS covering the whole planet with the consequent possibility for the terrorists to join one side of the surface of the Earth to the other side. This may well have been the first phenomenon of global percolation in the history of terrorism. Within such a percolating situation, the majority of the potential targets are not only accessible but they are also reachable via several different paths.

3.8 There exists no military solution

Given a territory, the distribution of passive supporters yields the range of terrorist action. If the passive support density p is less than the percolation threshold ($p < p_c$), most of the territory is safe with only one limited area under terrorist threat as shown in Figure 3.9. In contrast, as soon as p becomes larger than the percolation threshold ($p > p_c$), all the territory falls under the terrorist threat, as illustrated in Figure 3.10.

In physical terms one has to determine the size of the territory for which the condition $p > p_c$ is satisfied. It is the region A where terrorism is a real threat. There the AOS is percolating. The only difference from one form of terrorism to another is the scale on which passive supporters percolate. The change of scale does not change the fundamental nature of the terrorist phenomenon, but it modifies the number of threatened people, which is not a negligible difference.

With respect to any military operation, the destruction of a terrorist cell has immediate advantages, but only for a finite period since as soon as a new terrorist group is formed, it can strike again immediately within the whole associated active open space, which remains unchangedly accessible to its members. A long term solution needs to bring the condition $p > p_c$ back to some new condition $\tilde{p} < p_c$ by lowering the density of passive supporters from p to $\tilde{p}$. Such a reduction will induce a shrinking of the whole territory accessible to terrorist action. The AOS could thereby eventually be reduced to a narrow geographical area. However, a physical implementation of such a veritable pogrom at the military level is simply terrifying and unacceptable for ethical and moral reasons as well as from the point of view of natural justice. Since the passive supporters are unknown and randomly spread, the outcome of such a military implementation would be the destruction of a large number of agents, many of them hostile to the terrorist group. An militarily efficient solution would lead to the destruction a good part of the planet, potentially increasing the of the number of terrorists. At the same time, any partial military solution appears to be useless since without an effect on the level of the terrorist threat (leading to fragmentation of their global AOS), it voids any hope in curbing terrorism. The current terrorist danger would thus remain unchanged at the present level of global threat.

3.9 From no feasible military solution to novel social perspectives

As described above, the physically-inspired scheme to suppress global percolation by decreasing the density of active sites has been shown to lead to a deadlock. However, taking into account the social nature of our problem opens an alternative. Since for humanitarian reasons nothing should be done directly to the passive supporters, intervention should be focused on modifying the value of the percolation threshold itself instead of changing the density of passive supporters. This scheme potentially provides very promising perspectives but is it achievable? To answer the question it should be noted that the actual value of the percolation threshold depends primarily on two independent parameters, the connectivity of the associated network q and the dimension of the space d which embeds the network.

A square lattice has $q = 4$ and $d = 2$, which yields $p_c = 0.59$. A cubic lattice has $d = 3$ and $q = 6$, giving $p_c = 0.31$. The four dimensional hypercube has $d = 4$ and $q = 8$, yielding $p_c = 0.20$. Increasing the dimension or the connectivity (or both) decreases the percolation threshold. the origin of this effect of lowering the value of the percolation threshold is the appearance of more possible paths to connect from one site to another as the dimension increases.

However, very few geometrical networks allow an exact calculation of their percolation threshold. Most of them are calculated numerically using large scale simulations. Moreover the geometric structure of a social network is not defined unambiguously. We might guess (from considering the number of people and organizations with which we, as individuals, are connected) that on average it could have a connectivity of the order of $q = 16$, with a dimension of a priori $d = 2$, the surface of the earth. This would correspond to an unknown network in physics. Thus its associated percolation threshold is unknown.

Fortunately a universal formula for all percolation thresholds was discovered few years ago (Galam and Mauger, 1996). It is, quite generally,

$$p_c = a[(d-1)(q-1)]^{-b} , \tag{3.1}$$

where $a = 1.2868$ and $b = 0.6160$. This *Galam-Mauger* formula yields, with less than 1% error, all known thresholds. In addition it can be used to predict the value of the threshold for any network given the values of both its connectivity q and dimension d. The formula is shown in three dimensions in Figure 3.17. It is seen that percolation thresholds decrease significantly with increasing q or d or both.

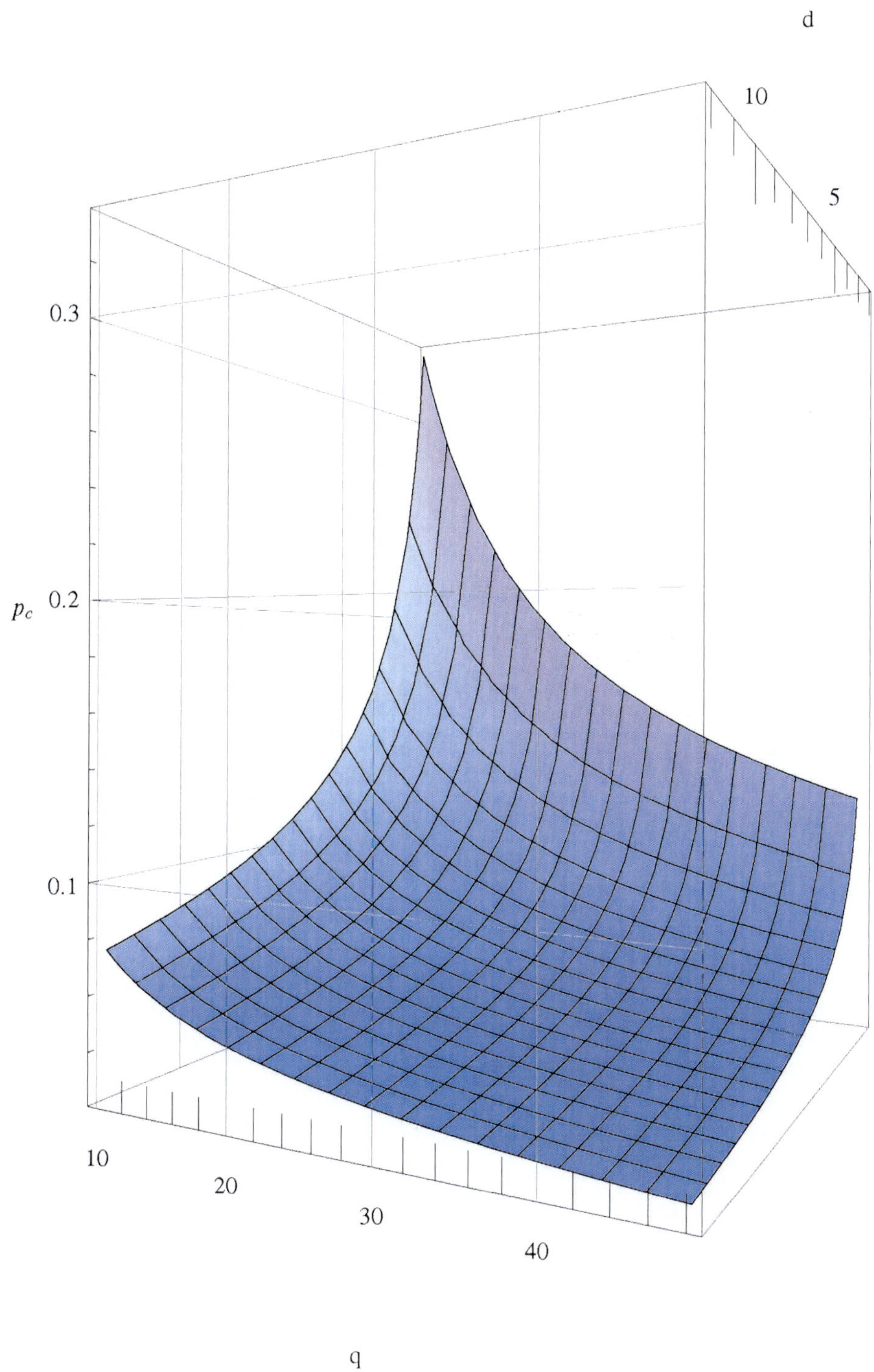

Figure 3.17: Three dimensional representation of the Galam-Mauger universal formula (Galam and Mauger, 1996) for percolation thresholds as a function of connectivity and dimension.

Putting $q = 16$ and $d = 2$ into equation (3.1) yields $p_c \approx 0.24$. This is too high a value to be realistic, since it is difficult to imagine that almost one fourth of the population passively supports a terrorist group. If that were the case, that very group would eventually take power to become the official government! The terrorism would therefore end by achieving its goal. We could accept this interpretation of events in the context of a so-called war of liberation, such as have occurred throughout the history of decolonization. But it would not appear to apply to either traditional terrorism or to the current global terrorism.

This unrealistic outcome underlines the fact that something is wrong in the estimate of either q, d or both. The value of p_c must be much lower to be in accord with common observation. To get a flavour of the dependence of p_c on q and d, Table 3.1 shows the value of p_c for a series of values of (q, d).

Table 3.1: Values of the critical threshold p_c for a series of values (q, d).

q	$d = 2$	$d = 3$	$d = 4$	$d = 5$	$d = 6$	$d = 8$	$d = 10$
10	0.33	0.22	0.17	0.14	0.12	0.10	0.09
15	0.25	0.16	0.13	0.11	0.09	0.08	0.06
20	0.21	0.14	0.11	0.09	0.08	0.06	0.05
25	0.18	0.12	0.09	0.08	0.07	0.06	0.05
30	0.16	0.10	0.08	0.07	0.06	0.05	0.04
40	0.14	0.09	0.07	0.06	0.05	0.04	0.03
50	0.12	0.08	0.06	0.05	0.04	0.03	0.03

While it is reasonable to evaluate q between 10 to 20, higher values should be justified by some investigation on the ground. The challenge is indeed to estimate the dimension of the terrorism social space. To get an estimation we could state that whereas a level of support of around 10% to 15% sounds right for traditional terrorism, support of only a few percent seems more realistic for the current global terrorism. From Table 3.1 this implies a value of $d = 4$ for traditional terrorism and $d = 10$ for the global one.

Both estimates yield values of d that are higher than 2. This prompts the hypothesis that in a social percolation phenomenon, extra dimensions are produced in which individuals may position themselves, in addition to the physical two dimensions of the Earth's surface. We identify these dimensions as social paradigms or the 'flags' around which terrorists articulate their fight (Galam and Mauger, 2003). People may then identify, with more or less support, to each one of these flags.

Typically for most terrorist groups, the first flag is a territorial claim based on either independence or autonomy. This flag constitutes the first social dimension independent of the two dimensions of geography. Moreover, as soon as a terrorist group starts to implement destruction, it induces some state repression against it, which in turn determines a new additional flag, since agents may approve or disapprove of the state repression and its severity. That gives already 4 dimensions. Thus, any terrorist social dimension seems to have a value of at least four.

These findings are consistent with traditional terrorism since we have $p_c = 0.13$ for $q = 15$ and $d = 4$. As soon as a terrorist cause gains support of 12% of a population, it is possible to move freely on all the associated territory. This value is not very high and is certainly reached in traditional terrorism. It may explain the continuous ongoing incapacity of the authorities concerned to quash these terrorist activities.

The problem is that not much freedom of action seems to be available to curb terrorism since a dimension of four is irreducible. It is the dimensional lower limit of any terrorist activity. Traditional terrorism of low dimension could thus remain active. It may however be useful to note that, within the present framework, the absence of repression would bring the dimension down to $d = 3$, taking the threshold up to 16%. Such numbers would suggest that for instance in the case where the popular Corsican support for independence ranges between 12% and 16%, it is indeed the repression which allows it to percolate all over Corsica. To do nothing would shrink the extent of the territory accessible to the terrorists.

With regard to the new international terrorism the situation seems to be qualitatively and quantitatively different. Indeed it is difficult to believe it has support of more than 10% of the world population. At the same time, it appears to be clearly successful in having its passive supporters percolate worldwide. On the other hand $p_c = 0.06$ at $q = 15$ and $d = 10$ is a reasonable guess. The problem is then to identify the extra six dimensions over and above the core of the four defined for traditional terrorism.

Actually, what characterizes the current international terrorism is the broad spectrum of flags on which it deploys its claims (Galam and Mauger, 2003). To start with, it strives for the independence of several different states and for the changing of the political system of a series of countries in order to establish religious states. In addition, it has a world religious dimension, an ethnic dimension, a bipolarizing dimension (the aim of partitioning the worldE, a social dimension, a regional dimension and a historical dimension. That brings its social dimension to at least 10. Accordingly, its percolation threshold is as low as 5%, which allows it to percolate at the global level.

3.10 Neutralizing flags to curb global terror

Current world terrorism has drastically increased its potential range of destruction by an increase in the number of its independent flags. That process has enabled a global percolation with only a few percent of passive supporters, to facilitate its terrorist activities. No military action could neutralize such a small fraction of essentially dormant passive supporters spread over the whole planet.

But the discovery of the features of global terrorism provides useful hints for finding some solution effective in curbing its threat of world-ranging action. In particular, contrary to low dimension terrorism for which a dimension reduction is impossible, here action on the number of flags becomes feasible. The strategic aim should therefore be the neutralization of some of these flags, whose eradication is not possible by the use of military means (Galam and Mauger, 2003; Galam, 2003).

It is within the political, economic and psychological realms that adequate measures could be elaborated to neutralize some of these flags. To lower the social dimension from $d = 10$ to $d = 6$ would make the threshold jump from 5% to 8%, see Figure 3.18. Such an increase would very likely immediately suppress the world percolation, and terrorism would at once shrinks to only one area of the world as for other traditional terrorism.

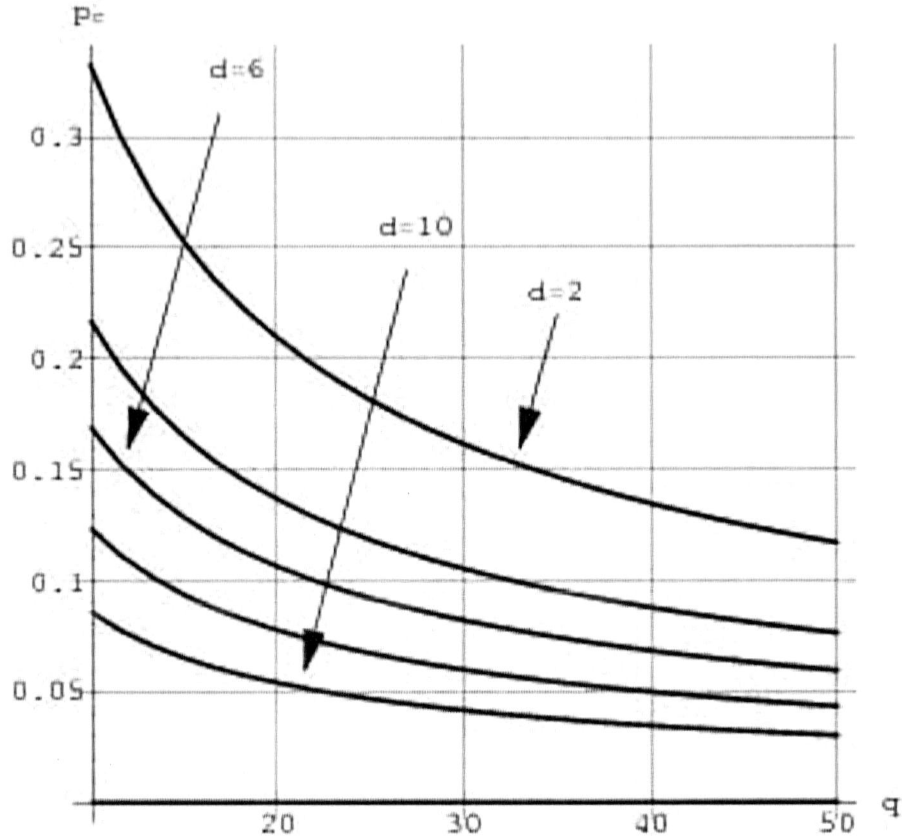

Figure 3.18: The percolation threshold p_c calculated from the universal Galam-Mauger formula (3.1) for fixed dimension as a function of the coördination q. It is seen that the threshold values drop with both q and d.

3.11 Conclusion

It should be emphasized that to determine how to put in action a flag neutralization process is beyond the scope of physicists working alone. It requires an interdisciplinary collaboration with specialists from a large spectrum of other disciplines. The possibility of an efficient non-destructive large scale war against international terrorism requires substantial interdisciplinary research. (Unfortunately, such a goal may turn out to be a challenge more difficult than the fight against terrorism.)

The approach presented here also applies to a range of clandestine activities, which similarly develop using the existence of networks of passive supporters. Such activities include tax evasion, corruption, illegal gambling, illegal prostitution, black markets, and others. Indeed, it is essentially a universal model of clandestine coöperation.

It would be fruitful to apply this general framework to real cases with data and facts, but clearly such a task is beyond the physicist working alone. It is also worth restating that this analysis does not claim to be an exact quantitative description of reality. It only aims to shed a new light on linking passive individual support to a terrorist cause and the associated range of action of the corresponding terrorist group.

3.12 References

E. Ahmed, A.S. Elgazzar and A.S. Hegazi, Phys. Lett. A 337 (2005) 127.

L. Francart and I. Dufour. Stratégies et décisions, la crise du 11 septembre. Paris: Economica (2002).

S. Galam, Y. Gefen and Y. Shapir. Math. J. Sociol. 9 (1982) 1.

S. Galam. Eur. Phys. J. B 25 (2002a) 403.

S. Galam. Eur. Phys. J. B 26 (2002b) 269.

S. Galam. Physica A 330 (2003) 139.

S. Galam. Physica A 336 (2004) 49.

S. Galam and A. Mauger. Phys. Rev. E 53 (1996) 2177.

S. Galam and A. Mauger. Physica A 323 (2003) 695.

T. Sandler, J.T. Tschirhart and J. Cauley. Am. Political Sci. Rev. 77 (1983) 36.

D. Stauffer and A. Aharony. Introduction to Percolation Theory. London: Taylor and Francis (1994).

Complexity and Security
J.J. Ramsden and P.J. Kervalishvili (Eds.)
IOS Press, 2008

Chapter 4

The 'What', 'Who' and 'How' of contemporary security

Trevor Taylor

Department of Defence Management and Security Analysis,
Cranfield University at Shrivenham, UK

4.1 Introduction

The aim of this essay is to outline the difficulties of addressing security in both a conceptual and a practical sense and to suggest that the number and dynamism of factors associated with security justify the conclusion that security in the contemporary world should be characterized as marked by complexity.

Social sciences struggle from being prisoners of popular language, from being compelled to use terms that often lack a clear meaning. Terms such as 'value', 'welfare' and 'interest' are central to our everyday discourse and to our sense of understanding of the world we inhabit, and yet they lack the precision of terms at the centre of most physical scientists' concerns. Tritium is a much more delineated material than happiness or democracy. In the realm of political science, security is among the most contested concepts. There are as many views of the domain of security as there are writers on the topic, indeed, given that some writers themselves are not entirely clear in their thinking, there are probably more views than there are writers. This chapter attempts to bring a degree of structure, order and perhaps insight to current debates.

A useful starting point is to recognize that at least three groups regularly use security as a key element in their activities. The first are government agencies, companies and other organizations that are seeking to generate some specific effects. Thus some governments have a national security policy and the USA, for instance, has a National Security Council. The EU has long sought to develop a Common Foreign and Security Policy, while many companies have a security manager and security staff.

The second group comprises academics and other researchers who write about security as part of an effort to understand the world. They are numerous, especially within the subject of international relations. Thus sitting on my bookshelf as I write are volumes of various ages with titles such as *British Security Policy* (Croft), *European Security in the Nuclear Age* (Wyllie), *Managing Security in Europe* (Algieri et al.), *Understanding Global Security* (Hough), *Security and International Relations* (Kolodziej), and *People, States and Fear: The National Security Problem in International Relations* (Buzan).

The third group comprises those whose prime motive behind writing is to sway the actions and thought of those in power and with authority. Many of these writers may be associated with a think tank or non-governmental institutions (such as Saferworld in the UK).

These groups are not entirely separate, with governments and companies feeling that they use security in a rigorous and coherent way, and with the academic authors of having books often at least an aspiration to influence the behaviour of governments, corporations and other powerful bodies by their writing.

Why does this matter and why does it happen? At least part of the answer here lies in the universal recognition that security matters are non-trivial; indeed they are highly significant. Thus if an issue is recognized by governments and other major controllers of money as falling within the security sphere it is almost certainly going to attract attention and resources. There is thus a temptation for any political campaigner (or government) to press for an issue to be 'securitized', i.e. placed within the security domain (Mandel, 1994). A stark illustration of the securitization of an issue occurred in 1979–1980 when, following the Soviet invasion of Afghanistan, the U.S. government sought to pressure the American Olympic movement into boycotting the Moscow Olympics. The Olympic movement was reluctant to comply but fell into line once President Carter linked any US participation in the Moscow Olympics to American U.S. national security. Over the past decades environmental factors have become widely recognized as being either closely linked to, or an intrinsic element of, the security domain, as is reflected both in political discourse and in the dozens of academic and other non-fiction works dealing with the environment and security. Thus many environmental questions can be said to have been securitized.

4.2 Perception and reality

Security is a two-sided coin. On the one side is reality, to which we have no absolutely reliable access. On the other side are the perceptions what drive our behaviour. People may feel secure and yet may be in great danger, as were the citizens of Lockerbie in Scotland in 1986, not realizing that a Boeing 747 had been destroyed in mid-air and its remains were falling on their village. When Donald Rumsfeld became U.S. Secretary of Defense in 2001, terrorism did not loom large in his thinking and was not a priority issue for him (Stevenson, 2007, p. 166).

On the other hand people may also feel insecure and yet in reality be in little peril. Agoraphobics have this problem every day. For more than four decades after 1945, the West and the USSR worried about a possible attack by the other, which it appears neither had much intention of launching. The perception-reality relationship is a constant and unavoidable feature of security issues, with an important element after 2001 being how governments in the West should present the realistic possibilities of terrorist attack. Too much stress could make people feel insecure and unconfident, not least about investment. Too little stress could lead to complacency and reluctance to commit resources to terrorist-related problems.

Linked to this point is that reactions to risk and even calculations of risk include many subjective features. The deployment of US cruise missiles in Europe after 1979 appears to have made a significant number of people feel more insecure (and they subsequently protested), whereas others either felt more secure or at least unaffected by the development. Similarly people around the world, in different personal circumstances, react differently to shared data on environmental damage and climate change.

The overlaps and contrasts between the world of perception and that of reality provoke major intellectual debates. The practitioner has to be aware that it is perceptions that shape behaviour, but that reality is also of enormous importance. Furedi (2002) has argued that contemporary Western societies are far too concerned with risk avoidance and safety, and that perceived risks are often greatly exaggerated. This introductory statement can do no more than recognize the problem.

4.3 The 'What' of security analysis—the domain of security from core to periphery

While people may disagree about the extreme extent of security, they concur that malicious acts of violence against the physical well-being and existence of human beings and against the integrity of property lie at the heart of security. Protection of citizens and organizations, including their assets, against the hostile intent of other states, and against the violence of criminals, are long-standing functions of the state and are accepted as definitely within the security domain. This has led to the close association of the state with security, and sometimes to the discussion of security only in state and governmental terms. Violent acts are most obviously associated with malicious intent but malice can be manifested in a range of ways. The launch of an economic embargo and, potentially of huge importance, cyber attacks, are both illustrations of deliberate but non-violent acts that would be seen by most as falling with the security sphere, either because they weaken military defences, or because they might cause loss of life. Press reports in the late summer of 2007 suggested that China was seeking to be able to disrupt U.S. and Western military capability and critical national infrastructure to a significant extent through hacking attacks on American computer systems,[1] although China denied the allegations (Hope and Blair, 2007; Coonan, 2007; Ward, Sevastopulu and Fidler, 2007; Pilkington and Johnson, 2007; McGregor and Sevastopulu, 2007; Thompson, 2007).

Since the end of the Cold War there has been significant debate in many states about the priority to be given to external defence and thus to armed forces, but historically the existence of a world of sovereign states has been accepted as presenting special security challenges. Because in the world of states there is no body with authority over individual countries, all countries have to be concerned with possible attack from other states. Moreover, because even defensive preparations to make any attack unsuccessful can be seen as threatening by others, and because such preparations can stimulate military responses by neighbours, the discipline of international relations has laid much stress on the 'security dilemma'—by strengthening its military forces, a state may provoke reactions in others that in the medium and long term actually weaken its security.

This focus on the protection of people and assets against hostile intent as the essence of security is reflected in governments' international as well as domestic policies. The UK's Department of International Development has long had a programme of security sector reform designed to strengthen 'security' institutions in the d eveloping world. These institutions always include the armed forces, the police, the judiciary, and the intelligence services. Where they exist, the gendarmerie and border control forces are also covered. Some of the reasoning behind this focus was summarized in a ministerial introduction to the British Government's *Security Sector Reform Brief* (2003):

> Most wars today are within states rather than between them. While the root causes often lie in the denial of rights, discrimination, and poverty, the risk of conflict is heightened when security forces are not subject to proper discipline or civilian control and where there is ready access to weapons. On the other hand, properly constituted security forces can be a force for good, thereby helping to reduce instability and contributing to a reduction in human suffering.

Moreover, away from government and in the domain of the individual and the organization, security is often thought of primarily in terms of protection against malicious acts. A guide to effective bodyguard action begins with a definition: 'The object of personal security is to reduce the risk of kidnap, assassination or criminal act by the application of certain principles and procedures to our normal daily life' (Consterdine, 1995, p. 5).

While hostile intent, malice and violence lie at the core of security, there is also an increasing tendency also to treat as security challenges those contingencies not directly linked to human malice, which could threaten the existence and property of significant numbers of people. Natural disasters such as floods and earthquakes, and large scale industrial accidents such as Chernobyl, are seen by many as within the security domain. It is important to recognise that 'accidents' often involve some human shortcoming and that 'natural' developments also may be affected by human behaviour. Thus there is extensive contemporary debate about the relationship between human behaviour and global warming (cf. Part III), and certainly deforestation and building can change the vulnerability of an area to floods. Things that may seriously harm people in significant numbers can be seen as candidates for the security label, which leads to the view that the spread of AIDS should be treated as a security challenge. The reality that environmental factors can threaten the existence of many people adds weight to the claim that security should not just cover physical violence. Consider the following extract from a New York Times report:[1]

> Public health is reeling. Pollution has made cancer China's leading cause of death, the Ministry of Health says. Ambient air pollu-

[1] Kahn and Yardley, 2007.

> tion alone is blamed for hundreds of thousands of deaths each year. Nearly 500 million people lack access to safe drinking water. Chinese cities often seem wrapped in a toxic grey shroud. Only one percent of the country's 560 million city dwellers breathe air considered safe by the European Union. Beijing is frantically searching for a magic formula, a meteorological deus ex machina, to clear its skies for the 2008 Olympics. Environmental woes that might be considered catastrophic in some countries can seem commonplace in China: industrial cities where people rarely see the sun; children killed or sickened by lead poisoning or other types of local pollution; a coastline so swamped by algal red tides that large sections of the ocean no longer sustain marine life. China is choking on its own success. The economy is on a historic run, posting a succession of double-digit growth rates. But the growth derives, now more than at any time in the recent past, from a staggering expansion of heavy industry and urbanization that requires colossal inputs of energy, almost all from coal, the most readily available, and dirtiest, source.

Sitting firmly alongside the view that 'natural events' and 'accidents' should be seen as in the security domain are the arguments of those such as Charles Perrow, who recognize that such events already kill on the scale of major wars, and who see the potential for even worse developments than we have seen so far. In the context of the U.S. nuclear power industry, it is argued that poor senior management practices, pursuing short term profit at the expense of safety arrangements, are at least as much a threat to the lives of millions of Americans as are terrorists (Perrow, 2004, p. 23).

Moving further from the core of security lie those who argue that anything threatening serious disruption to a society's way of life and central values should be viewed as a security challenge. From a human perspective, it may reasonably be thought desirable that people should be able to live as free from fear as possible. Freedom from fear arguably should go beyond the absence of worry about violence or major damage to property, but into wider areas including economic well-being and even cultural practices. Mandel (1964) saw that "national security entails the pursuit of psychological and physical safety, which is largely the responsibility of national governments, to prevent direct threats primarily from abroad from endangering the survival of the régime, their citizenry *or their ways of life* (this author's emphasis)."

This widest conception of security presents both intellectual and practical challenges, not least because, as Furedi stresses, society's concern with safety is growing and 'one of the unfortunate consequences of the culture of fear is that any problem or new challenge is likely to be transformed into an issue of survival' (Furedi, 2002, p. lxiii). Intellectually it is hard to see how anyone could develop expertise in a topic as broad as security might be. In the broadest conceptions of security, it becomes synonymous with importance, i.e. it becomes virtually impossible for something to constitute a really important problem without it also being classifiable as a security problem. Practically and ethically, there should be awareness that a very broad concept of security has been used by oppressive régimes to justify practices such as the involvement of state security agencies in the tight vetting of the reading, viewing and audio material of populations. From the 1950s, there was a significant Soviet effort to restrict the access of

USSR and Eastern European populations to Western thought and culture as a matter of state security.

The argument so far, summarized in Figure 4.1, is that the 'what' of security can been seen as having a core an inner area and a more distant zone. This needs to needs to be qualified by pointing out that, even if an effort is made to constrain the security domain to the hostile/malicious acts category, issues in the outer areas still need careful attention because events there can lead people to threaten or use violence. Environmental factors and change can lead to mass migration and then violence. There is a particular concern about the potential of water supply issues to lead to wars (Klare and Thomas, 1994; Homer-Dixon, 1994; Barnet, 2007; Dent, 2007). Some ethnic groups maintain their identity in part by practising extreme violence against those women who seek to marry outside the group. In this sense, 'very important' issues are at least potentially in the security domain, because it is just such issues that can inspire large-scale violence.

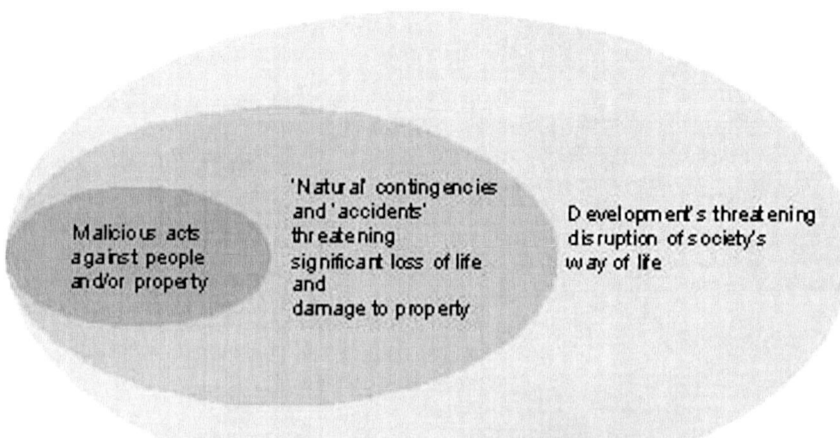

Figure 4.1: The 'What' of security analysis.

Thus it is possible to re-present Figure 4.1 in terms of phenomena and their causes (Figure 4.2). On the security front line are matters that pose a threat to people's physical existence, through malicious human action. In the second echelon, close by, are matters that pose a threat to people's physical existence because of environmental damage, accidents and natural disasters. In the third, fourth and fifth echelons are those matters what indirectly or directly could feed the first two echelons because they loom large in people's consciousness. . These are many and varied, such as concerns about the authority of tribal leaders in Afghanistan, religious and linguistic freedoms, the viability of a Basque way of life in Spain, and the failure of Islamic societies to match the economic and technological advances of more secular societies. As Jonathan Swift reminded us more than two hundred years ago with his description of the 'Big Enders' and the 'Little Enders', it is hard to put limits on the things that people are on occasions ready to fight about.

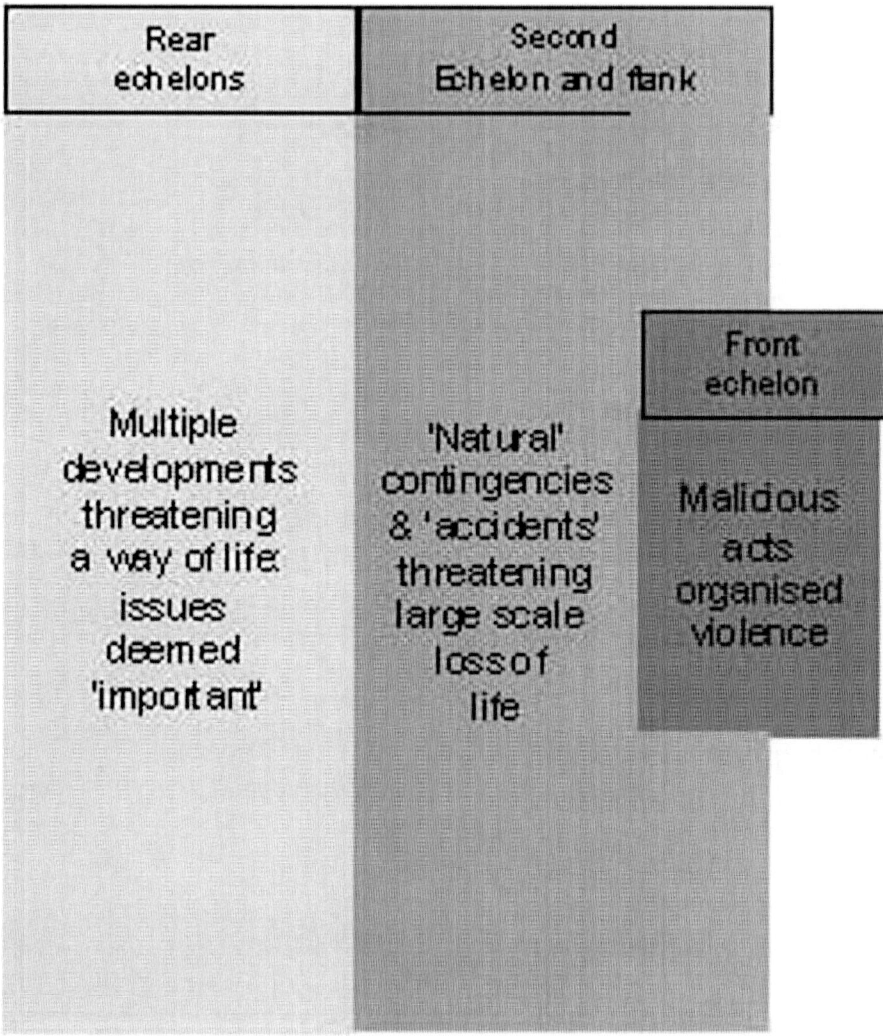

Figure 4.2: The 'What' of security analysis (another view).

4.4 The 'Who' of security analysis

There is a need to go further and to address two additional areas, the first of which is whose security is of concern? The second area involves a discussion of the management of security in the light of the causes of insecurity. This can be thought of as the 'how' of security.

There are many units whose security could and should be a focus of concern. Most obvious is the individual, with concerns about individuals likely to be strongest in cultures where the expectation of a good society is that it should enable the individual to live a full and rich life. Intriguingly the state's use of the term 'social security' in terms of unemployment and other benefits indicates a conception of security that is in the largest area in Figure 4.1, since it is meant to protect the individual from the worst disruptions to a way of life arising from illness, loss of employment, old age and so on.

There can be concern with the security of groups within societies, such as religious groups, ethnic groups or people from a specific region. Group insecurities were central to the dissolution of Yugoslavia and the subsequent violence of the 1990s in that area.

Society as a whole can be the subject of security concerns, as the threat to many people in the developing world from AIDS demonstrates.

Individual organizations such as companies and even universities have to be concerned with security. By the start of the 21st century, even in the West there were few clear principles delineating what companies could expect the state to do to protect their assets against crime (including terrorism) and what companies were expected to do for themselves. Arguably the state is doing less as the private 'security' industry continues to grow and all major companies employ a significant number of 'security' staff as well as placing sub-contracts for services such as guarding.

At the highest organizational level are the security of the state and of the government, the distinctions between the two do not always receive the prominence they merit. The state is a legal entity and the government is the organization entrusted to act on its behalf. Forms of government can change with the state remaining in continuous existence, as was the case with Poland after 1990. The individuals holding office in a government, that we can collectively call the régime, can change without the form of government changing, most obviously through an electoral process. On the other hand, the failure of a government can lead to the collapse of the state as a meaningful actor, as happened in Somalia in the early 1990s. Thus there is a need to think separately about the security of the state, of the governmental system, and of the régime of the moment. Extreme behaviour can often be justified by claims that, if a particular régime falls, the governmental system will collapse, as will the state as a whole. Criticism of the state President then becomes a threat to the State.

Outside formal organizations, there is a strong case for not overlooking the security of people in general, wherever they live (the human security dimension) (Human Security Centre 2006). As wider concern about the rights of women and rights of children has grown (cf. Chapter 26), as the distinction between combatants and non-combatants has often evaporated in wars, and as women have often been specifically targeted in conflicts, women and security has become a focus for concern (Kennedy-Pipe 2007; Brocklehurst, 2007, pp. 75–90 and 367–382). Addressing humanity in general, especially during the Cold War there was the possibility of a large-scale nuclear war that could have generated a nuclear winter, affecting everyone on the planet. Global warming too appears to justify near universal concern. Finally there is a case for taking serious account of challenges to the well-being of the bio-sphere as a whole, i.e. the whole range of living things on the planet (Figure 4.3).

The key point to be made about this list is that there is no automatic compatibility between the security of the different elements in the 'who' of security. Most obviously, to protect society as a whole and the individuals and organizations within it, it may be necessary to require certain individuals,

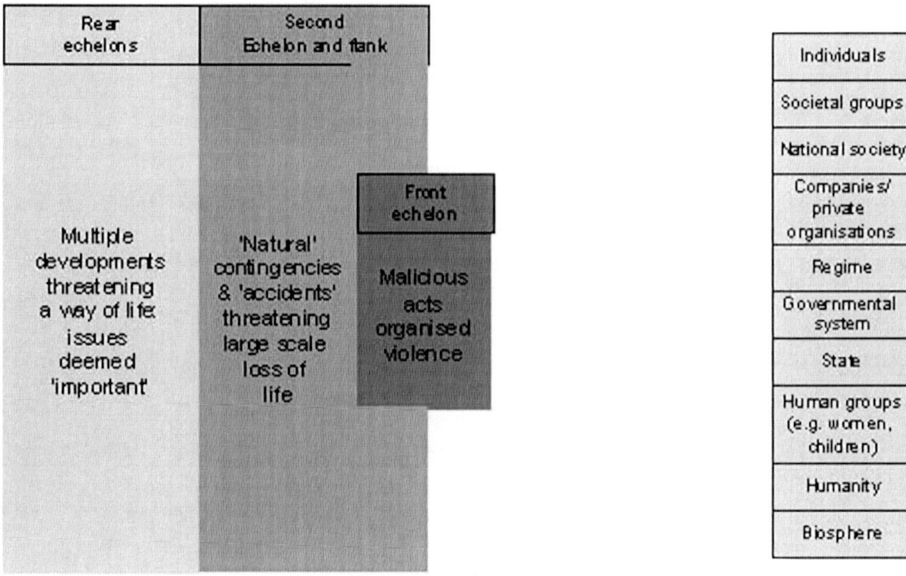

Figure 4.3: The 'What' and 'Who' of security analysis.

notably members of the armed forces and often the police, to expose themselves to physical danger and their families to the possibility of economic disruption. In the developing world it is not uncommon for the regíme to make itself feel secure by making large sections of society feel insecure. A government may feel tensions between maintaining economic security and enhancing environmental security, and so on. In a world of clashing objectives, politics and prioritization are key factors. Arguably, efforts to generate economic growth and thus perhaps enhance state military security are damaging the environmental security of the planet as a whole.

4.5 Management: the 'How' of security

Finally there is a need to address the 'how' of security—how different aspects of security are to be delivered and reassurance provided.

Beginning with the core of security (the domain of hostility, malice and violence), it might be reasonably concluded that some types of malicious act can only be deterred or defeated by forces capable of organized violence, such as the armed forces, the police and similar uniformed services. Certainly when actual violence is threatened, all save pacifist thinkers believe that armed force must normally play a significant part in the response.

But, as is widely recognized, it is not normally enough simply to deal with the m anifestations of actual or potential violent action. There are needs to address the issues, frustrations, and even injustices that lead individuals, groups and even national societies to resort to violence. Domestic and international peace cannot be built by armed forces, police and other 'security sectors' alone, as was demonstrated after 2003 by Western experiences in Iraq. Action across a range of issues, organized and implemented by a range of agencies, is often needed to take place, and in a coordinated manner.

This is now formally recognized in British governmental and military thought and the thinking of many foreign militaries. Britain's 'conflict prevention' expenditure has been allocated by a combination of the Department of International Development, the Foreign and Commonwealth Office and the Ministry of Defence since soon after the Labour Party came into office in 1997. While the degree of actual coordination and cohesion achieved can be queried, the high-level recognition of the limitations of single department approaches cannot. The UK military now recognize that they need to be focused on 'effects' and outcome generation, that many important effects (including stabilization) do not rest solely on military action, and that therefore comprehensive cross-governmental coordinated strategies and action are needed (Development, Concepts and Doctrine Centre webpage). Such action may also bring in non-governmental actors, including foreign aid charities. Experience in Iraq has also led to criticisms in the U.S. of failure to reflect this thinking in its policy towards Iraq, in particular its reluctance to think in detail before the spring of 2003 about what needed to be arranged for Iraq after Saddam Hussein's overthrow, its disbandment of extant Iraqi security forces, and its allocation of overall responsibility for Iraq's political and economic development to the Pentagon.

Similar considerations apply with regard to dealing with many of the important sources of violence at home, especially violent crime, and including terrorism. The police and the judicial system may detect and even deter much crime but they need other help, from educational authorities, from social workers, from national strategies on cultural cohesion and diversity and so on, to reduce the *disposition* to crime and terrorism. Thus, dealing with violent civil and international conflict at the individual, group and social levels needs many players to work together, which is usually not easily achieved even in a single government, which historically has felt it best to allocate a defined range of tasks to a particular ministry.

The complexity of the security domain is clearly emerging. A related consideration is that many of the 'accidents' and 'natural disasters' that threaten the existence of large numbers of people have the same feature as violent conflict, i.e. they need a multi-dimensional, multi-agency approach for their mitigation if they occur. They also often need a multi-agency approach to prevention, for instance minimizing the chances of flood damage needs coordination of those responsible for urban planning and development as well as civil engineers.

These arguments can be summarized by means of a "butterfly diagram" (Figure 4.4, with a number of security challenges and their multiple causes on the left side, and means of dealing with them and the multiple organizations involved on the right. Patterns of activity vary as to whether the immediate focus is on prevention, damage limitation after shock, or rapid recovery.

To populate this model in an illustrative way, consider the challenge of Islamic terrorism in the UK as Security Challenge 1. Its causes are contested but certainly numerous, including the limited cultural and political integration of immigrants, declining parental authority even in immigrant groups, immigration as a reflection of global wealth disparities, the alienation of some indigenous British people from today's society, Western foreign policies towards the Middle East and the Islamic world, the global and local place of Islamic fundamentalism, and the availability of explosives and detonators. Significantly also, most

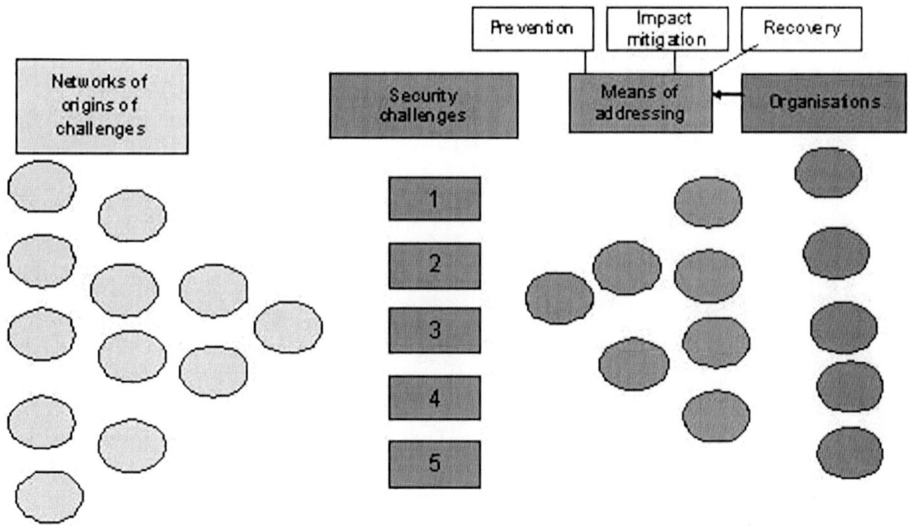

Figure 4.4: The 'How' of security: conspicuous complexity?

of these factors are not static, but are changing either suddenly or steadily, and often not in an overt fashion.

Preventing terrorism should address all the perceived causes, perhaps through education campaigns, enhanced foreign aid, surveillance and monitoring of suspicious groups and individuals, detection devices in potential target areas, strikes against terrorist command centres outside the UK, and the monitoring of sales of potentially dangerous products such as fertilizers, gas canisters, mobile phones and so on. Such measures involve (coordinated) action by a whole range of government departments, some non-governmental organizations including companies, and even foreign governments and international organizations.

Formal British doctrine for addressing terrorism, the CONTEST structure, with its four element structure (Prevent, Pursue, Protect and Prepare) clearly addresses these points (H.M. Government, 2006) and the breadth of the Prevent domain can be recognized by noting the collective finding of seven working groups arranged by the government. They found that "the responsibility for tackling extremism and radicalization in all its forms was the responsibility of society as a whole." The Working Groups are united in the view that whilst their remit was to tackle extremism and radicalisation, most if not all the strands see that the solutions lie in the medium to longer term issue of tackling inequality, discrimination, deprivation and inconsistent Government policy, and in particular foreign policy (Preventing Extremism Together, 2005, p. 3)

Consideration of mitigating the impact of any shock that should occur, and of rapid recovery after that shock, would involve a somewhat different list of means and organizations, with the emergency services having a potentially big role in mitigating impact. It should also never be forgotten that the ability of the population as a whole to deal with danger and damage is a significant element in resilience, the reaction of the people of London and other cities to the Blitz bombing campaign in World War 2 being a clear example. The determinants of resilience at the societal level are likely to be numerous and difficult to grasp with confidence.

4.6 Conclusion

This short piece represents at best an outline sketch of a reality. It has noted that security is an elastic term in academic and political discourse whose meaning cannot be controlled in the wider world.

The argument here is that the core of security should probably lie in matters that threaten the physical existence of human beings on a large scale, especially from malicious acts, but even such a narrow definition throws up a huge territory for the security expert (academic or professional) to understand and requires the integrated action of many players. The reason for this is that the origins of so many things that threaten human life on a large scale are varied, as are the dimensions of the strategies and tactics that are needed to address them effectively. Moreover, the factors associated with security are constantly evolving, generating interrelationships that are highly dynamic.

While the essence of complexity is itself a matter of debate (cf. Part II), there is not doubt that the security domain is marked by three features commonly associated with complexity. First the individual systems and subsystems that generate security and insecurity are marked by continuous change, both in terms of their environment and the internal components of the systems. To quote Stacey, Griffin and Shaw (2000, p. 17) writing about the features of complexity, "the whole is more than the sum of the parts, with both the whole and the parts following iterative, nonlinear laws". Thus 'irregular' patterns of behaviour emerge that "cannot be reduced in any simple way to the parts of which any of them are composed." When we consider the world of security, it is marked by organizations and groups that constantly evolve, and security systems are thus much more akin to biological rather than mechanical systems.

A second aspect of complexity is that small changes should have the potential to generate significant and difficult-to-predict changes. The world of terrorism seems certainly to be marked by this feature, with the personal experiences of an individual perhaps leading to a terrorist act causing massive loss of life.

A second feature of security is that, as illustrated by even brief consideration of how it is to be addressed, it is a very broad, multidisciplinary field. Stephen Johnson (in Prencipe et al., 2005) observed:

> We can define a complex system as a set of humans and technologies united to perform a specific function, which are collectively incomprehensible (in total) to any individual person. Examples are legion but include nuclear power plants, modern jet aircraft and ballistic missiles, computerized command and control systems, etc.

Even narrow conceptions of security challenge the security expert to address a huge number of considerations, and hence the argument is made here that the focus for security studies should be issues that contribute to the threat and use of violence and malicious acts. There should be scope for some issues to be of extraordinary importance and yet not to fall almost automatically under the security umbrella.

4.7 References

Algieri, F., Janning J. and Rumberg, D. (eds), Managing Security in Europe. Gütersloh: Bertelsmann (1996).

Barnet, J., Environmental Security, in: Collins, A. (ed.), Contemporary Security Studies. Oxford: University Press (2007).

Brocklehurst, H., Children and War, in: Collins, A. (ed.), Contemporary Security Studies. Oxford: University Press (2007).

Buzan, B., People, States and Fear: the National Security Problem in International Relations. Brighton: Wheatsheaf (1983).

Collins, A. (ed.), Contemporary Security Studies. Oxford: University Press (2007).

Consterdine, P., The Modern Bodyguard. Leeds: Protection Publications (1995).

Coonan, C., The Internet Balance of Power'. The Independent, 6 September 2007.

Croft, S., British Security Policy. London: Harper Collins (1991).

Department for International Development, Foreign and Commonwealth office and Ministry of Defence. Security Sector Reform Policy Brief. London: DfID (2003).

Dent, C., Environmental Security, in: Collins, A. (ed.), Contemporary Security Studies. Oxford: University Press (2007).

Development, Concepts and Doctrine Centre of the UK Ministry of Defence.[2]

Furedi, F., Culture of Fear. London: Continuum (2002).

H.M. Government. Countering International Terrorism: the UK's Strategy. London, July 2006.[3]

Homer-Dixon, T., Environment, Scarcity and Violence. Princeton: University Press (1999).

Hope, C. and Blair, D., Chinese hackers "hit 10 Whitehall departments". Daily Telegraph, 6 September 2006.

Hough, P., Understanding Global Security. London: Routledge (2004).

Human Security Centre, Human Security Briefing 2006, University of British Columbia, 2006

Kahn, J. and Yardley, J. 'Choking on growth, as China roars, pollution reaches deadly extremes'. New York Times, 26 August 2007.

Kennedy-Pipe, C., Women and Conflict, in: Collins, A. (ed.), Contemporary Security Studies. Oxford: University Press (2007).

Klare, M.T. and Thomas, D.C., World Security: Challenges for a New Century, chs 12–18. New York: St Martins Press (1994).

Kolodziej, E.A., Security and International Relations. Cambridge: University Press (2005).

McGregor, R. and Sevastopulo, D., China denies hacking into Pentagon computer network. Financial Times, 5 September 2007.

Mandel., P., The Changing Face of National Security: A Conceptual Analysis. London: Greenwood (1989).

Perrow, C., The Next Catastrophe. Princeton: University Press (2004).

---

[2] `http://www.mod.uk/DefenceInternet/AboutDefence/WhatWeDo/DoctrineOperationsandDiplomacy/DCDC/DcdcMultiagencyOperationsTeamProfile.htm`

[3] `http://www.intelligence.gov.uk/upload/assets/www.intelligence.gov.uk/countering.pdf`

Pilkington, E., and Johnson, B., China flexes muscles of its 'informationalised' army. The Guardian, 5 September 2005.

Prencipe, A., Davies, A. and Hobday, M. (eds), The Business of Systems Integration. Oxford: University Press (2003).

Preventing Extremism Together. Working Groups Report, p. 3. August–October 2005.[4]

Stacey, R.D., Griffin, D. and Shaw, P. Complexity and Management. London: Routledge (2000).

Stevenson, C.A., SecDef: the Nearly Impossible Job of Secretary of Defense. Washington, D.C.: Potomac Books (2007).

Thompson, D., China flexes its limited muscles. Financial Times, 5 September 2007.

Ward, A., Sevastopulo, D. and Fidler, S., U.S. concedes danger of cyber attack. Financial Times, 6 September 2007.

Wyllie, J., European Security in the Nuclear Age. Oxford: Blackwell (1986).

[4] `http://www.communities.gov.uk/documents/communities/pdf/152164`, accessed 11 September 2007.

Part II

What is Complexity?

Complexity and Security
J.J. Ramsden and P.J. Kervalishvili (Eds.)
IOS Press, 2008

Chapter 5

An introduction to complexity

Jeremy J. Ramsden

Cranfield University, Bedfordshire, MK43 0AL, UK

Complexity is indubitably all around us. Even what might appear to be a very simple, static landscape such as a patch of desert contains vast numbers of entities (grains of sand) of different shapes, colours and sizes, that interact with each other with a variety of forces. And as soon as life is present, we seem to require a higher stage for complexity in order to capture the richness of even a single living cell: Gerhard Michal's well-known diagram "Biochemical Pathways", frequently found on the walls of biochemical laboratories, is an intricate diagram covering several square metres with a dense network of nodes and vertices. Even the most primitive cells contain several hundred different types of proteins, each of which can interact to some degree at least with each of the others. Multicellular organisms increase their complexity by constructing internal organs such as the kidney,[1] and the brain; the most complex exemplar being the human brain, containing of the order of 10^{10} neurons, each of which may be connected to 10^3–10^4 others. And when we consider a society of human beings, thanks to the Internet connected to such a degree that in principle every one of the 6×10^9 inhabitants of the planet can communicate with every other one, we appear to be confronted with a system that possibly transcends human understanding in its complexity.

Appreciation of the complexity of human society and its institutions is by no means new. For example,[2]

> In place of a cumbrous engine controlled by a single will, an army had become a vast living mass instinct with vivid life throughout

[1] A good source for appreciating the immense complexity of the kidney is S.R. Thomas's paper "Modelling and simulation of the kidney" (J. Biol. Phys. Chem. (2005) 70–83).

[2] From 'Moltke', in: Lord Sydenham of Combe, Studies of an Imperialist, p. 58. London: Chapman and Hall (1928). The essay was first published in the Edinburgh Review in October 1891.

> its whole being, an instrument of extreme complexity, flexible to the last degree, a loose aggregate of men or a weapon of terrific power according to the perfection of its parts and the spirit and intelligence of its thousands of subordinate leaders.

By this, Sydenham rightly implies that a modern army is more complex than its classical predecessors, just as a modern multicellular organism is more complex than its amoeboid precursor. Perhaps the military sphere is the first one in which complexity had to be explicitly considered—with, of course, the powerful motivation that while mastery of complexity could lead to victory, failure to master it could lead to defeat and death.

Armies of course evolved mechanisms for dealing with their complexity without having to ponder on its meaning, and without that presently rather amorphous and ill-defined body of knowledge known as "complexity science". One could draw a parallel with the early history of the steam engine, developed without the science that we now call thermodynamics, but once that science had become established—drawing heavily on the results from practical experience—it played a most important role in further improving that type of engine. In the same way, the motivation for studying complexity (i.e. the complex systems all around us) and for attempting to construct a theory of complexity is that it may as a result be possible to improve the management of these complex systems.

The question then immediately arises, "What is a complex system?" The phrase appears at first sight to be a tautology, since a system is defined as an assemblage of parts that cannot be meaningfully studied independently of the whole, and it is precisely because of the existence of such interactions that complexity arises. Every system should therefore in principle be complex, and if it is not, it is not a system. But still, we might wish to know, when confronted with what is apparently a system, whether it really is, and whether it is therefore complex. Typical characteristics making something into a system would include a large number of components and different types of components, and interactions between them, which might themselves be of different types. Hence liquid argon would not qualify as a (complex) system; although its components (the individual argon atoms) are numerous, they are all the same and their interaction with each other is minimal (scarcely going beyond the constraint that two atoms cannot be in the same place). At present complex systems are chiefly identified ostensively—we know (or think we do) what complexity is when we see it. Very characteristic of the kinds of human complex systems that are relevant to this Workshop are groups of warring tribes possibly fighting an enemy loosely perceived as being common to all, as one sees in Darfur today and which was frequently encountered throughout the British empire. Note however that such systems not only have numerous, different and interacting components, but the components themselves are evolving (their number, type or interactions, or all three of them) in an open-ended, indeterminate fashion. Whereas complex *physical* systems typically show extreme sensitivity to starting conditions (the "butterfly effect"), and are rich in nonlinear interactions and delayed feedbacks, frequently resulting in chaotic behaviour,[3] the evolving nature of the system is rather typical of *biological and social* systems. Therein lies the essential distinction between the system—for which the well-developed "general systems theory" exists (it will be outlined later on in this chapter)—

[3] A.B. Pippard, Response and Stability. Cambridge: University Press (1985).

and the complex system, the latter containing evolving elements and therefore being indeterminate. The presently rather loosely defined body of knowledge known as "complexity science" is aimed at providing a theoretical framework for handling complex systems comparable to the framework provided by general systems theory for systems.

Identifying a system as complex is valuable because it immediately indicates that its behaviour will transcend the framework of general systems theory. The identification should then give some indication of the degree of effort that may need to be expended on understanding the system, e.g. by constructing a model. Simple models based on linear interactions and equilibria are almost certainly not going to work if the system is truly complex, i.e. they will not succeed in capturing the essential behaviour of the system. There are of course situations in which a simple deterministic description is sufficient, and because such a description is much easier and cheaper to realize, there is no point in going beyond it if it is not actually needed. In a sense complexity science is at present privative, serving above all to define what is inadequate. Models will need to be evolvable and open-ended (indeterminate), but there is no general recipe for constructing them. Measuring the degree of complexity of a system—one approach to which will be explored in Chapter 7—should allow the resources necessary to properly understand and deal with it to be more carefully determined than otherwise.

In many cases, where the system is human-made, we presumably have some degree of control over its degree of complexity, and it could be very important to determine the optimal level of complexity at which to aim, in order to (for example) maximize security. Is there a direct relationship between complexity and security? This is one of the core questions that this Workshop seeks to answer. In the military sphere, complexity would generally appear to be advantageous. It is clearly associated with variety: the greater the variety, the greater the power and flexibility of effective responses to danger, implying more safety and more security (and less of both for the enemy). For example, highly sophisticated, complex technologies such as the submarine induce a certain insecurity for the enemy because of the mysterious, unknown aspects of their presence. On the other hand, complex weapons may be more prone to breakdown (there are more things to go wrong), and they are more difficult to test reliably, since the sheer vastness of their parameter space usually makes it impracticable to cover all possibilities (cf. Part IV). Although in this particular sense complexity would appear to engender fragility, that weakness is, at root, only due to our current inability to truly master complexity. Complexity emerges as an inevitable consequence of the growth of variety required to cope with a richer and more varied environment. Moreover, exaptations may vastly increase the scope of the initial diversification. A very good illustration of this course of events (as has been analysed by V.V. Maximov[4]) is the growth in brain power of predatory marine organisms living in very shallow waters, which needed to track moving prey against a background of moving shadows generated by the sun shining through wind-generated ripples in the water. The necessary growth of information processing sophistication (needed for actual survival) was likely the key to the subsequent emergence of colour vision (as an exaptation of that

[4]Environmental factors which may have led to the appearance of colour vision. Phil. Trans. R. Soc. Lond. B 355 (2000) 1239–1242.

expanded brain sophistication). To the modern human being, who has inherited this enhancement, colour vision is an extremely useful aid to survival in a complex world—and yet at the same time new aberrations of sight are peculiar to that enhanced sophistication. Nevertheless the very fact of its persistence suggests that there is a net gain to its possessor.

In general (not least from observing the continuing existence of bacteria, still the most numerous form of life) we might suppose that we could survive at a simpler level by discarding richness and sophistication. But is it really possible to go back? Once we (as individuals, and as a species) have tasted, seen and moved into the new world of complex possibilities, the transition becomes irremeable. Therefore we as humans *must* learn to cope with and master complexity.

What is complexity?

The elements of complexity are clear enough to state: a large number of constituent elements, a large variety of elements, and a large interconnectivity. We might then define a "simple system" as a system that can be meaningfully (in the sense that useful predictions of its behaviour can then be made) decomposed into its constituent parts (and is therefore not a true system at all); a "system" is one in which such decomposition is not possible for all practical purposes; and a "complex system" is one in which the elements and their interconnexions are evolving, and which therefore transcends the framework of general systems theory.[5]

In other words, while all systems are in principle complex, practically speaking it may be possible to model and predict the system sufficiently accurately for the purpose in hand by neglecting the complexity, i.e. neglecting one or more of the number, variety and interconnectedness of the constituent parts.[6] In fact, science has been doing this even before it was called science. Newton's model of the solar system reduces the planets to massy points without extension, differing only in their masses and the positions of their orbits; all other details such as mineral composition (and the presence of life) are neglected—and this model is highly successful at predicting the positions of the planets sufficiently accurately for a huge number of applications.

That is why *systemicity* is a prerequisite of complexity: we can have quasi-simple systems that are not complex (according to the heuristic criterion introduced above, i.e. practically speaking the complexity, or a sufficient part of it, can be neglected for the purposes of prediction), which leads to the following definition of complexity: *the attribute of a system that for practical purposes cannot be approximated quasi-simply* (which means treating its elements largely independently).

[5] It may occur to the reader that there are certain cellular automata (see S. Wolfram, Cellular automata as models of complexity. Nature (Lond.) 311 (1984) 419–424) whose rules (encompassing the elements and interconnexions) are fixed, but whose results are evolving and unpredictable (class IV automata in Wolfram's scheme). This suggests that cellular automata might be a fruitful direction to take in the search for a general theoretical framework for complex systems.

[6] At the risk of being pedantic, it should again be mentioned that the concept of 'part' or 'element' is really alien to the notion of system, since by definition the system cannot be decomposed into such constituents. All we can concede is that the parts are the *precursors* of the system, but as soon as they are incorporated, they lose their full individuality.

This definition suggests two important corollaries: one is that the boundary between complex and not-complex is not particularly sharp; there is no critical point to be traversed; it is not even a crossover, but rather a gradual transition. The other corollary is that complexity is relative to the context. For example, in planning an interplanetary voyage, a more detailed description of the planets is necessary than that of Newton's original model, even though the fundamental laws describing their motions are unchanged.

Other practical definitions along these lines suggest themselves. Objects often acquire complexity by virtue of their internal structure. For the viewpoint of an interior designer, it may be sufficient to consider a telephone as a block of basalt or ebony with a particular shape. An object that is complex when its full functionality is taken into account is therefore reduced to something that is simple indeed in the particular context of interior design.[7] An object might be considered to be simple if its function is apparent from its form, although in order to make this a proper definition the prior knowledge possessed by the viewer would need to be carefully specified.

5.1 The relation of complexity to systems theory

According to general systems theory, which in many ways can be considered as a precursor to complexity science, systems can be characterized by

- number
- type
- relationships

of or between their parts. Let us recall R.L. Ackoff's four criteria for deciding whether a collection of objects constitutes a system:

1. One can talk meaningfully of the behaviour of the whole of which they are the only parts.
2. The behaviour of each part can affect the behaviour of the whole.
3. The way each part behaves and the way its behaviour affects the whole depends on the behaviour of at least one other part.
4. No matter how one subgroups the parts, the behaviour of each subgroup will affect the whole and depends on the behaviour of at least one other subgroup.

[7]To the geologist or botanist, basalt or ebony might be anything but simple—but this is again a different context.

The temporal evolution of the system is given by the equation

$$\begin{aligned}\frac{dg_1}{dt} &= \mathcal{G}_1(g_1, g_2, \ldots, g_n)\\ \frac{dg_2}{dt} &= \mathcal{G}_2(g_1, g_2, \ldots, g_n)\\ &\vdots\\ \frac{dg_n}{dt} &= \mathcal{G}_n(g_1, g_2, \ldots, g_n)\end{aligned} \tag{5.1}$$

where the functions $\mathcal{G}$ include terms proportional to $g_1, g_1^2, g_1^3, \ldots, g_1g_2, g_1g_2g_3, \ldots$ etc. In practice, many of the coefficients of these terms will be close or equal to zero.

If we only consider one variable, we have

$$\frac{dg_1}{dt} = \mathcal{G}_1(g_1)\ , \tag{5.2}$$

expansion of which gives a polynomial

$$\frac{dg_1}{dt} = rg_1 - \frac{r}{K}g_1^2 + \cdots \tag{5.3}$$

where $r > 0$ and $K > 0$ are constants. Retaining terms up to g_1 gives simple exponential growth,

$$g_1(t) = g_1(0)e^{rt}\ , \tag{5.4}$$

where $g_1(0)$ is the quantity of g_1 at $t = 0$. Retaining terms up to g_1^2 gives

$$g_1(t) = K/[1 + e^{-r(t-m)}]\ , \tag{5.5}$$

the so-called logistic equation, which is sigmoidal with a unique point of inflexion (at $t = m, g_1 = K/2$), at which the tangent to the curve is r, and with asymptotes $g_1 = 0$ and $g_1 = K$. In ecology, r is called the growth rate and K is called the carrying capacity.

Consider now two objects,

$$\left.\begin{aligned}dg_1/dt &= a_{11}g_1 + a_{12}g_2 + a_{111}g_1^2 + \cdots\\ dg_2/dt &= a_{21}g_1 + a_{22}g_2 + a_{211}g_1^2 + \cdots\end{aligned}\right\} \tag{5.6}$$

in which the functions $\mathcal{G}$ are now given explicitly in terms of their coefficients a (a_{11}, for example, gives the time in which an isolated G_1 returns to equilibrium after a perturbation). The solution is

$$\left.\begin{aligned}g_1(t) &= g_1^* - h_{11}e^{\lambda_1 t} - h_{12}e^{\lambda_2 t} - h_{111}e^{2\lambda_1 t} - \cdots\\ g_2(t) &= g_2^* - h_{21}e^{\lambda_1 t} - h_{22}e^{\lambda_2 t} - h_{211}e^{2\lambda_1 t} - \cdots\end{aligned}\right\} \tag{5.7}$$

where the starred quantities are the stationary values, obtained by setting $dg_1/dt = dg_2/dt = 0$, and the λs are the roots of the characteristic equation, which is (ignoring all but the first two terms of the right hand side of equation 5.6)

$$\begin{vmatrix} a_{11} - \lambda & a_{12} \\ a_{21} & a_{11} - \lambda \end{vmatrix} = 0\ . \tag{5.8}$$

Depending on the values of the a coefficients, the phase diagram (i.e. a plot of g_1 vs g_2) will tend to a point (all λ are negative), or a limit cycle (the λ are imaginary, hence periodic terms appear), or there is no stationary state (the λ are positive). Regarding the last case, it should be noted that however large the system, a single positive λ will make one of the terms in (5.7) grow exponentially and hence rapidly dominate all the other terms.

Although this approach can readily be generalized to any number of variables, the equations can no longer be solved analytically; indeed the difficulties become forbidding. Hence one must turn to statistical properties of the system. Equation (5.6) can be written compactly as

$$\dot{\mathbf{g}} = A\mathbf{g} \tag{5.9}$$

where $\mathbf{g}$ is the vector $(g_1, g_2, \ldots)$, $\dot{\mathbf{g}}$ its time differential, and A the matrix of the coefficients a_{11}, a_{12} etc. connecting the elements of the vector. The binary connectivity C_2 of A is defined as the proportion of nonzero coefficients.[8] In order to decide whether the system is stable or unstable, we merely need to ascertain that none of the roots of the characteristic equation are positive, for which the Routh-Hurwitz criterion can be used without having to solve the equation. Gardner and Ashby determined the dependence of the probability of stability on C_2 by distributing nonzero coefficients at random in the matrix A for various values of the number of variables n.[9] They found a sharp transition between stability and instability: for $C < 0.13$, a system will almost certainly be stable; and for $C > 0.13$, almost certainly unstable. For very small n the transition was rather gradual, viz. for $n = 7$ the probability of stability is 0.5 at $C_2 \approx 0.3$, and for $n = 7$, at $C_2 \approx 0.7$.

As will be emphasized in Chapter 6, the effort of solving equation (5.1) is hardly warranted, due to its lack of relevance to most real systems (especially those being considered in this book), that are characterized by open-endedness and indeterminacy.

As Gardner and Ashby have elegantly showed,[9] it is not even necessary to solve this equation in order to assess the likelihood of a system being stable: instability means that at least one of the roots of its characteristic equation are positive (meaning that one of the elements of the system will start to grow exponentially), and the presence of a positive root can be determined via the Routh-Hurwitz criterion. Numerical investigations have shown that if more than 13% of the elements of the binary interaction matrix A are nonzero, then the system constituted by those elements will almost certainly be unstable. For any large system, the transition from stability to instability is rather abrupt.

5.2 Frustration

When a large number of interactions are present, while local regions may be able to organize themselves into an optimal state while more or less isolated from their neighbours (in a purely physical system, this would probably mean that they reach an energy minimum), when interactions with adjacent local agents

[8]The ternary connectivity takes into account connexions between three elements, i.e. it contains coefficients like a_{123}, etc.

[9]M.R. Gardner and W.R. Ashby, Connectance of large dynamic (cybernetic) systems: critical values for stability. Nature 228 (1970) 784.

are allowed to operate, it becomes impossible to optimize those interactions while retaining the optimal internal arrangement.

The "spin glass" is a real physical system showing this kind of behaviour.[10] Suppose we have an initial sample of N sites subdivided into m smaller blocks, ℓ atoms to a side, where ℓ is defined by $N/m = \ell^d$, d being the dimensionality of the system. Two neighbouring blocks will touch along the surface with area $A \approx \ell^{d-1}$, and this is the number of bonds that have to be cut to make the subdivision. For the spins S in two neighbouring blocks B_1 and B_2, with sites i and bonds ij, the Hamiltonian is

$$H = \sum_{i}^{N} \sum_{jnn(i)} J_{ij} S_i S_j \ . \tag{5.10}$$

If the interaction J_{ij} is now replaced by i in B_1 and j in B_2, the total energy (without allowing any spins to relax) is

$$E = \sum_{i \text{ in } B_1} \sum_{j \text{ in } B_2} J_{ij} S_i^0 S_j^0 \ , \tag{5.11}$$

the superscript 0 denoting the values set in the lowest energy states in the separated blocks. As A becomes large, $E \propto AJ$; the ordering in two blocks will be either maximally compatible or incompatible with each other, and the variance $\bar{E}^2 \sim A^2$. Hence $\lim_{A\to\infty} \bar{E}^2/A$ cannot approach zero (as the system would like to do) and is typically around unity, i.e. $\bar{E} \propto \sqrt{A}$. Note that the energy coupling the structure together is therefore not an intensive variable, implying that the transition to order must be dynamical rather than structural.[11] The result is typically a large number of globally quasi-optimal states: there is no global optimum to which all can strive. The state is described as 'frustrated' (because the desire of the system to reach its optimum cannot be achieved). This is a feature of every human society, and suggests that Hobbes' social contract can only be a provisional concept, because it implies that there is a uniquely optimal state for society.

Examples of "social frustration" are ubiquitous. Motoring permits individual freedom, but is also rather dangerous, hence motorists are allowed, but motor-cars are subjected to speed limits and a host of other regulations. The value of many consumer goods is enhanced in the eyes of their purchasers if they are durable; on the other hand few companies could survive if all their products last a lifetime. Degradation leading to the need for replacement of goods is moreover crucial for the ongoing development of design. Pharmaceutical companies develop drugs to diagnose and treat disease; if the treatment is successful then the need for further diagnosis (and treatment) is eliminated; hence a permanent cure might not be in the best interests of the industry. The British Broadcasting Corporation (BBC) is funded by a television licence fee in order to allow it to broadcast programmes that might not have mass appeal, yet because they perceive that if too few people watch their programmes, the

[10] A spin glass is an alloy of a spin-possessing atom in a matrix of spinless atoms, such as manganese in copper (S.F. Edwards and P.W. Anderson, Theory of spin glasses. J. Phys. F 5 (1975) 965–974).

[11] See G. Toulouse, Theory of the frustration effect in spin glasses: I. Comm. Phys. 2 (1977) 115–119; P.W. Anderson, The concept of frustration in spin glasses. J. Less-common Metals 62 (1978) 291–294.

government might withdraw the television licence income, they feel obliged to imitate the commercial television stations by broadcasting programmes with mass appeal (which, unfortunately, seems to mean programmes with minimal intellectual content and artistic value).

Hence while path dependence (nonergodicity) of complex processes means that their outcomes are unpredictable, frustration (wherever present) ensures not only a multiplicity of paths, but also a multiplicity of endpoints.

5.3 Regulation

Regulation refers to the ability of the system to maintain itself in a viable state. Figure 5.1 shows the classical scheme.[12] The type of regulation (or survival strategy) depends on which information channels are open. Information is in any case passing from D to T. Only in the case of some very cumbersome systems, such as the tortoise, is this channel virtually blocked. The most perfect type of protection from danger is achieved if the regulator R can independently receive information about the dangers, and direct those parts of the system under its control (T) to act accordingly. A human being operates on this principle: the eyes and other senses can directly perceive what is happening, and after processing these sensory inputs, the brain directs the muscles to fight or run. Primitive systems, such as a thermostatted waterbath, operate through "regulation by error": some danger, which for the thermostat might be a cold draught, or someone bringing a flask of cold water to immerse in the bath, firstly causes the relevant essential variable, the temperature, to sink, which is then noticed by the regulator, which then switches on the heater accordingly. Much unconscious regulation in living systems doubtless operates on this principle.

Two aspects are particularly important regarding the ability of the system to ensure its survival. One is that the channels through which information flows have sufficient capacity (typically measured in terms of bits of information per unit time). The other is that the system (comprising T and R in 5.1) possesses a sufficiently varied repertoire to respond appropriately.[13] This is formalized as Ashby's *Law of Requisite Variety.*[14] Examples will be easy for the reader to construct. A dueller (with pistols) who could only fire straight ahead and horizontally or after raising his arm by 15° would have a poor chance of survival, etc.

The thermostatted waterbath, although clearly a system, might qualify as "simple", since evidently it could be described and modelled rather straightforwardly. Even though the same basic scheme (Figure 5.1) could describe a living creature, complexification occurs because of the huge repertoire of possible responses (that have evidently arisen as a generalized response to the huge variety of possible dangers). This underlines the fact that *variety is crucial to complexity.* Furthermore, even a 'simple' bacterium (which is, of course, not at all simple in the sense that we have defined) has a potentially open-ended response to environmental danger.[15] Living organisms seem to have sufficient plasticity, either

[12] W.R. Ashby, An Introduction to Cybernetics. London: Chapman & Hall (1956).

[13] Appropriateness is defined more formally in §5.4 below.

[14] W.R. Ashby, Requisite variety and its implications for the control of complex systems. Cybernetica 1 (1958) 83–99.

[15] This has been shown quantitatively by J. Vohradský and J.J. Ramsden, Genome resource

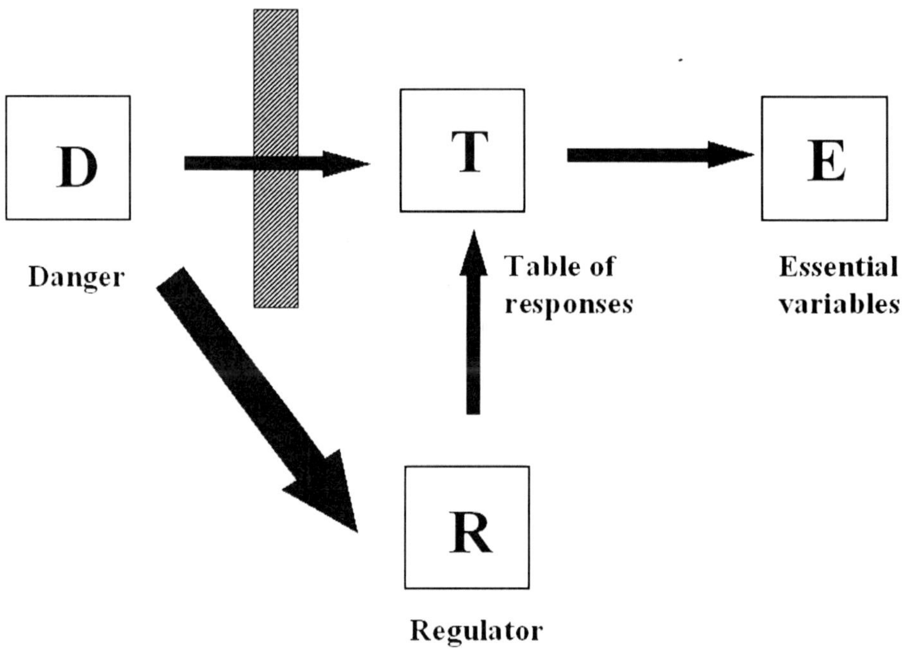

Figure 5.1: Diagram showing information flows (arrows) in a general responsive system. Environmental dangers impacting on the system are denoted by D. The machinery of the system, except that part of it expressly devoted to regulation, is denoted by T, because it can be written as a table of responses to stimuli. Those responses can be controlled by the regulator R. The output E, denoting essential variables, is what keeps the system viable. The lines joining these components represent channels along which information can flow.

genetic or epigenetic, in order to extend their responses into hitherto unexplored regions. Among bacteria, mutations and exchange of genetic material appears to be an important mechanism; in eucaryotes, alternative ways of assembling the RNA precursors to proteins from the DNA could be significant, and there is of course the vast reservoir of unexpressed DNA whose precise function has not yet been unambiguously demonstrated.[16] Therefore, not only do living systems rank as complex through their vast repertoires, but also because the repertoire can be and is being constantly extended, both throughout the lifetime of an individual member of the species, for the species as a whole.

5.4 Directive correlation

Sommerhoff ascribes a key position to the concept of adaptation.[17] The concept is rooted in the appropriateness of the responses of a (living) system to the needs of the organism, brought about by the exigencies of its environment, and to the effectiveness of those responses. It is almost self-evident that a response must be to something (an event that occurs in the environment). Appropriateness implies an effective contribution to some future state (the 'goal' of the response). There are four primary elements in the overall process:

1. The environmental circumstances (events) that evoke the response, called the coenetic variable.
2. The response itself.
3. The environmental circumstances during the execution of the response, relevant to its success.
4. The outcome ('goal') of the response, called the focal condition.

The spatio-temporal relations between these primary elements are sketched out in Figure 5.2.

It is important to note that 'appropriateness' transcends 'efficacy' insofar as the latter covers a particular set of circumstances, whereas the former implies that had the circumstances been somewhat different, a related response would have been equally effective. In other words, one imagines a set of alternative environmental circumstances $E'_{t_1}, E''_{t_1}, E'''_{t_1}, \ldots$, and a correlated set of effective responses $R'_{t_1}, R''_{t_1}, R'''_{t_1}, \ldots$, the members of the two sets standing in a one-to-one correspondence such that corresponding pairs will lead to achievement of the goal G_{t_2}. Adaptation (adaptedness) transcends appropriateness in that it implies the inerrant selection of an appropriate response corresponding to essentially any environmental circumstances that may be encountered by the system, for a fixed focal condition (which, for a living organism, might be "survival").

We can state that the response R is *effective* with respect to the environmental circumstances E and the focal condition G if the joint occurrence of R and E causes the subsequent occurrence of G, *appropriate* if E is a member of an ensemble of Es, and R is the corresponding member of an ensemble of Rs,

utilization during procaryotic development. FASEB J. 15 (2001) 2054–2056.

[16]For a concise introduction these topics, see J.J. Ramsden, Bioinformatics. Dordrecht: Kluwer (2004).

[17]G. Sommerhoff, Analytical Biology. Clarendon: Oxford (1950).

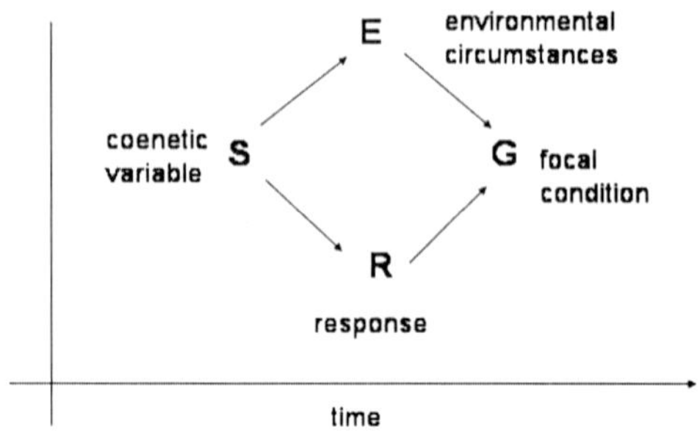

Figure 5.2: Sketch of directive correlation (after Sommerhoff).

each of which is effective in bringing about G if and only if it occurs in conjunction with the corresponding member of the ensemble of Es, and R is *adapted* to G if the system "causally determining R is objectively so conditioned that if certain changed initial circumstances [S] had caused the occurrence of any alternative member of the ensemble of Es it would in each case also have caused the occurrence of the corresponding member of the correlated ensemble of R" (appropriate responses).[18]

The coenetic variable S therefore appears as a common causal determinant of E and R. This constitutes the *directive correlation* of R and E, which underlies adaptation, which therefore appears as a tetradic relationship.

5.5 Delayed feedback

Feedback is typically pervasive in a system. Even in such a simple example as the thermostatted waterbath, however, if there is any feedback delay—essentially in the transmission of information along a channel—chaotic, unpredictable behaviour is likely to ensue.[19] The implications of this are immense, yet surprisingly they do not seem to be as widely appreciated as that immensity would lead one to suppose.[20] As Pippard points out, "The advantage of strong negative feedback in hastening the approach to equilibrium is soon lost once delay is introduced. Long before β (defined by the equation for $z(t)$, $\tau \mathrm{d}z(t)/\mathrm{d}t + z(t) + \beta[z(t - t_0)]$, where τ is a decay time constant and t_0 is the delay) has reached its critical value the decay of oscillation has become slow enough to offset the initial quick response, and the longer the delay, the less the tolerable feedback."

[18] Sommerhoff, loc. cit.

[19] A.B. Pippard, Response and Stability. Cambridge: University Press (1985).

[20] Clausewitz recognized such delay as one of the major difficulties in conducting a war.

5.6 Implications of complexity

Sensitive dependence on initial conditions is well known in dynamical systems theory (it explains the difficulty of balancing a pencil on its point so perfectly that it remains vertical). Bifurcations (small perturbations leading to qualitative changes in the model's state) is also well known from chaos theory. In what ways does complexity go beyond these results? One is the demonstration of the necessary effect of innovations (mutations), in enabling a new behaviour to invade an existing system,[21] The take-home message of the following paper of this Part (Chapter 6) that deals with modelling complex systems is that complexity science is essentially the realization that not only the extreme inherent nonlinearities and the extreme sensitivity to initial conditions familiar from general systems theory, but also the fact that not only the coefficients of the variables, but even the equations themselves may be constantly changing, all need to be taken into account in modelling the system under consideration, which may be the closest that one can get to understanding it, and that underlying all this is the ever present possibility of novelty (innovation) appearing (emerging?), which ensures the ultimate open-endedness of the complex system.

One corollary of this realization is that one may in the end have to be satisfied with mastery of the system in the absence of precise specifications of all parts of it. Some remarks of Szentágothai may be relevant here: "Whenever he is looking at any piece of neural tissue, the investigator becomes immediately confronted with the choice between two conflicting issues: the question of how much of the intricate wiring of the neuropil is strictly determined by some genetically prescribed blueprint, and how much freedom is left to chance within some framework of statistical probabilities or some secondary mechanisms of trial and error, or selecting connexions according to necessities or the individual history of the animal. Even on brief reflexion one has to arrive at the conclusion that the case may not rest on either extreme."[22] In the case of the brain, that most marvellously complex system, it appears that the individual neurons and their connexions cannot be specified genetically—the required amount of information far exceeds what can be stored in our genome; hence at most an algorithm for selecting the best connexions could be provided by our genes, and it may even be "merely" an algorithm for constructing the algorithm that makes the selection. Mastery of complexity may ultimately rest on the ability to generate it in this fashion, without insisting on the precise specifications of every detail.[23]

The attainment of extreme variety is known as vastification. Although on its own vastification does not automatically imply complexity, the proliferation of individual objects is generally accompanied by the proliferation of interconnexions between them, much as the proliferation of cells (as in, for example, cancer) is swiftly followed by angiogenesis. Certainly the vastification of the brain during the evolution of higher life forms has led to an enormous increase in the complexity of the organ.

Complexification, in one way or another, can have some quite direct and

---

[21] P.M. Allen, Evolution, population dynamics, and stability. Proc. Natl Acad. Sci. USA 73 (1976) 665–668).

[22] J. Szentágothai, Specificity versus (quasi-)randomness in cortical connectivity. In: Architectonics of the Cerebral Cortex. M.A.B. Brazier and H. Petsche, eds. New York: Raven Press (1978).

[23] The expression "complexity ceiling" is used to denote the highest complexity that can be mastered with explicit specification of every detail.

straightforward impacts on security. The more possible combinations a lock has, the more secure it is. Similar considerations apply to encryptation technology. The newest ten dollar Hong Kong bank notes are composites of various materials including a transparent window, which certainly would appear to make them very difficult to forge.

5.6.1 Emergence

Emergence is often considered to be one of the hallmarks of complexity. The term is derived from the sudden appearance of a butterfly from the chrysalis. Emergent phenomena are those that could not have been predicted from a lower, "more fundamental" level of theory. Thus, much of condensed matter physics (e.g. superconductivity) cannot be predicted from elementary particle theory (even if those theories were perfect); chemistry is not just applied physics; biology is not just applied chemistry; and so forth.[24] Emergence demonstrates that reductionism, in the sense of a programme to reduce all phenomena to laws at the level of the elementary particle, can never be successful. In biology, we cannot predict an organism from its DNA sequence.

Emergence is sometimes described as "the whole is more than the sum of its parts". It denotes the appearance of some quality, not inherent in the parts individually, when they are brought together. Thus, two tasteless gases, oxygen and hydrogen, and a black tasteless solid, carbon, can be brought together to make a solid that is colourless and sweet, namely sugar; two gases, hydrogen chloride and ammonia, can be brought together to make a solid, ammonium chloride;[25] two electrically insulating substances, pure water and pure salt, can be mixed to make a conducting solution. Other examples abound. You would not be able to reconstruct this text if I simply gave you a list of the numbers of its different letters and punctuation marks. If you are reading it on paper, you would not be able to reconstruct it from a mass of cellulose fibres and a bottle of ink. Seen in this light, emergence appears as something rather trivial and obvious. There is no "theory of emergence". At best, one must simply be aware of it. Given its evident ubiquity, it is surprising that anyone could have ever believed that all these phenomena could be reduced to laws of the elementary particle level. The notion of the "whole being greater than the sum of its parts" is however somewhat misleading. Why summation? Evidently the way the parts are combined is typically far more complex than summation. In some cases, it might just be multiplication, giving us the aphorism "the whole is greater than the product of its parts", but in general we need to capture the notion that the structure of an object is essential to its essence: "things depend on how their parts are arranged". With very complex objects, even this might be insufficient: for example the structural difference between an animal dead and one living might be infinitesimal just after its death; in this case it is the sequence of successive structures that provides the clue to the true nature of its state.

In the context of the topic of this book, one of the most pertinent exemplars of emergence is Hobbes' Leviathan—the emergence of a social contract from a set of individual citizens.

[24] P.W. Anderson, More is different. Science 177 (1972) 393–396.

[25] These examples were constructed by W.R. Ashby.

5.6.2 Innovation

It has been a moot point whether the appearance of novelty deserves to be called emergent. In some way, of course, the imago can be predicted from the pupa—we simply do not understand the incredible complexity of the system sufficiently well, but in the sense that the pupa always results in the imago (except in the case of some usually understood aberration), the laws that govern the formation of the imago are immanent not only within the pupa but also within the caterpillar. The fact that the DNA of a creature is necessary but insufficient for us to predict its structure, much less its behaviour, is again a reflexion of our very limited understanding of the system (including any necessary parts of the environment of the proto-creature bearing the DNA). Novelty, however, seems to be in a different category. It implies the "leap of faith" involved in inductive reasoning. To be sure, emergence lies beyond deduction, but innovation lies beyond emergence.[26]

A microcosm of this debate is contained within the bacterial world—whether bacteria can 'direct' their own mutations, that is, select which ones occur rather than rely on sufficient variety of random mutations occurring within the population to provide the necessary means of survival under changed circumstances;[27] perhaps at the time of Cairns et al.'s experimental work, when the interpretation of directed mutation was put forward, the plasticity of the repertoire of bacterial responses was not sufficiently well appreciated.[28]

* * *

As already mentioned, the chapter (6) that follows deals with mutable, evolving (open-ended) complex systems. The quantitative measurement of the complexity of descriptions of entities is dealt with in Chapter 7. The final two chapters of this Part (8 and 9) deal respectively with soil and condensed matter as paradigms for complex systems; the world's climate (Chapters 10 and 12) could also have been appropriately included here as a paradigm of complexity.

It should be borne in mind that throughout this book the systems under consideration nearly always involve that most complex of entities, the human being (even as an individual), and it is sobering to recall that "We cannot trace any uniformity in the operations of a human being by merely looking at the actions themselves, as we can in the fall of a stone or the course of a planet.

---

[26]The word innovation is used here to signify true novelty, in accord with its dictionary definition (e.g., "the making of changes in something already existing, as by introducing new methods, ideas, or products"—from the Concise Oxford Dictionary, 10th edn). However, one should note that other, more restrictive, usages also exist, such as those of the UK government's Department for Business, Enterprise and Regulatory Reform (BERR)—"innovation is the successful exploitation of new ideas". Such usages are presumably influenced by the appreciation of the role of innovation in economics, as was particularly emphasized by J.A. Schumpeter (Capitalism, Socialism and Democracy. New York: Harper & Row, 1942).

[27]J. Cairns, J. Overbaugh and S. Miller, The origin of mutants. Nature (Lond.) 335 (1988) 142–145.

[28]Cf. J. Vohradský and J.J. Ramsden, Genome resource utilization during procaryotic development. FASEB J. 15 (2001) 2054–2056.

It is the unseen feelings that furnish the key to the vast complication of man's works and ways",[29] a topic that will be further taken up in Part V.

[29]Bain, A., The Senses and Intellect, p. 3. London: Parker (1855).

Complexity and Security
J.J. Ramsden and P.J. Kervalishvili (Eds.)
IOS Press, 2008

Chapter 6

Complexity, stability and crises

Peter M. Allen and Mark Strathern

Complex Systems Management Centre,
School of Management, Cranfield University,
Bedfordshire, MK43 0AL, UK

6.1 Introduction

Complexity science tells us that natural and human systems evolve over time in a series of discontinuities, successive transformations and crises, rather than in a smooth and continuous process of growth. These ideas therefore provide us with a new basis on which to understand both the occurrence of instabilities and crises, and also the way to design systems that can survive such events—a new basis for the development of resilient systems. The first point to be made are that complex systems science began when it was realized that open systems could exhibit spontaneous self-organization—a qualitative evolutionary process—in which structural change could occur as new system configurations and variables emerged. This was clearly shown in the original work on 'dissipative structures' (Nicolis and Prigogine, 1977) and on synergetics (Haken, 1977). Since then, enormous developments have occurred by many researchers around the world, but the fundamental point remains true that structural change occurs through a series of instabilities which arise from a dialogue between the 'system' as viewed at a given time, and the fluctuations and disturbances present within it, or in its environment (Allen, 1976; Allen and McGlade, 1987; Allen, 1990, 1994; Allen and Strathern, 2004; Allen, Strathern and Baldwin, 2006).

Clearly then, structural instability is not always bad—because in reality it marks the process of evolutionary development, and this is never a question simply of quantitative growth, but is also that of an increase in functional types, and in the linkages and interactions that constitute a society or organization. However, evolution does mean the appearance of new things—innovations—as well as the disappearance and extinction of others. Therefore in itself it cannot be said to be good or bad, but will be thought 'good' by the winners and 'bad' by the losers. By adding processes of adaptation and flexibility into a system, we can make failure less painful and allow the movement along an evolutionary trajectory with less conflict and pain, just as the invention of the limited liability company allowed a much faster growth of economic activities, and exploration of possible innovations.

If 'progress' in one group's eyes is 'total destruction' for the lives of another group, then they will fight to avoid the change. This points to the fact that in order to explore security and indeed complexity, we must adopt a 'multi-agent' perspective, and examine potential conflicts through the eyes of the different 'stakeholders'. In reality, different qualitative structures of role and function have emerged in different societies, with different accompanying belief systems and social rules and norms, and there is no scientific basis for saying that one society is 'better' than another. All depends on values and ethics of a society and these have co-evolved with their respective physical, biological, geographical and technological systems (Allen, Strathern and Baldwin, 2006).

Security is therefore a problem of defending an existing system either against accidental or natural events, or against attacks made by people who see the existing system as a threat to themselves. If improving security means that we decrease the frequency and damage caused by accidents, natural events or by attack, then we can:

- decrease the tension or pressure that underlies the events;
- anticipate them and if possible avoid them;
- reduce the damage of events once they occur.

This chapter will try to show how complexity and complex systems science is involved in all of these.

6.2 Complexity and crises

What is a crisis? It has to correspond to a situation in which some pre-existing system is threatened with radical change of some kind, which for at least some of the participants is seen as negative. But how do we come to see the world in terms of a particular system or situation, and one that we wish to preserve and maintain? We understand situations by making creative, but simplifying assumptions. We define the domain in question (the boundary) and by establishing some rules of classification (a dictionary) that allow us to say what things were present when. This means that we describe things strategically in terms of words that stand for classes of object. The 'evolutionary tree' is an abstraction concerning types of things rather than things themselves. In order to go further in our thinking, and get more information about an actual situation, we then consider only the present, and say, what is this system made of NOW, and how is it operating NOW. This is operational not strategic. It therefore assumes structural stability and takes us away from open, evolutionary change, to the effects of running a fixed set of processes. If the events considered are discreet, then the running is according to a probabilistic dynamics, and we have what is called stochastic non-linear dynamics, where different régimes of operation are possible, but the underlying elements never change nor learn, nor tire of their behaviours. If we assume that we can use average rates instead of probabilities for the events, then we arrive at deterministic, system dynamics. This is in general non-linear dynamics, and may be cycles or chaos or at equilibrium, but what happens is certain, simple and easy to understand.

In Figure 6.1 we show how successive assumptions are made in order to approach an understanding of the real situation. On the left hand side we have the cloud of reality and practice. Here, we are in the realm of non-science, in which people try to sum their experiences informally, and come up with heuristic rules and folklore of various kinds to deal with the problems of the real world.

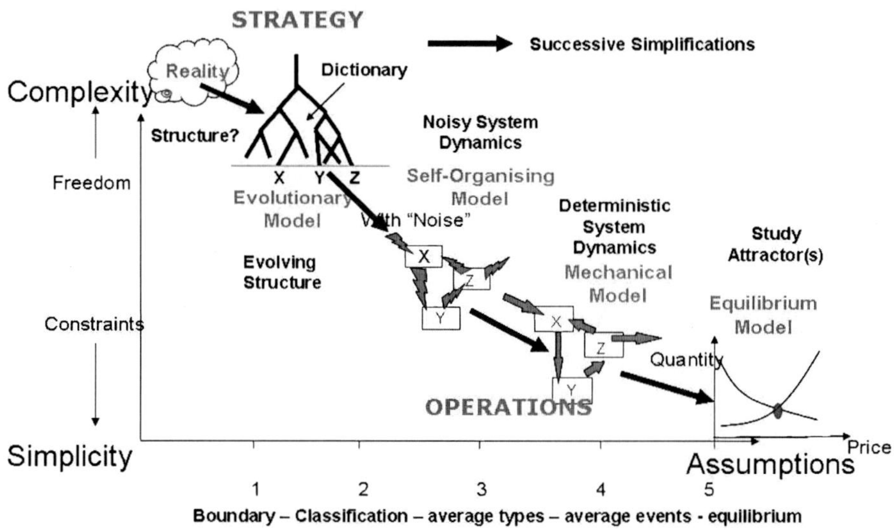

Figure 6.1: The choice of successive assumptions that lead to 'scientific' understanding of a situation (see also Table 6.1).

Science begins by deciding on a boundary within which explanation will be attempted, in the context of the environment outside. The second assumption is that of classification. The elements present within the boundary are classified into types, so that potentially, previously established behaviour and responses of similar types can be used to predict behaviour. In examining any system of interest over some long time, however, it will be found that the components and elements present in the system have in fact changed over time. Qualitative evolution has occurred in which some types of component have disappeared, others have changed and transformed, and others still have appeared in the system initially as innovations and novelties.

This possibility of transformation and evolution is the domain of the complex system that co-evolves with its environment and we see that evolutionary systems are more natural than mechanical systems, and that therefore it is completely natural that time should be marked by successive moments of transformation and change. Normality consists of periods of relative stability separated by successive instabilities.

During a period of apparent structural stability, we can simply represent the system as being made up of the existing components and their interactions.

The description of the behaviour of the system will be one of the probabilities of different possible transitions and events, leading to probabilistic, coupled equations. These will be probabilistic in the absence of detailed knowledge of the precise, mechanical operation of the parts. This probabilistic dynamics will give rise to a dynamics of average values, of variances and higher moments, as a result of the coupled processes. Because there may be different possible attractor basis for the dynamics, the statistical distribution will reflect this, being multi-modal and not a simple Gaussian or normal distribution. The dynamical system will itself generate the distribution and all its moments, including the variance, and so there will be no simplification into an average dynamic with a given random variance. Instead, all the moments will really play a role in defining the evolution of the distribution. These kind of probabilistic dynamics are described by what are called the 'master' or Kolmogorov equations governing the change in the probability distributions as a result of the non-linear interactions.

There are now two different routes to simplification:

- Assume that sufficient time has occurred that we can look at the stationary solution of the probabilistic non-linear dynamics which leads to 'self-organized criticality' (SOC) exemplified by the growing sandpiles and earthquakes of Bak (2000);[1]

- Assume that the system is still changing dynamically and that it can be represented by the dynamics of the average, mean or first moment of the probability distribution, assuming in addition that this can be uncoupled from the higher moments. This leads to deterministic system dynamics—a mechanical representation of the system. We can then study the attractors of this simplified system, and find either point, cyclic or chaotic attractor dynamics as the long term outcome.

This succession of models arises from making successive, simplifying assumptions, and therefore models on the right are increasingly easy to understand and picture, but increasingly far from reality. The operation of a mechanical system may be easy to understand but that simplicity has assumed away the more complex sources of its ability to adapt and change. A mechanical model is more like a description of the system at a particular moment, but does not contain the magic ingredient of microdiversity that constitutes evolutionary drive. The capacity to evolve is generated by the behaviours that are averaged by assumptions 3 and 4 (Table 6.1)—average types and average events—and therefore organisations or individuals that can adapt and transform themselves, do so as a result of the generation of micro-diversity and the interactions with micro-contextualities. This tells us the difference between a reality that is "becoming" and our simplified understanding of this that is merely "being" (Prigogine 1981; Prigogine and Stengers, 1983).

Ultimately then, our perception of change in terms of successive crises arises from our habit of seeing any present situation as a stable mechanical system, rather than as just part of a historical process of continuous change and transformation (cf. Figure 6.2).

In reality, complex systems thinking offers us a new, integrative paradigm, in which we retain the fact of multiple subjectivities, and of differing perceptions

[1] Real sandpiles may not display SOC.

Table 6.1: The general complexity framework.

Number	Assumption made	Resulting model
1	Boundary assumed	Some local sense-making possible—no structure supposed.
2	Classification assumed	Strategic, open-ended evolutionary—structural change occurs Statistical distributions part of evolutionary process, can be multimodal
3	Average types	Operational, probabilistic, non-Linear equations, master equations, Kolmogorov equations—assumed structurally stable. Statistical distributions can be multimodal or power laws
4	Statistical attractors	Self-organized criticality, power law distributions

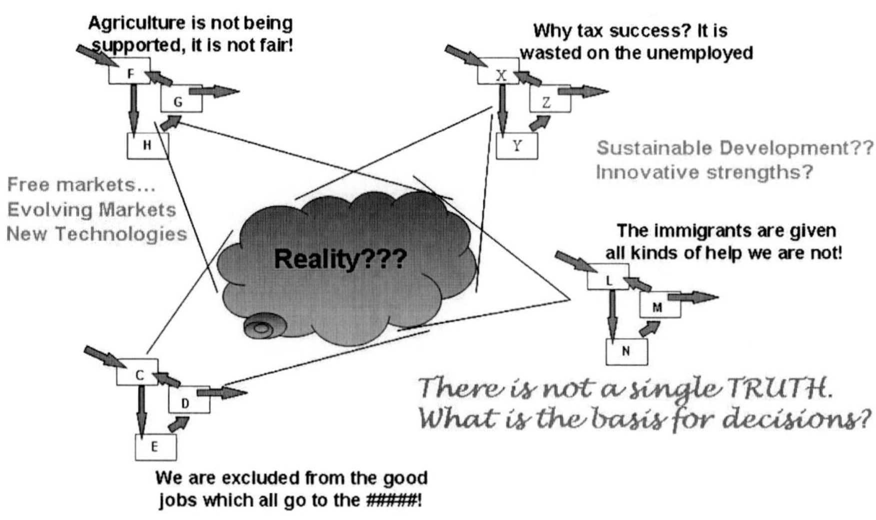

Figure 6.2: Different people see the same system in different ways. Each can however be rational and consistent, whilst advocating quite different actions or policies.

and views, and indeed see this as part of the complexity, and a source of creative interaction and of innovation and change. The underlying paradox is that the knowledge of any particular agent will necessarily imply a *lack of knowledge* of other points of view. But history, society and the market result from the interaction of multiple agents and their different perspectives and understanding, as well as those of players entering or leaving the system. Actions based on any particular domain of knowledge, although seemingly rational and consistent, will necessarily be inadequate. In such a world, crises are inevitable, and so the more important question is that of the ability of systems and individuals to cope with the unexpected and to adapt and evolve successfully.

What matters over time is the expansion of any system into new dimensions and conceptual spaces, as a result of successive instabilities involving dimensions additional to those the current system appears to occupy.

This idea of evolution as a question of invadability, with respect to what was not yet in the system, was the subject of a very early paper by the author (Allen, 1976). Essentially then, systems have to be seen as temporary, emergent structures that result from the self-reinforcing non-linear interactions that result from successive invasions. History is written not only by some process of rational improvement in its internal structure but more fundamentally by its dialogue with elements that are not yet in the system—successive experimental linkages that either are rejected by the system, or which "take off" and modify the system irreversibly.

Rational improvement of internal structure, the traditional domain of systems thinking, supposes that the system has a purpose, and known measures of performance that can indicate the direction of improvements. But, a more fundamental structural evolution of complex systems results from the successive take-offs, invasions, within the system by new ideas, elements and entities, characterized by emergent properties and effects, leading to new attributes, purposes and performance measures. So for example, the adoption of a new technology may transform society and people's needs, but equally a fundamentalist revolution that occurs in a country will lead to new measures of performance and success on the part of society. This kind of structural evolution does not occur randomly, as successful transformations of systems only occur through the revelation of positive feedback and synergy, leading to new, internally coherent, structures that exhibit some temporary degree of stability.

An overarching view of the origins and implications of possible changes in the technical, social and economic system can be obtained by looking at cities and regions as evolving, self-transforming systems in which behaviour, decisions and the value systems underlying these all evolve over time. This leads to a view of a city or region as a complex evolution of spatially distributed learning reflecting local stresses, opportunities and exploratory responses such that people not only change what they do, but also their knowledge of what they could do, and what they want to do. Qualitative, structural changes occur both in the macroscopic forms of the collective structure, and also in the microscopic structures within individuals' brains that govern their trade-offs and decision making, which in turn govern the future structural evolution of the collective system and of the individuals that inhabit it.

6.3 Urban and regional complexity

In reality then a city is a complex system, as is a neighbourhood, a block, a household, and an individual which all sit within an ecosystem, pulling in natural resources and pumping out wastes. These represent nested levels of description, and we can develop mathematical models that can explore different possible evolutionary pathways and possible futures under the assumptions of different possible interventions. This work started in 1976 when the US Department of Transportation commissioned our early research on developing such models, initially only dynamic, but later on developing fully complex, learning, multi-agent models. The essence of these models is shown in Figure 6.3 in which the patterns of location of people, jobs, transport and of infrastructure are coupled together, so that their combined evolution can be explored under different interventions and plans. Such models represent the system as a pattern of locations and flows representing the patterns of activity and leisure of a population.

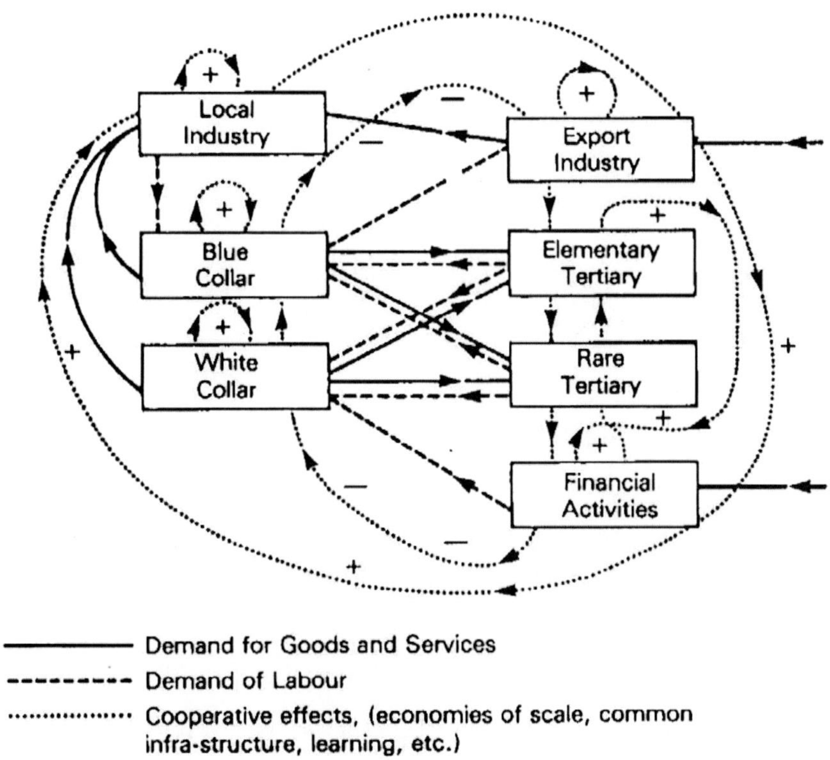

Figure 6.3: The interaction diagram spatially distributed multiple agents of different kinds.

This is an evolving pattern of flows which both reflect and affect the spatial distribution of diverse activities and opportunities:

- Dispersed distribution of affordable/desirable housing;
- Concentrated distributions of employment;

- Concentrated distributions of retail opportunities;
- Dispersed distributions of leisure and cultural facilities.

Since the 1970s, work has been going on developing computer models that take into account the complex interactions of linked responses that lead to a co-evolution of urban structure (patterns of retail, commercial and manufacturing employment, and different qualities of residence) with transportation infrastructure. These models describe (cf. Figure 6.4):

- Different types of actor at each zone, with characteristic needs to be fulfilled;
- These characteristic needs are stable, but the behaviour of actors depends on the changing circumstances;
- The spatial distributions of the different types of job and different kinds of people affect each other as the potential for housing demands, commercial activities and for travel affect and are affected by transportation and land-use.

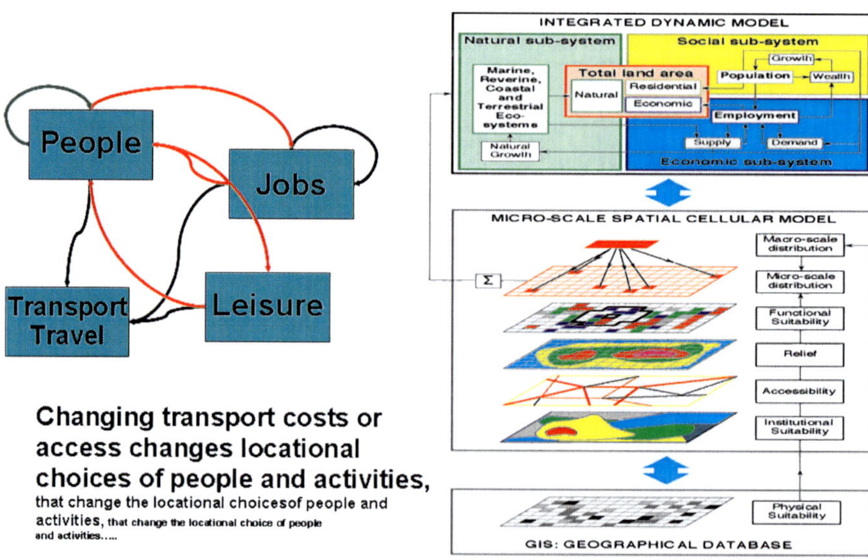

Figure 6.4: Software systems have been developed that allow the interacting spatial distributions of people, jobs, leisure facilities and transportation can be studied (White and Engelen, 2001).

The development of these models has been described in "Cities and Regions as Self-Organizing Systems" (Allen, 1997a and 1997b). These kind of model demonstrate how change over time concerns the interacting behaviours of the actors in region, as they all struggle to respond to changing opportunities and pressures, each pursuing their own goals, as a function of their cultural, religious and ethnic identities as well as of their financial means. In many ways the issues of security and complexity arise from the conflicts of interest that arise in this evolution as different groups succeed or fail in the on-going struggle, and grievances either grow or dissolve over time.

The model above can be used to explore the spatial evolution of a European city such as Brussels, showing how the model can capture qualitative changes in the spatial structure of the city. Although, theoretically all the variables could potentially be spread homogeneously across the whole space of the city, the interactions between the different types of actor lead to the instability of this 'solution' and the emergence of clusters of business, industrial, retail and residential concentrations. In principle, this pattern could have followed many possible paths and come out quite differently from the Brussels we observe today, but the precise history that occurred there actually resulted in the observed pattern, and the model allows us to generate this as one of the possible futures of a simple "proto-Brussels". However, the qualitative changes that occur as the industrial centres get pushed to the periphery, and blue-collar workers must re-align their daily travel and places of residence, are evidence of periods of tension when the preferences of particular types of agent are not well served by the structure in which they find themselves. This pressure, in the case of Brussels, is resolved when the planning laws and building sector respond to the potential demand for new industrial locations and housing leading to a spatial change in the patterns of location and transport flows that resolve these pressures (Figure 6.5).

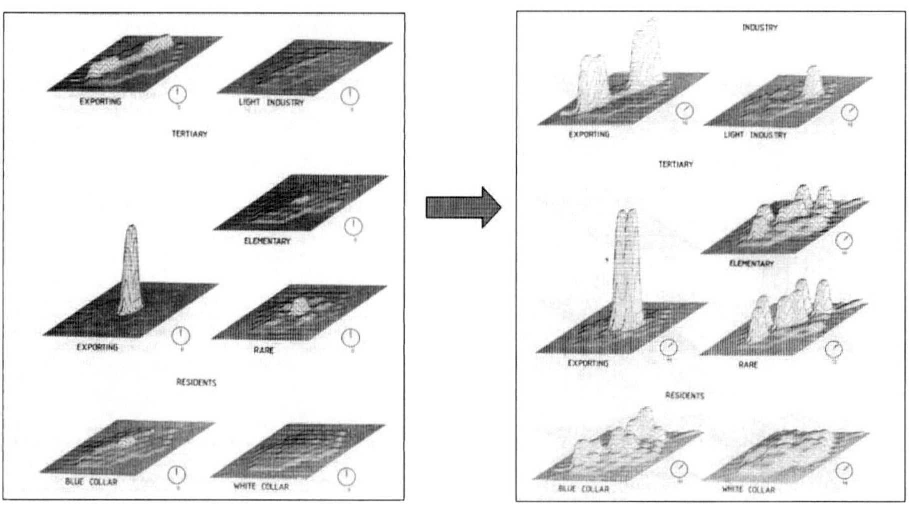

Figure 6.5: Emergence of complementary spatial structure of the 7 variables.

Clearly then, our model supposes a political and economic system that can detect and respond to poorly satisfied demand and is able and willing to do so. When this is not so, then the model will generate dissatisfaction that if not addressed would lead, in real life, to problems of public security and eventually terrorism. Clearly, there are therefore real sources of conflict of interest and genuine grievance that will occur as part of the normal evolution of a society, and of course between different societies and groups living in the same region where territorial conflicts will occur. Without some mechanisms of response that can give rise to what is perceived by many as being fair resolutions, then the conflict can grow and lead to a crisis.

Complexity plays an important role in generating conflict because people cannot necessarily see the longer term or wider consequences of their actions. In this way the actual systems that can emerge may be the result of self-reinforcing loops of self-justification and the complementary ones of blame and enmity, and so eventually to hostilities and violence. In our model of Brussels the background that was not made explicit was of a well-functioning society with channels of communication and political response that allowed the pressures of evolution of the spatial structure of the city to evolve qualitatively over time in response to the emergent pressures arising for different agents in the system. This is an interesting example of course, because very recently, the linguistic tensions in Belgium have led to the possibility that there will be a political break-up of Belgium into its two basic cultural components—the (Dutch-speaking) Flemish and the (French-speaking) Walloons. This leaves Brussels in a complicated situation at the meeting point of the two. In fact the spatial dynamics of Brussels has been affected for many years by this latent conflict, and the continual compromises that have been created in order to avoid confrontations have led to some quite absurd situations in which 80% of the residents in an officially Flemish commune may speak actually speak French. The linguistic definitions of Communes around Brussels have for a long time affected migration out of the city to the suburbs. Without going into too much detail, it is clear that some solutions of the Brussels problem could lead to clashes between marching crowds and this could rapidly degenerate way beyond the real reasons for the conflict. Of course, Belgium is a highly civilized country and so such events seem impossible, but what the recent wars in the Balkans and elsewhere have shown is just how rapidly and devastatingly a relatively prosperous and civilized area with ethnic fragmentation can degenerate into violence. In reality, language, culture, religion and identity really matter to most people, and when they play a role in the economic and social mechanisms of a region can then lead to disaster even though nobody really wishes it.[2]

6.4 Anticipating crises

In the section above, we discussed how modern computing power can allow us to build complex, evolving models of urban and regional systems, and to use them to explore possible futures, including both the planning and policies of social and economic development and also the possible crises arising from conflicts of interest of opinion. We can make different possible simulations and explore the probability of crises occurring. However, these models require considerable data and take some time to develop, and so it is important to ask whether there is a simpler and faster approach coming from complex systems thinking that could be used for the anticipation of possible crises in different possible countries.

The answer is affirmative. Instead of attempting a quantitative mathematical model with the different types of agents in different zones, and considering all their interactions, we use a qualitative approach. In this we consider the different active agents within a region—the different ethnic, linguistic, political, religious and geographical zones and the interactions and impacts each group

[2] Cf. S. Brams, Paradoxes in Politics. New York: Free Press (1976), for examples of this in the context of voting systems *(Editor's note)*.

on the others, and also of the way that each of them will experience the changes occurring in the system as it runs. The reflexion is therefore two-fold: the agents of different kinds that make up the system and the economic, demographic, political, environmental and natural trends and events that are driving the system as a whole.

The idea behind the approach is to explore the kinds of instabilities that could possibly occur in the system as it evolves forward in time under different possible scenarios. In particular we are looking for the *loops of positive feedback* that if not interrupted can only end in crisis and instability. In general there will be slowly changing variables such as demography, pollution, economic growth or decline, changing technology of communications and social changes in political rights but as they run forward they can either make the system more stable or on the contrary more vulnerable to instability. So, either the slow changes can lead to greater stability, reduced tensions and greater prosperity or, on the contrary can lead to various "tipping points" of political or economic instability. The direction in which the system evolves will depend on whether loops of negative feedback dominate those of positive self-reinforcement. In the latter case, if loops of positive feedback are not defused by some economic or political actions, then the system can only move inexorably towards some crisis as instability eventually occurs.

In addition of course, the system may be relatively far from a tipping point, but be pushed over it by the occurrence of some external event—perhaps an earthquake, flood or drought—which pushes the system off its previous trajectory and off towards a different dynamic attractor. A further consideration for the anticipation and analysis of possible crises in whether or not it would be possible, having perceived a growing risk of instability, to intervene in the system and guide it away from a potential tipping point and towards more acceptable behaviour. By reflecting on different possible crises, and the factors that would seem to inflame or diminish an economic crisis, social collapse, external attack, environmental crisis, inter-communal conflict—we can arrive at strategies to intervene in the development and defusing of crises.

Processes that correspond to slow changes are made explicit and their interaction taken into account. From a spreadsheet representation the connexions can be seen between different processes, which would increase or decrease their strength. Obviously over time this shows how the processes affect each other, and how the complex pattern of effects may change through time. In particular, by multiplying this matrix of interacting processes by its transpose, we are able to show the loops of self-reinforcement, which will indicate where processes are self-regulating, or more importantly, self-reinforcing.

Possible triggers and events can then be classified into two basic categories: Independent Events, and Dependent Events. The former refer to possible events whose probability is not affected by any of the processes running in the system, for example, an earthquake, while the latter refer to events such as political instability that do depend on what is going on in the country. For each dependent event, the processes that increase or decrease its probability need to be made explicit, providing a basis upon which the probability of the different dependent events can be updated regularly. The output of the processes is therefore a method by which the probability of various events can be updated over time. This would include the changing pattern of strength of the contributing processes as they interact and self-reinforce, and would also include any changes in external circumstances that might have occurred. We will need to look for indicators of these processes to indicate how near a crisis may be and also decide whether a particular type of crisis is amenable to intervention and which possible actions could be taken.

6.4.1 The drivers of change

The factors at work in any society that frame the interactions of the different agents, political, economic, ethnic and religious groups within a society are for example:

- Demography
- Environmental change
- Economy
- Coherence in politics
- Alliances and alignments
- Access to knowledge
- Science and technology
- Public perceptions and attitudes

Each of these domains can be considered in turn, and possible events imagined, together with a notion as to whether they will destabilize or stabilize the society being studied, which processes and events they may affect in turn, and whether or not they are amenable to any kind of intervention. Examples of such events might include:

- Economic crisis
- Social breakdown
- External attack
- Environmental crisis
- Inter-communal conflict

We may also decide whether they are fast or slow variables and also place contingent processes or events into scenarios, exclude those items with no clear meaning or direction for effects and separate processes and events. We can also consider two different types of event: (a) those independent of processes inside the system; and (b) those whose probability of occurrence is affected by what happens in the system (potential "tipping points").

6.4.2 The output

In order to illustrate how this would work, let us take an imaginary example, which is based on an exercise carried out some years ago in the context of the Caucasus. Based on reflexions concerning the 8 drivers of change, we find the following, summarized in Tables 6.2 to 6.9.

Table 6.2: Processes related to demographic change.

No	Process	$\mathfrak{S}^a$	$\mathfrak{A}^b$	$\mathfrak{V}^c$	Outcome
1	Stable or declining population	S	N	S	Pressure on resources, land, water, infrastructure, housing
2	Declining public health (alcohol)	D	N	S	High costs, lack of productive work,
3	Increasing ethnic consolidation and tension	D	N	S	Increasing hatred, conflict, lack of shared interests
4	Increasing adaptability of women to new economic opportunities	S	N	S	Higher household incomes, and diversity of income
5	Growing population in the capital city (rural emigration)	D	N	S	Strain on resources in capital city, pollution, unemployment, congestion costs, food flows, housing problems, water
6	Growing instability of refugees	D	N	S	People want to return home, injustice, call for action

[a] Stabilizing (S) or destabilizing (D)?
[b] Amenable (A) or not (N)?
[c] Slow (S) or fast (F)?

Table 6.3: Processes related to environmental change.

No	Process	$\mathfrak{S}^a$	$\mathfrak{A}^b$	$\mathfrak{V}^c$	Outcome
7	Impact of pollution, heavy metals, smog,..	D	A	S	Public health, anger where pollution occurs
8	Increasing drought in *External Country 2*	D	N	S	Economic costs, unemployment Food prices
9	Improved crop yield due to biotechnology	S	A	S	Economic gain, lower food prices,
10	Declining pollution from closure of industry	S	N	S	Less health problems, lower costs
11	Declining crop yields due to salination and chemical pollution	D	A	S	Food prices, conflict, anger
12	Rising sea level	D	N	S	Decreasing agricultural production, conflicts
13	Increasing over-fishing, caviar shortage	D	A	F?	Food prices, conflict. The over-fishing is in fact slow, but the effect could be fast.
14	Increasing deforestation	D	A	S	Conflicts

Table 6.4: Processes related to economic change.

No	Process	$\mathfrak{S}^a$	$\mathfrak{A}^b$	$\mathfrak{V}^c$	Outcome
15	Increasing dissatisfaction of countries without oil boom	D	A	S	Conflict
16	Continuing lack of development of economic diversity	D	N	S	Economy not adaptive, vulnerable
17	Impacts of environmental pollution	D	A	S	Conflicts, economic costs of health
18	Dwindling prospect of subregional or regional economic integration and security issues	D	N	S	Conflicts, high prices
19	Growing US/Iranian tensions block/distort economic links	D	N	S	Conflicts, costs
20	Growing anxiety over security reduces external investments	D	N	F	Lack of investment, costs, prices

Table 6.5: Coherence of politics.

No	Process	$\mathfrak{S}^a$	$\mathfrak{A}^b$	$\mathfrak{V}^c$	Outcome
		External Country 1			
21	Increasing development of civil society (e.g. impact of NGOs	S	A	S	
		Country studied			
22	Rising corruption	D	N	S	Build-up of tension
23	Decreasing centralization of the political system	S	N	S	More local autonomy
24	Increasing Instability in occupied territory	D	N	F	Conflict, a break point again
25	Increasing development of civil society (impact of NGOs)	S	A	S	
		External Country 2			
26	Increasing regional tensions, local and central élites in conflict	D	N	S	Conflicts
27	Increasing politicization of refugees	D	A	S	Conflict
28	Developing free press	S	A	S	Better information
29	Increasing development of civil society (impact of NGOs)	S	A	S	

Table 6.6: Alliances and alignments.

No	Process	$\mathfrak{S}^a$	$\mathfrak{A}^b$	$\mathfrak{V}^c$	Outcome
30	Growing importance of the disputed region	D	N	S	Conflicts
31	Growing geopolitical advantages to *Country studied* in territorial dispute with neighbour	D	N	S	Conflict, but *Country studied* has oil, so has friends
32	Increasingly entrenched zero-sum perspective on regional competition locally	D	N	S	
33	Increasingly entrenched zero-sum perspective on regional competition in US/Russian dialogue	D	A	S	

Table 6.7: Access to knowledge.

No	Process	$\mathfrak{S}^a$	$\mathfrak{A}^b$	$\mathfrak{V}^c$	Outcome
34	Growing de-modernization (qualitative collapse)	D	A	S	Breakdown of infrastructure, cCosts, prices, conflicts
35	Growing generational gap in thirst for/access to knowledge	D	A	S	
36	Growing diversion of energies into introspective nationalism	D	N	S	Values related to historical identities, to historical conflicts

Table 6.8: Science and technology.

No	Process	$\mathfrak{S}^a$	$\mathfrak{A}^b$	$\mathfrak{V}^c$	Outcome
37	Declining opportunities and prestige for top scientists—increasing brain drain	D	A	S	
38	Improving expertise in product development techniques	S	A	S	Helps get business rolling
39	Continuing failure to diversify from military technology	D	A	S	Vulnerable economy

Table 6.9: Perceptions and attitudes.

No	Process	$\mathfrak{S}^a$	$\mathfrak{A}^b$	$\mathfrak{V}^c$	Outcome
40	Growing nationalism	D	A	S	Conflicts
41	Growing importance of informal economy and organized crime as the way ahead	D	N	S	Social breakdown
42	Decreasing respect for or trust of political classes	D	N	S	Social breakdown

6.5 Analysing the structure

The important point is the establishment of two types of term in the changes that are occurring: *processes* (assumed to be running fairly continuously but not too rapidly) and *events* (assumed to be rapid changes that have a certain probability of occurring). This distinction then allows us to treat the processes and events separately, and to attempt to demonstrate how the processes, over time, affect the probability of the possible events. The changing probability of these events therefore becomes part of the output of the method. This is summarized in (Figure 6.6).

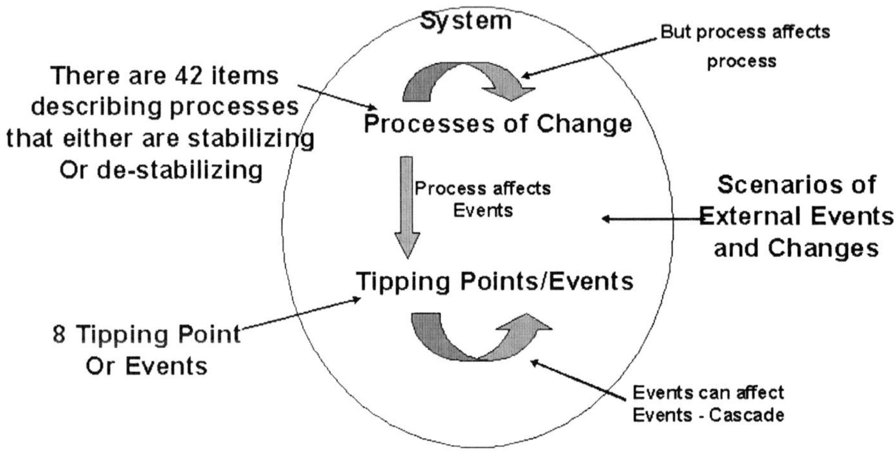

Figure 6.6: Diagram showing how one can separate the processes and the events and consider the interactions between and within them.

The next phase of the analysis is to take the processes that have been stated to be operating within the system, and see how they interact with each other. To do this we simply read off the each process in turn and consider its possible effect on all the 42 processes that have been defined.

So, for example, taking the first 10 processes from "Stable or declining population", to "Declining pollution as industry closes", then we can insert numbers corresponding to no effect, weak or strong effect either positive or negative (Figure 6.7).

Causes Effect of row on column

	1	2	3	4	5	6	7	8	9	10
1	1	0	-1	0	-1	1	1	0	0	1
2	2	1	1	0	0	1	2	1	0	0
3	1	1	2	0	1	2	2	0	0	0
4	1	1	0	1	0	0	0	0	0	0
5	-1	1	1	1	1	1	1	0	0	0
6	1	1	2	0	-1	1	1	0	0	1
7	2	3	2	0	2	1	1	0	-1	0
8	1	2	1	0	0	0	1	0	-1	0
9	0	-1	0	0	1	0	1	1	1	0
10	0	-1	0	0	0	0	-2	0	1	1

Figure 6.7: Matrix for the interaction between the first 10 processes of the modified matrix.

Obviously, this process can be continued for all 42 processes and we can arrive at a representation of the manner in which each process affects and is affected by the others. By including the specification of whether a process was in fact stabilizing or destabilizing, we can arrive at a picture of how the processes related to stability and instability are reinforced or diminished by the others. We can relate destabilizing to 'bad' and stabilizing to 'good' in that they either tend towards or away from crises. We can then show the complete interaction matrix between the 42 processes (Figure 6.8).

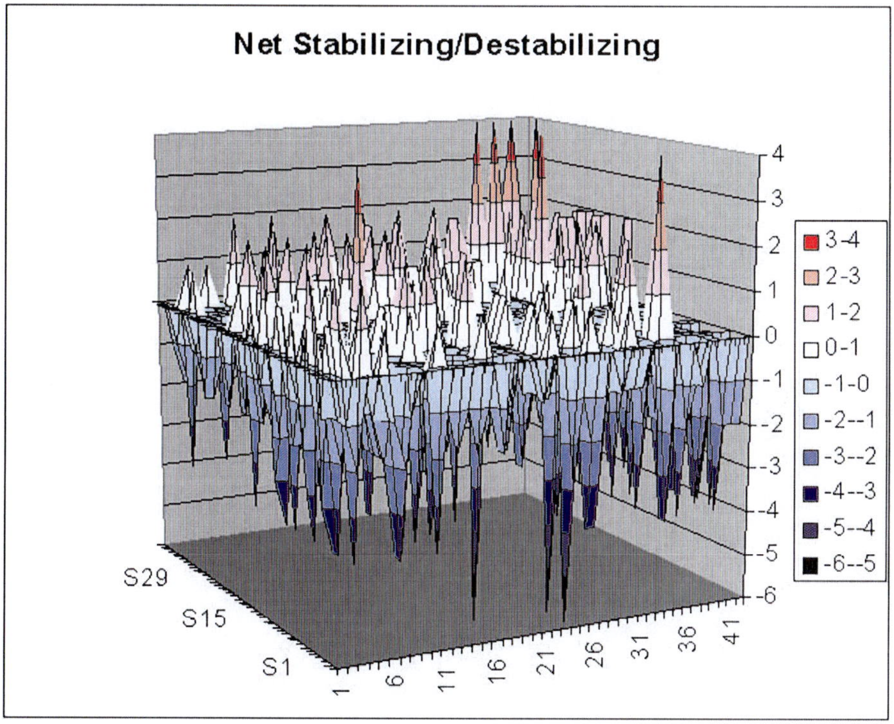

Figure 6.8: The interaction matrix for the processes retained as characterizing the imaginary case study.

This shows us that there are considerable interactions between the different processes of change identified in the exercise. It also shows us that any particular process of change will feed on to those that it affects. In turn, the modified changes in these will feed on to all the processes that each of these affects too. In this way, the whole complex pattern of linkage between processes is captured by the matrix. The importance of each process is therefore being modified through time as the system runs, and this will in turn affect the probability of occurrence of possible break points of the system.

We can now carry our analysis a stage further by considering how processes interact with other processes, and how these may in turn interact back on the original one. Let us consider 2 processes, X and Y. Several possibilities arise:

- X increases Y and Y increases X
- X increases Y, Y decreases X or X decreases Y and Y increases X

- X decreases Y and Y decreases X

In the first case, there is a risk of runaway growth of X and Y. If X and Y are destabilizing this will be a pressure towards a crisis. In case 2, we have a self-regulation where Y helps to reduce changes in X. The exact result would depend on the relative sizes of the two effects. In the third case we again have a self-reinforcement as the reduction in Y could then decrease X and lead to the collapse of X and Y.

We can study this question of self-reinforcement (Figure 6.9), and possible run-away effects by considering the matrix of interaction between processes multiplied by its transpose.[3]

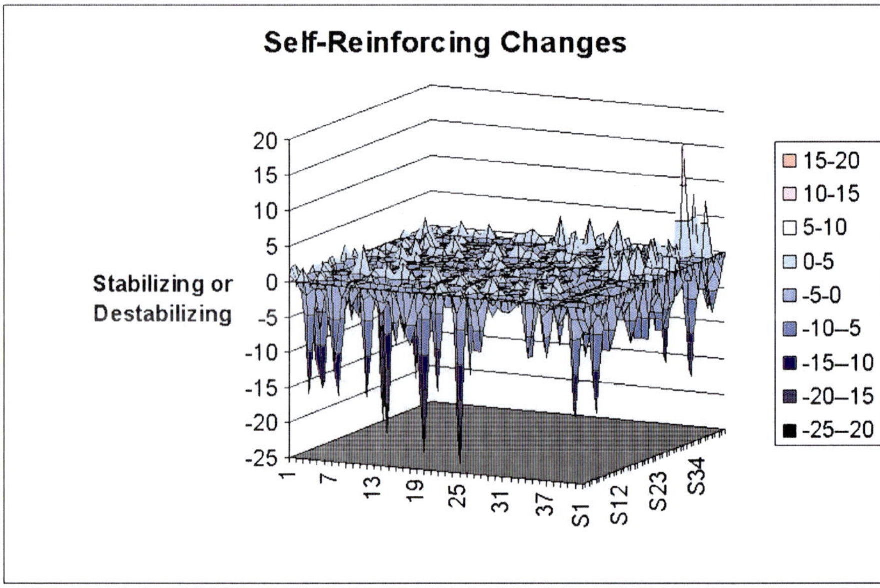

Figure 6.9: The pattern of self-reinforcement of processes 1–42.

This allows us to spot the self-reinforcing loops which, if not countered, can only end in crisis. Let us check out the self-reinforcement. The strongest of these are:

- 3 and 24
- 7 and 17

These correspond to:

- 3. Increasing ethnic consolidation and tension
- 24. Increasing instability in occupied territory (NK)

and

- 7. Impact of pollution, heavy metals, smog, ...

[3]The transpose is the matrix constructed from the original one by swapping the rows and columns.

- 17. Impacts of environmental pollution

The next strongest are 3 and 11, 17 and 11, 3 and 7, 34 and 41, 3 and 40, 7 and 41:

- 3. Increasing ethnic consolidation and tension
- 11. Declining crop yields due to salination and chemical pollution
- 17. Impacts of environmental pollution
- 40. Growing nationalism
- 34. Growing de-modernization (qualitative collapse)
- 41. Growing importance of the informal economy and organized crime as the way ahead
- 7. Impact of pollution, heavy metals, smog, ...
- 41. Growing importance of the informal economy and organized crime as the way ahead

Good: 38 and 41:

- 38. Improving expertise in product development techniques decreases a bad thing.
- 41. Growing importance of the informal economy and organized crime as the way ahead

These show us that our method does demonstrate the connexions:

- 4 is good for 34
- 25 is good for 34
- 38 is good for 41, 36, 34
- 3 is bad for 24
- 6 is bad for 24 and 7 is bad for 17.

6.6 Scenarios

In considering the context for the analysis we start from the global, and move in to the scale of the region. Each scenario should correspond to a set of processes (e.g. world economy grows) and events with a given probability. Some of these probabilities will change over time, while some, like the probability of an earthquake, may remain constant. However, in general, over time these probabilities will change, and so if the strategic analysis is repeated annually for example, the probabilities of these events will be expected to change.

We should really reflect on a series of scenarios starting from the global and moving inwards. For example:

- World economy grows or slows

- Regional economy slows or grows
- Oil markets are good/bad and this gives/not gives an oil boom in the region
- Ethnic problems arise in the region

These could be considered to be simply additive possibilities rather than nested (Figure 6.10).

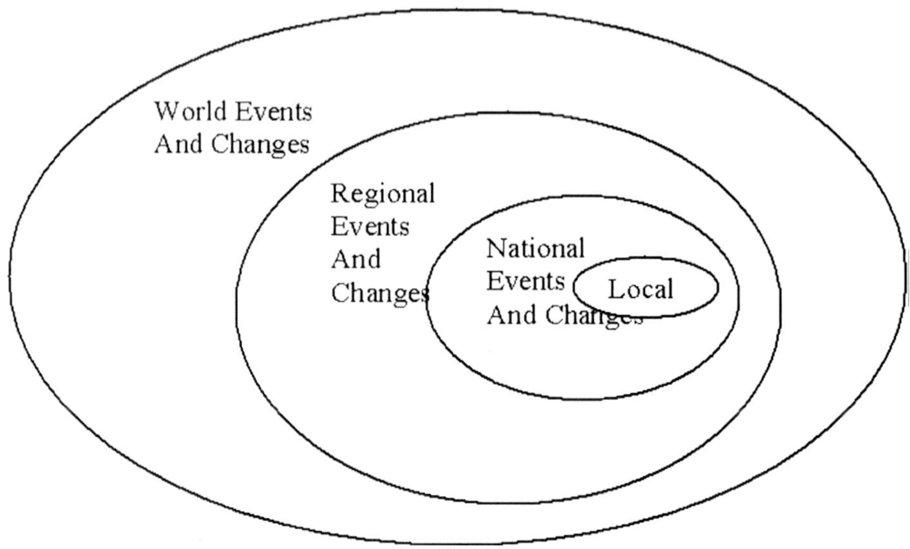

Figure 6.10: The scenarios need to reflect successive layers of spatial context.

From these scenarios for external events we need now to consider the possible scenario of internal events. These should of course be considered within the context of the different possible scenarios described above.

We can consider first the internal events whose probability is not affected by the (42) processes of change identified in §6.4.2. This might include externally inspired terrorism, earthquakes, the unexpected death of a leading politician, or a nuclear power station disaster. All these are internal events that are not the result of any of the processes that are running in the system.

For events that are made more or less probable by the 42 processes running in the system, these may correspond to steady changes in the probability of events, or in some case to the occurrence of "break points" when the system suddenly becomes potentially unstable and vulnerable to the triggering of a rapid change.

We have now classified the possible changes in our system as being processes, independent and contingent events. These run under a selection of external scenarios concerning processes and perhaps events occurring outside the region of study.

We have shown that processes affect processes, and so the relative importance of different processes may change over time. Processes also affect contingent events in the system, and therefore this is the initial output of our analysis. However, it is also true that events affect events, and so once an event has been triggered there may be a cascade that is set in motion. This cascade may be what gives the overall behaviour the appearance of one particular type of crisis as opposed to another.

A cascade may result both from independent or contingent events. So, we need to consider the probabilities of both types. We can build an event/event interaction matrix and explore how one event may trigger others.

6.7 Implications

Having described how one might develop a qualitative view of the factors and linkages that may exacerbate or reduce crises, we can now turn to thoughts on how security might be increased and the system kept away from a disastrous "tipping point". A main aim would be to reduce the driving forces behind the pressures and drivers that motivate terrorism (cf. Chapter 3). In reality of course, it is not necessarily the actual terrorists that can be eliminated, since there will always be extreme views in any group, but the support offered to them by some host population. Without the tacit support of a population terrorists are much less dangerous. What we must address is how it is possible that virtually a whole population can feel that violence and terrorism are justified. The new reality suggested by complexity science forces us to recognize the fact that there is not one objective reality but many different perspectives and views, each conditioned by history and subjective experience. If a people suffer economic and physical hardship and consider that it has been treated unjustly, then eventually it will support almost any actions that might change the situation, and maintaining security may become almost impossible. Trying to increase controls, vigilance, and the suppression of freedom will only exacerbate the problem and simply increase the resistance and the recruitment of terrorists.

Returning to our complex systems view of regional and urban evolution, we see that there are collective outcomes that nobody can really be aware of. The emergence of communities and of identities while positive in many ways, can also lead to different subjective experiences concerning opportunities and deprivation. If the collective processes of the complex system—its governance—cannot detect and respond to emerging discontent and difficulties, then these may grow and lead on to runaway processes of violence and insecurity. The very diversity that complexity tells us is an adaptive force which can create a better future can also lead to conflict if the different experiences become too unequal. The whole history of civilization is associated with towns and cities where people have come together and cooperated in the evolutionary development of skills and knowledge that has transformed their lives. However, where groups are excluded or identified as "the enemy", or where nationalist bigotry, religious intolerance and cultural imperialism dominate, then unless these forces are reduced the outcome can only be conflict.

Security, in the long term, therefore depends not just on finding new ways of tapping into secret communications, or developing highly trained intervention and intelligence gathering systems, it depends on reducing the injustices that lie behind insurgency.

The methods outlined here coming from complexity recognize the reality of multiple viewpoints, and the potential legitimacy of multiple perspectives. In the end open systems will outcompete closed ones, and just as it is the lack of adaptation in firms that leads them all to fail sooner or later, it is the locking in of identity and the assumption that ones own views are the 'real' and 'correct' ones that leads to conflict. The fundamental rule of the successful evolution and co-evolution of complex systems is their openness and willingness to consider other points of view. Instead of objective truth, complexity tells us of multiple subjectivities and perspectives, and the power they have when networked together.

6.8 References

Allen, P.M., 1976. Evolution, population dynamics and stability. Proc. Natl Acad. Sci. 73, 665–668.

Allen, P.M., 1990. Why the future is not what it was. Futures 22, 555–569.

Allen, P.M., 1994. Coherence, chaos and evolution in the social context. Futures 26, 583–597.

Allen, P.M., 1997a. Cities and regions as evolutionary complex systems. Geographical Systems 4, 103–130.

Allen, P.M., 1997b. Cities and Regions as Self-Organizing Systems: Models of Complexity. Taylor and Francis.

Allen, P.M. and McGlade, J.M., 1987. Evolutionary drive: the effect of microscopic diversity, error making and noise. Foundations Phys. 17, 723–728.

Allen P.M., Datta, P. and Christopher, M., 2006. Improving the resilience and performance of organizations using multi-agent modelling of a complex production-distribution system. J. Risk Management 8 (Special Issue: Complexity, Risk and Emergence), 294–309.

Allen, P.M. and Strathern, M., 2004. Evolution, emergence and learning in complex systems. Emergence 5, 8–33.

Allen, P.M., Strathern, M. and Baldwin, J., 2006. Evolutionary drive: new understanding of change in socio-economic systems. Emergence Complexity Organization 8, 2–19.

Bak, P., 1996. How Nature Works: The Science of Self-Organized Criticality. New York: Copernicus.

Foster, R. and Kaplan, S., 2001. Creative Destruction: Why Companies that are Built to Last Underperform the Market—And How to Successfully Transform Them. New York: Random House.

Haken, H., 1977. Synergetics. Springer.

Nicolis, G. and Prigogine, I., 1977. Self-Organization in Non-Equilibrium Systems. New York: Wiley Interscience.

Mandelbrot, B.B., 1982. The Fractal Geometry of Nature. New York: Freeman.

Mandelbrot, B.B., and Hudson, R.L., 2004. The (Mis)Behavior of Markets: a Fractal View of Risk, Ruin and Reward. London: Profile.

Prigogine, I., 1981. From Being to Becoming: Time and Complexity in the Physical Sciences. San Francisco: W.H. Freeman & Co.

Complexity and Security
J.J. Ramsden and P.J. Kervalishvili (Eds.)
IOS Press, 2008

Chapter 7

The representation of complexity

Jeremy J. Ramsden

Cranfield University, Bedfordshire, MK43 0AL, UK

The introduction to this Part (Chapter 5) has outlined some general features of complexity. In this brief chapter the focus is especially on those aspects complementary to the complex modelling described in Chapter 6, in particular how to quantify complexity.

7.1 Types of complexity

Complexity can be classified into ontological, semiotic, and interpretative. Giving an account of the ontological complexity of a material object requires that it be fully known, but this we can obviously never achieve: even a single atom may have existed for millions of years and we would have to know its entire history in order to fully know it. An approach towards ontological complexity is the intrinsic complexity, which is deduced from direct consideration of the object. Semiotic complexity is the complexity of the *description* of the object.[1] Finally, interpretative or i-complexity is the complexity of the interpretation of the object. This can sometimes follow straightforwardly from the d-complexity, but sometimes (as in the case of riddles and paradoxes), it may be much more difficult to obtain.[2]

Most of this chapter will be concerned with d-complexity. The first problem is then to obtain the description, in other words, devise a suitable coding scheme for capturing the description. Whereas the (in practice unattainable) ontological complexity is absolute, semiotic complexity is relative to the framework selected

[1] Called d-complexity by L. Löfgren.

[2] For example, consider the statement, "the least set that is a member of itself". Very complex objects, such as a Beethoven piano sonata, may have extraordinarily expansive interpretations, which far transcend any description in terms of the musical notes alone.

for constructing its description. It is therefore the first duty of the investigator to ensure that this framework is relevant and adequate.

Before launching into the more formal aspects, let us start by looking at what we intuitively feel to be complex.

7.2 Intuitive notions of complexity

Most of us seem to have quite a good idea of what is complex and what is not. Contrapuntal music (e.g. most of the compositions of J.S. Bach) seems to be more complex than romantic sonatas (e.g. compositions of W. A. Mozart).[3] The poems of John Keats seem to be more complex than those of William McGonagall. Complexity undoubtedly has considerable aesthetic allure. Indeed, Beardsley included complexity (along with unity and intensity) in his criteria for artistic value.[4] A complex work will almost certainly require a longer description than a simple one if we have to explain the work in prose. Therefore, we already have the beginnings of a formal measure, namely, that complexity is proportional to the length of the description of the object. We might wish to refine this statement slightly by saying that it is proportional to the amount of information needed to describe it concisely, thereby avoiding the possibility of increasing the d-complexity of an object merely by 'padding'.

We also note the perception that certain games are more complex than others. Bridge is more complex than snap; it would take great deal longer to explain the intricacies of bridge than to explain snap, presupposing that the listener was interested (if the listener had no interest in bridge, then its complexity (to him) would be zero). On the other hand, although gō (wei ch'i) it more complex than chess, which in turn is more complex than draughts, the rules of gō are simpler to explain than those of chess. We shall return to this point later. As for the i-complexity, this ranges from zero (no meaningful interpretation can be associated with the game) to extremely high, in the minds of those who see the games as microcosms of life and the universe.

7.3 Intrinsic complexity

The intrinsic or structural complexity of an object may be deduced from a direct consideration of its physical features. Generally speaking, complexity increases with the number of states of a system (which also implies an increase in the number of interconnexions). Many systems are networks, in other words they have nodes and connexions between the nodes (in graph theory, the connexions are usually called edges or vertices). Simply counting the number of nodes and the number of connexions can be used as a rudimentary measure of complexity. This is essentially the basis of Halstead's complexity measure of an algorithm.[5] Given an implementation of an algorithm in any programming language, it is possible to identify all the operators and operands. Their total number is the complexity. McCabe's measure associates a directed graph (called the program

---

[3]Cf. R. Tureck, Cells, functions, relationships in musical structure and performance. Proc. R. Inst. 67 (1995) 277–318.

[4]M.C. Beardsley, Aesthetics: Problems in the Philosophy of Criticism. Indianapolis: Hackett Publishing Company (1958).

[5]M.H. Halstead, Elements of Software Science. New York: Elsevier North Holland (1977).

control graph) with a computer program.[6] Each node in the graph corresponds to a block of code in the program where the flow is sequential, and the edges correspond to the program branches (it is assumed that each node can be reached by the unique entry node, and that each node can reach the unique exit node). The complexity is then given by the cyclomatic number $V(G)$ of the graph G with n nodes, e edges, and p connected components:[7]

$$V(G) = e - n + p \tag{7.1}$$

Complexity measures are of considerable practical importance in software engineering, since experience has shown that the time needed to write a functioning program is proportional to the complexity of the algorithm. Empirically, these measures shown good correlation with actual programming times.

If the network (graph) has the form of a tree,[8] which is typical for hierarchical systems, in which elements are clustered according to the strength or importance of their interactions, a very convenient procedure can be used to compute its complexity. The complexity $\mathfrak{C}$ of a tree T consisting of b subtrees $\mathrm{T}_1, \ldots, \mathrm{T}_b$ (i.e., b is the number of branches at the root), of which k are not isomorphic, is defined as[9]

$$\mathfrak{C} = \mathfrak{D} - 1 \tag{7.2}$$

where the diversity measure $\mathfrak{D}$ counts both interactions between subtrees and within them, and is given by

$$\mathfrak{D} = (2^k - 1) \prod_{j=1}^{k} \mathfrak{D}(\mathrm{T}_j^{(i)}) \,. \tag{7.3}$$

If a tree has no subtrees, $\mathfrak{D} = 1$; the complexity of this, the simplest kind of tree, is zero (hence equation 7.2). Any tree with a constant branching ratio at each mode will also have $\mathfrak{D} = 1$ (and hence zero complexity). This complexity measure satisfies the intuitive notion that the most complex structures are intermediate between regular and random ones.

Structural complexity is often defined simply as the number of parameters needed to define the graph. Other measures of network complexity include the number of different spanning trees of the network; the variability of the second shortest path between two nodes (edge complexity); and network or β-complexity, given by the ratio $\mathcal{C}/L$, where $\mathcal{C}$ is the clustering coefficient (the mean number of edges per node) and L is the network diameter (the smallest number of edges connecting a randomly chosen pair of nodes).

Thermodynamic depth is a more physically oriented concept that attempts to measure the process of constructing an object (a complex object is difficult

[6] T.J. McCabe, A complexity measure. IEEE Trans. Software Engineering (SE-2) 4 (1976) 308–320.

[7] It has been proved that V is equal to the maximum number of linearly independent circuits in a strongly connected graph. A linear graph is strongly connected if for any two edges r and s, there exist paths from r to s and from s to r.

[8] A tree is a graph in which each pair of vertices is joined by a unique edge; there is exactly one more vertex than the number of edges. In a binary tree each vertex has either one or three edges connected to it; a rooted tree has one particular node called the root. Their space is ultrametric, satisfying the strong triangle inequality $d(x, z) \leq \max\{d(x, y), d(y, z)\}$ where x, y and z are any three nodes and d is distance between a pair of nodes.

[9] B.A. Huberman & T. Hogg, Complexity and adaptation. Physica D 22 (1986) 376–384.

to construct).[10] According to this notion, the average complexity of a state is the Shannon entropy of the set of trajectories leading to that state (i.e., $-\sum p_i \log p_i$, where p_i is the probability that the system has arrived at that state by the ith trajectory).

From the viewpoint of the designer and maker of an object, its complexity comprises the number n of independent dimensions (in parameter space) that must be specified to describe its shape (for example, $n = 1$ for a sphere, since its radius uniquely specifies it); the precisions $\Delta\ell_i, i = 1, \ldots, n$ with which the dimensions must be specified; and its (a)symmetry, harder to quantify and dependent on current manufacturing technology (for example, it is easy to make round shapes on a lathe, but irregular shapes should pose no problems for molecular manufacturing).[11]

Leaving aside the symmetry, the complexity C is then given by

$$C = n \log_2 \bar{\ell}/\bar{\Delta\ell} \tag{7.4}$$

where $\bar{\ell}$ and $\bar{\Delta\ell}$ are respectively the geometric mean length and the geometric mean precision (tolerance), defined as

$$\bar{\ell} = (\prod_{i=1}^{n} \ell_i)^{1/n} \tag{7.5}$$

and

$$\bar{\Delta\ell} = (\prod_{i=1}^{n} \Delta\ell_i)^{1/n} . \tag{7.6}$$

The rationale for this is to consider C (measured in the bits) to be related to the probability S of success in making the specified object, as $C = -\log_2 S$, and the success depends on the product of the ratios of the lengths and their tolerances, i.e. $S = \prod_{i=1}^{n}(\ell_i/\Delta\ell_i)$.

7.4 Encoding an object

In this digital age, when almost everything is stored in the memory of a computer, we should be very familiar with encoding objects as symbolic strings. It is a prerequisite for carrying out any kind of formal manipulation of the object, including computations such as extracting its complexity. Regardless of the sophistication (or otherwise) of the algorithm used subsequently to compute the complexity, the manner of encoding the object of interest is absolutely crucial to the success and relevance of all subsequent operations, such as the computation of the complexity.

Consider that the object of interest is a pencil. A verbal description could take many forms. It could simply describe what is visible ("a cylinder of graphite encased in wood")—adding further details as required ("painted red", "sharpened to a point at one end", etc.). It could capture its function ("an object used for drawing lines on paper"). It could describe how it is manufactured, or it could describe its history. The description chosen (which could encompass

[10] Sketched out in S. Lloyd and H. Pagels, Complexity as thermodynamic depth. Ann. Phys. 188 (1988) 186-213.

[11] M.F. Ashby, Materials Selection in Mechanical Design. Oxford: Pergamon Press (1992).

all four, and could indeed fill a thick book) depends on the interest of the person making the description (or the person for whom the description is made). Alternatively, we might choose a picture, which could be a stylized drawing, including cross sections parallel and perpendicular to the long axis of the pencil, or a photograph, which might be in monochrome or colour, at different resolutions. Perhaps it would be appropriate to describe the pencil by recording the actions of someone writing with it on cinematographic film. In a sense, the most exact description would be the coördinates and identities of all its constituent atoms—but from this it would be very difficult to discern the function of the pencil. [12]

Once the relevant description has been chosen, if we wish to determine its complexity by a numerical procedure, it must then be encoded in some form permitting computations to be carried out with it. Essentially this means encoding it as a string of symbols. These symbols could be numbers of any base, and without loss of generality we can suppose that the numbers are in base two, that is, the string would consist solely of zeros and ones (binary encoding). Again depending on our interest and purpose, we might specify a very simple encoding, such as zero for vowels and one for consonants if the description is a piece of text; or captured by each successive letter of the alphabet being represented as a number from 1 to 26, extra numbers being used forand punctuation marks, and each number being written with three digits. Alternatively, each complete word in the description could be assigned a unique number, and so forth. If a picture has been chosen as the representation, it could be digitized by placing a grid over it (the choice of grid fineness depending on the desired resolution), and each pixel (grid square) assigned a number corresponding to its grey level—or simply below (0) or above (1) a threshold. The final encoding is then a long string of numbers corresponding to successive pixels, line by line; a line drawing might be more usefully vectorized, i.e. encoded as a series of line segments identified by their starting points, slopes and lengths.

We denote the string encoding the object whose complexity is to be calculated by s.

7.5 Regularity and randomness

Intuitively, one would state that highly regular objects are simple, therefore not complex. The text that consists of the letter 'A' repeated one million times is extremely simple, as is anything that can be represented by such a text (e.g. a polyalanine molecule). Its description is extremely short, namely "repeat 'A' one million times". Similarly with spatial patterns: a checkerboard is very simple to describe ("rows of alternating black-and-white squares"), but the detailed irregular patterning of the fur of a tabby cat would require a very lengthy description. As the regularity diminishes, the description becomes longer and longer, until eventually one reaches a completely random sequence of letters or spatial arrangement of objects, in which case no compression at all of the description is possible, but each letter must be given separately. Randomness may, indeed, be considered to be the antithesis of regularity.

[12]Note the similarity of these procedures with feature extraction in pattern recognition. Pattern indeed is sometimes considered to be synonymous with structure.

The approach of searching for repeated sequences and replacing them with a shorter piece of text is used in data compression algorithms.[13] For example, the phrase "business manager" might be replaced by the code B12 together with information about the positions of its occurrences in the text.

7.6 Information

Information removes uncertainty. Consider the simple example of measuring the length of a piece of wood, known to be at most one foot long. The gain in information achievable by measuring the piece of wood is simply the information available before the measurement, subtracted from the information available after the measurement. The gain presupposes the existence of a world of objects and knowledge, including the ruler itself and its calibration in accepted units of measurement. The 'information' has therefore two parts, a prior part embodied by the physical apparatus, the knowledge required to carry out the experiment or observation etc., and a posterior part equal to the loss in uncertainty about the system due to having made the observation.[14] The prior part (K) can be thought of as specifying the set of possible values from which the observed value must come; in a physical measurement, it is related to the structure of the experiment and the instrument it employs. The posterior part (I) is sometimes called "missing information' ', because once the prior part is specified, the system still has the freedom, quantified by I, to adopt different microstates; in some circumstances the magnitude of I corresponds to the degree of logical indeterminacy inhering in the system, i.e. that part of its description that cannot be formulated within itself; it is the amount of *selective* information lacking.

I can often be calculated according to the Shannon index,[15]

$$I = -\sum_{I=1}^{n} p_i \log p_i \ , \tag{7.7}$$

where the p_i are the probabilities of the n possible outcomes of the measurement. If, in the example at the beginning of this section, the ruler is marked in 1 inch segments, and their probabilities are equal (that is, there is no prior information about the length of the piece of wood, other than that it does not exceed 1 foot), then for all i, $p_i = \frac{1}{12}$. The computed value of i is also a measure of the surprise upon receiving the result of the measurement. K can be quantified using the concept of algorithmic information content (AIC), i.e. the length of the most concise description of what is known about the system, i.e. Kolmogorov information (see §7.7). Hence the total information is the sum of the ensemble (Shannon) entropy I and the physical (Kolmogorov) information (or entropy) K:

$$\mathcal{I} = I + K \ . \tag{7.8}$$

[13] Such as the Ziv-Lempel algorithm (J. Ziv and A. Lempel, A universal algorithm for sequential data compression. IEEE Trans. Inform. Theory 23 (1977) 327–343).

[14] Mackay has proposed the terms 'logon' for the structural (prior) information, and 'metron' for the metrical (posterior) measurement (D.M. Mackay, Quantal aspects of scientific information. Phil. Mag. (Ser. 7) 41 (1950) 289–311).

[15] C.E. Shannon, A mathematical theory of communication. Bell System Tech. J. 27 (1948) 379–423.

To summarize, the Kolmogorov information K can be used to define the structure of information, and is calculated by considering the system used to make a measurement.[16] The result of the measurement is macroscopic, remembered information, quantified by the Shannon index I.[17] The gain in information equals (final – initial information), i.e.

$$I = (I_f + K) - (I_i + K) = I_f - I_i . \tag{7.9}$$

It is unexceptionable to assume that the measurement procedure does not change the structural information, although this must only be regarded as a provisional statement: presumably any measurement, or series of measurements, which overthrows the theoretical framework within which a measurement was made does actually lead to a change in K.[18]

7.7 Algorithmic information content (AIC)

AIC, also called algorithmic or Kolmogorov complexity, and with which the name of Chaitin is also associated, is essentially a formalized version of the "length of description" measure of complexity. The formal definition of the AIC of a symbolic string (encoding the object being described) is "the length of the smallest (shortest) programme P that will cause the standard universal computer (a Turing machine T) to print out the symbolic string and then halt".

Symbolically (but only exact for infinite strings), denoting the AIC by K,

$$K(s) = \min\{|P| : s = C_T(P)\} \tag{7.10}$$

where $|P|$ is the length of the program (in bits) and C_T is the result of running the program on a Turing machine.

The determination of AIC is essentially one of pattern recognition.[19] The pattern of the string "1111111111111111111111111111..." is extremely simple to discern. "0101010101010101010101010101010..." is marginally more difficult, and "001001001001001001001001001001..." slightly more so. But what about "1111000100011110110011010101 1..."? From such a short piece of the string it

[16] K coincides with the term 'ontology' used in some fields such as bioinformatics. Roughly speaking it means "a common understanding across a community"; in bioinformatics this would comprise the general circulation and acceptance of terms used to label the objects in the field (in bioinformatics, these might be cellular components, molecular functions, and biological processes), as well as synonyms and abbreviations, and (in bioinformatics) the association of phenotypic objects with genes.

[17] The information quantified by the index is only the first step, as was fully recognized by its inventor (see C.E. Shannon and W. Weaver, The Mathematical Theory of Communication. Urbana: University of Illinois Press, 1949). The next step is to concern oneself with the meaning of that information (semantics). Sometimes, indeed, the word 'data' is used to mean the raw results of measurements, that is facts or "unconditional information", and 'information' is used to signify data endowed with meaning—in other words incorporating a conditional element, for meaning depends on theories and other constructs of the human imagination. The final step is to consider the action to which the meaningful information leads. Action may include concepts such as understanding (linked to explanation), and the term 'knowledge' is sometimes used to denote the totality of available information together with the understanding that enables one to carry out fresh measurements (in the context of scientific research, for example) and extend the boundaries of available information.

[18] In due course. In all the above, we have not given any consideration to the duration of the physical processes of making the measurements and transmitting their results.

[19] See footnotes 12 and 13.

is of course difficult to discern whether there is any regularity at all (although one can already note that some groups of symbols are repeated). In long strings, such as those encoding a photograph, which might comprise several million pixels, it is very difficult to be certain that one has discovered all patterns. Pattern recognition works by comparing the unknown object with known prototypes; there is no algorithm for discovering patterns de novo. Our brains are extremely good at picking out patterns, presumably because we have accumulated vast numbers of prototypes ever since infancy; it is not however actually known how the brain processes its visual input to ascertain the presence of patterns.

The maximum value of the AIC is equal to the length of the string in the absence of any internal correlations, that is, considering the string as random, viz.,

$$K_{\text{max}} = |s| \ . \tag{7.11}$$

Any regularities, i.e. constraints in the choice of successive symbols, will diminish the value of K from K_{max}, which is the unconditional complexity.

The joint algorithmic complexity, $K(s,t)$, is the length of the smallest program required to print out the two strings s and t:

$$K(s,t) \approx K(t,s) \lesssim K(s) + K(t) \ ; \tag{7.12}$$

the mutual algorithmic information is

$$K(s:t) = K(s) + K(t) - K(s,t) \tag{7.13}$$

(which reflects the ability of a string to share information with another string); and the conditional algorithmic information (or conditional complexity) is

$$K(s|t) = K(s,t) - K(t) \ , \tag{7.14}$$

which is the length of the smallest program that can compute s from t.

The main drawback of AIC as a measure of complexity is that it increases monotonically from a perfectly regular sequence to a perfectly random one, which not only is in disaccord with our intuitive notion of complexity, but also with the fact a random sequence conveys minimal information (possibly even less than a perfectly regular sequence), because it is statistically regular. Although we are unable to predict which symbol will occur next, hence formally the surprise in transmitting each symbol is maximal, no interpretation (and hence meaning) can be ascribed to the sequence of symbols thus accumulated, and in the sense that an equivalent sequence could be generated from a lookup table of random numbers, the information content is zero. Space precludes discussion of such phenomena as the digits of π, which approximates very well to a random sequence, it can be computed by a fairly simple program. A sequence of symbols written in a language that we do not know is essentially random.

Therefore, efforts to find other measures of complexity have continued.

7.8 Effective complexity (EC)

EC was introduced by Gell-Mann[20] in an effort to overcome the problem of AIC increasing monotonically with increasing randomness. EC is defined as

[20] M. Gell-Mann and S. Lloyd, Information measures, effective complexity, and total information. Complexity 2 (1996) 44–52.

the length of a concise description of the set of regularities of the description. The length of the concise description can be computed in the same way as the AIC. This construct does indeed overcome the problem: a very regular symbolic sequence will have few distinct regularities, and therefore a short description of them; a random sequence will have no regularities, and therefore an even shorter description. There will be some intermediate descriptions with many different regularities, which will yield a large EC, which hopefully corresponds to a large complexity. Essentially,

$$\mathrm{EC} = \mathrm{AIC} - \mathrm{RIC} \tag{7.15}$$

where RIC is the random information content.

EC measures knowledge—our knowledge about the object being described, in the sense that it quantifies the extent to which the object is regular (nonrandom), and hence predictable.

EC shares the same technical difficulties of AIC: that of finding regularities, both in compiling an initial list of them, and then in finding the regularities of the regularities. Like any measure of d-complexity, EC is observer-dependent in the sense that the choice of relevant prototypes to use when searching for regularities is made by the observer.

7.9 Physical complexity (PC)

A highly complex text written in a language of which the reader has no knowledge is likely to convey no meaning. A symbolic sequence can therefore be random and meaningless with respect to one observer (environment), yet highly meaningful with respect to another.

In order to capture this feature, and explicitly incorporate the context-dependency of complexity, Adami and Cerf have introduced another notion, that of physical (or mutual) complexity (PC), that attempts to quantify correlations between the symbolic sequence corresponding to the description of the object and the object that it describes,[21] in contrast to the AIC, or EC, which attempt to quantify correlations within the string.

Every object exists in a physical environment e.[22] The environment is all-important: it determines whether a description is meaningful or meaningless. For example, in an environment of Japanese speakers, a description in English is meaningless. It is a common fault of academic lectures that they are delivered without giving the context, the lack of which makes them often largely meaningless. If no environment is specified, then every string is random, conveying no information.

The conditional complexity (cf. equation 7.14) is actually the length $|P|$ of the smallest program that can compute s from e, namely

$$K(s|e) = \min\{|P| : s = C_{\mathrm{T}}(P, e)\} \; . \tag{7.16}$$

It represents those bits in s that are unrelated to e, i.e. they are random with respect to e. If no environment is specified, then all strings have the maximum complexity, $K_{\max}$. The the mutual algorithmic information defined above

[21] C. Adami and N.J. Cerf, Physical complexity of symbolic sequences. Physica D 137 (2000) 62–69.

[22] Kolmogorov-Chaitin AIC is implicitly embedded in the environment of the usual rules of mathematics.

(equation 7.13) is now

$$K(s:e) = K_{\max} - K(s|e)\ . \tag{7.17}$$

It represents the number of meaningful elements in string s. Note, however, that it might not be practically possible to compute $K(s|e)$ unless one is aware of the coding scheme whereby some of e is encapsulated in s.

7.10 Bibliography

J. Crutchfield, The calculi of emergence. Physica D 75 (1994) 11–54.

J. Crutchfield and K. Young, Inferring statistical complexity. Phys. Rev. Lett. 63 (1989) 105–108.

N.R. Hall and S. Preiser, Combined network complexity measures. IBM J. Res. Develop. 28 (1984) 15–27.

G. Kampis and V. Csányi, Notes on order and complexity. J. Theor. Biol. 124 (1987) 111–121.

J.J. Ramsden, Bioinformatics. Dordrecht: Kluwer (2004).

Complexity and Security
J.J. Ramsden and P.J. Kervalishvili (Eds.)
IOS Press, 2008

Chapter 8

Soil as a paradigm of a complex system

Karl Ritz

National Soil Resources Institute, School of Applied Sciences, Cranfield University, Bedfordshire, MK43 0AL, UK

Abstract. Soils form a relatively thin layer that acts as the interface between the atmosphere and lithosphere. They provide a wide range of ecosystem goods and services that both support human societies and underpin the terrestrial components of the earth system. Soils are important to local, national and global security since they protect food and fuel supply, underwrite environmental quality and enhance culture. Furthermore soils have security-related roles in environmental and criminal forensics and as a potential receiver compartment of hostile biological or chemical agents. Soils are remarkably complex in terms of their biological, chemical and physical constitution. Biodiversity belowground always exceeds that found above the surface, particularly at the microbial scale. Soil organic matter is extremely varied, being comprised of the simplest organic molecules through to vast randomly arranged structured polymers that can be highly resistant to decomposition. The physical structure of soils is manifest as pore networks of highly complex geometry that are connected across spatial scales that typically span several orders-of-magnitude. Soils exhibit properties characteristic of complex systems including indeterminacy, non-linearity, emergent behaviour and self-organization. They can serve as a potent model to study complexity, with the added incentive that there is an imperative to manage them sustainably.

8.1 Context: soil and security

Soil forms the outermost layer of the terrestrial system. It is the interface between the atmospheric and subsurface zones and connects to the hydrosphere. It is misleading to refer to soil in the singular sense. In reality, soil is an extremely heterogeneous material in space and time and there are many different soils distributed across the planet, with a concomitantly wide range of properties—the extent of this variability will be considered in more detail below. Although soils, often described in geological terms as 'loose superficial deposits', are generally manifest as relatively very thin layers some tens of centimetres thick, they play absolutely crucial functional roles in driving and governing the Earth system. These roles are myriad and revolve around the provision and delivery of key ecosystem goods and services. Indeed, soils are implicated in the majority of such factors as documented by the Millennium Ecosystem Assessment (2005). The services include:

Supporting. Soils are intimately involved in terrestrial primary production by virtue of the vegetation they support. They are heavily involved in the cycling of all biotically pertinent nutrients, acting as a source, reservoir and transforming matrix for all the major nutrient cycles, particularly nitrogen, phosphorus, potassium and sulphur. Soils provide a supporting platform for terrestrial life: vegetation grows on and in soils, and humankind builds its civilizations upon them.

Provisioning. Soils underpin the provision of fixed carbon via primary production, and hence energy to the bottom of the terrestrial food chain. By virtue of their porous nature, soils store water and modulate hydrological cycles. They underpin the provision of the food, fibre and fuel, upon which humankind relies for its survival.

Regulating. By virtue of their connexion to the Earth system and predominantly via key roles in carbon and nitrogen cycling, soils regulate the climate at a planetary level. In hydrological terms, they regulate flooding and purify water as it passes through them. They also regulate plant, animal and human disease by acting as a reservoir for such agents as well as attenuating pathogen populations.

Cultural. Soils provide a wide range of cultural services to humankind in aesthetic, spiritual, educational and recreational terms. They also provide a significant heritage function in that soils are the repository for the majority of archaeological material.

These ecosystem goods and services map strongly to the breath of constituents of well-being to humankind (Figure 8.1).

In relation to other aspects of local, national and international security, soils play important roles that in contemporary terms are perhaps under-appreciated. Primarily, there is an obvious and crucial requirement for security in the provision of all ecosystem goods and services at a local, national and global scale to protect food and fuel supply, underwrite environmental quality and enhance culture. Of paramount importance is food production, and the security of producing sufficient calories to sustain populations is unquestionably and irrevocably bound to soil. Awareness of this fact is increasingly being lost in those societies where urbanization of the population is dominant or increasing, and there is an associated dislocation of the population from rural environments and the origin of their foodstuffs.

Hence soils represent a fundamentally important resource, and notably one that is grounded in local and national terms yet is connected to the global

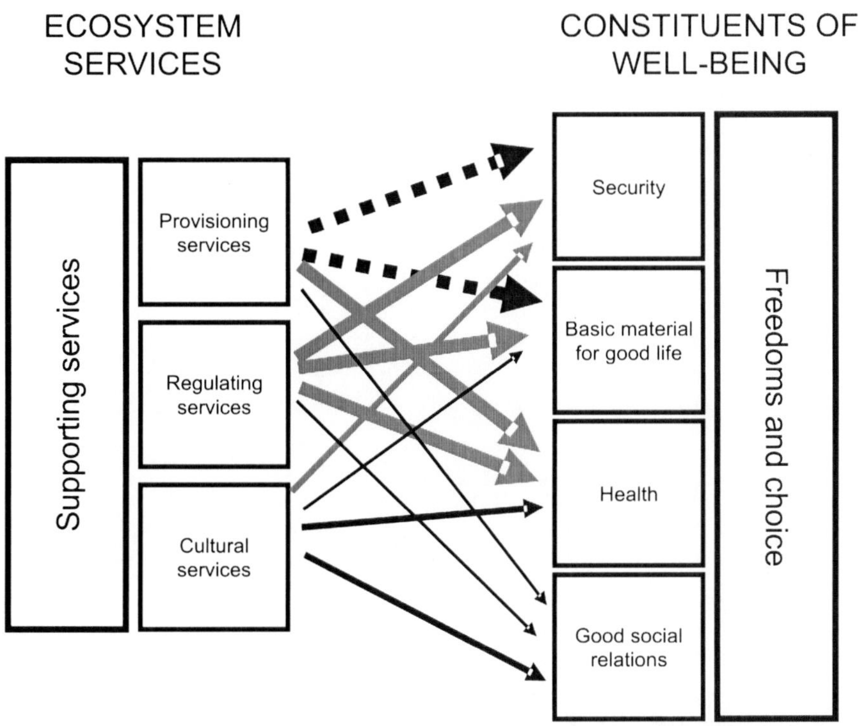

Figure 8.1: Linkages and associated strengths between categories of ecosystem services and components of human well-being that are commonly encountered. Width of arrow indicates strength of linkage, pattern indicates the extent to which the linkage can be mediated by socio-economic factors (black = low; grey = medium; broken = high). Derived with permission from Millenium Ecosystem Assessment (2005).

system. Their relationship to security pervades all these scales. At a local level, soils represent an immediately available and exploitable resource. Nationally, the 'land' prevails within geopolitical boundaries. Globally, the ecosystem and earthsystem services provided by soils connect, and hence local or national management of soils will have consequences for other nations. Such national resources must be conserved and protected, particularly so since soil is effectively a non-renewable resource on the timescale of human societies. Historically, civilizations which have not duly conserved their soils have not prevailed.

The need for soil protection is being increasingly realised at governmental levels nationally an internationally. For example, within the UK, a "Soil Action Plan" has been articulated (Defra, 2004). At a European level, there is the ongoing development of a Draft Soil Framework Directive, underwritten by a strategy which takes a threat-based approach to protecting soil resources (Commission of the European Communities, 2006). The acknowledged threats to soils in this initiative are identified as a decline in organic matter, soil contamination (local and diffuse), sealing, compaction, decline in biodiversity, salination, floods and landslides. Global initiatives also acknowledge such threats and particularly relate to desertification and erosion, such as the United Nations Convention to Combat Desertification (UNCCD).

Two more subtle relationships between soils and security relate to biosecurity and forensics. An acknowledged ecosystem service provided by soils includes modulation of disease, but in intensive systems, or under epidemic circumstances, soils can harbour as well as attenuate contagia. Soils will also inevitably act as receptor compartments for biowarfare or bioterrorism agents to some extent, and in this context urban soils may be particularly pertinent. Similarly, they will act as receptors and potential attenuators for noxious chemical agents. Soils can act as forensic material in a range of criminal or environmental contexts, from two perspectives: (i) sample origination, where the provenance of a soil is important, for example to narrow search areas to locate victims, artefacts or other evidence, and (ii) sample matching where it is necessary to confirm or deny that soil samples are related, for example in matching a suspect to other evidence.

Soils have played crucial roles in the development of the Earth as a planet, life has evolved in the context of soil systems and civilizations have risen and fallen by virtue of their exploitation and management of the earth they have inhabited. Soils continue to support the needs of contemporary societies, and this requirement will unquestionably prevail. Hence soils are closely linked to security and there is a clear need to audit, monitor and manage the resources they represent. The snag is that these tasks are particularly challenging due to the remarkable complexity of soils.

8.2 Soils and complexity

Soils are arguably unique in the way they show extreme structural heterogeneity across many orders-of-magnitude of scale, from mega- to micrometres (Figure 8.2). There is a long heritage of describing soils according to their intrinsic properties, their geological and biogeochemical origins and the ecosystems they support (Krupenikov, 1992). There are a number of extant systems used to classify soils, adopted to varying degrees in different countries, and there has been ongoing debate, which aspires to formulate the most effective systems. A current incarnation that particularly seeks to provide a comprehensive and universal system is the Word Reference Base for Soils (FAO, 2007). Clearly, one of the consequences of soil complexity is the challenge it presents to describe such systems in a coherent manner. There appears to have been relatively little discussion of the inherent diversity (or complexity) of soil constituents, particularly between soils. Indeed, soils are rarely described as being more or less 'complex'. The exception is in relation to the biology, and cataloguing the extent of soil biodiversity is something of a preoccupation in contemporary soil science (Ritz, 2005).

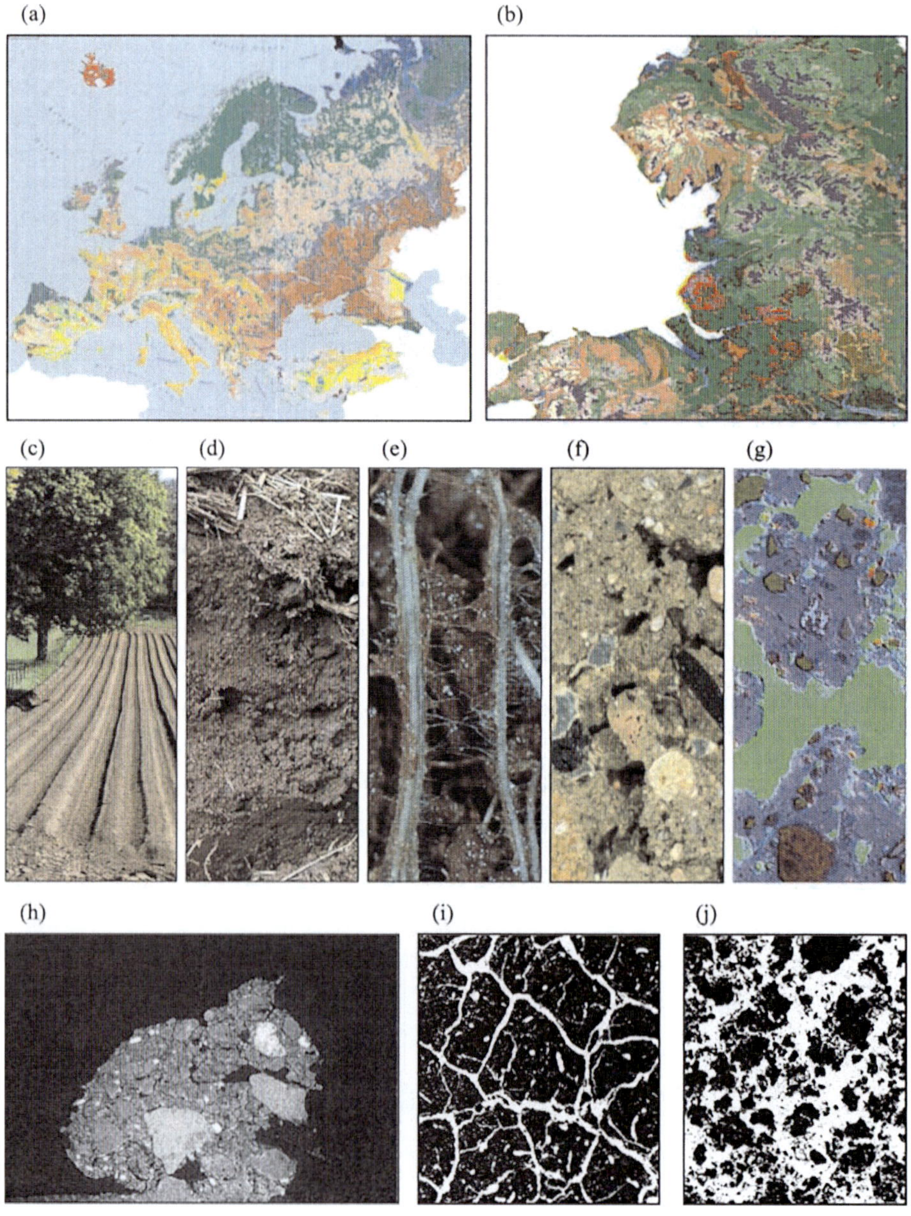

Figure 8.2: Soil structure across nine orders of spatial magnitude. For key see next page.

Key to Figure 8.2

(a) Soil map of Europe; map width *ca* 3000 km. (b) Soil map of part of Wales, U.K.; map width *ca* 150 km. (c) Field-scale view of potato field and adjacent pasture; view width (front) *ca* 5 m. (d) Surface view of arable soil; image width *ca* 15 cm. (e) magnified view of roots and bridging fungal hyphae in grassland soil; image width *ca* 15 mm. (f) Polished surface view of resin-impregnated block of undisturbed arable soil; image width *ca* 500 μm. (g) Thin-section of mineral soil viewed with transmitted light, false colour; image width *ca* 50 μm. (h) X-ray computed tomographic image slice of soil aggregate, grey scale relates to density of material such that black is lowest density; image width *ca* 3 mm, digital resolution = 4.4 μm. (i) and (j) thin-sections of mineral soil image-processed to show pores as white and solid as black; image width *ca* 1 mm. Sources: (a) Joint Research Centre, Ispra; (b) National Soil Resources Institute, Cranfield University; (c)–(h) the author; (i), (j) Crawford et al. (1993), with permission.

The basic principles of soil science are well expounded in a variety of textbooks (e.g. Wild, 1988; Brady and Weil, 2002; White, 2005). A brief overview of some of the major features important to the theme of this chapter is given below. From a physico-chemical perspective, soils are comprised of a wide variety of solid, semi-solid, liquid and gaseous constituents. The solid phases are inorganic ('mineral') or organic (carbonaceous), with the latter often divided into living (the biota) and non-living organic matter. The mineral phases of soils are fundamentally derived from the underlying geology, unless modified by mineral supplements added as a management factor. The development of soils from such parent material is termed pedogenesis. Processes of gradual transformation by biogeochemical actions including 'weathering', which involve heating-cooling (in extremis as freeze-thaw) and wetting-drying cycles, and the action of other chemical, physical and biological erosive mechanisms result in the production of a variably-sized population of mineral particles. These are conventionally classified, more or less arbitrarily, from small (nanometre) to large (millimetre), as clay, silt and sand fractions. These mineral constituents combine with organic components, usually originating in the first instance from biological autotrophic fixation of CO_2 (predominantly plants) to eventually form a relatively thin 'topsoil'. The variety of mineral components in soils can be very large, contingent on the associated geology. Soil minerals provide a solid phase that underwrites the physical structure of the soil, but also plays very significant and consequential roles in terms of providing reactive surfaces in relation to adsorption (and hence sequestration) and desorption of elements, nutrients and organisms. These are often electrostatic charge-based and both positively and negatively charged surfaces are always present in soils, albeit to different degrees and at different spatial scales, depending on the nature of the soil and its history. Clay minerals play particularly important roles in charge-based phenomena in soils and impart many properties which strongly affect soil functions. There are many subtleties in the properties associated with the sand-silt-clay fractionation of soils that are beyond the scope of this chapter.

Soil organic matter originates from primary production, and in large part from terrestrial vegetation. Carbon fixed by photoautotrophs enters the soil via belowground deposition in roots, rapidly in the form of soluble exudates that emanate from growing roots, and more slowly as deposition of cells and tissues. Aboveground plant parts are deposited on the soil surface, in large masses and periodically in temperate systems where the deciduous or annual habit is common, and more consistently over time in other biomes. Such material is incorporated into the soil largely by the action of the soil biota in natural systems. In agricultural systems, particularly intensive-industrially based ones, plant residues were traditionally incorporated into soils by mechanical means. However, there is now an increasing trend away from this approach with the advent of increasing production costs to achieve this (particularly fuel) and the increasing demonstration that such incorporation is not generally necessary to maintain a productive system. Hence zero-till and conservation tillage systems are increasing in their adoption by farmers. Organic material in soils is more or less continuously transformed by a very wide variety of chemical and biochemical mechanisms into a remarkably diverse range of compounds. These span the gamut of organic chemistry from methane (CH_4), through a vast range of intermediates, to huge randomly-structured polymers that are essentially indeterminate and hence have no specific molecular weight or structure. Furthermore, the rates of such transformations apparently range from seconds to centuries, as do the residence times of the compounds. Many of these are biochemicals and often have functional, regulatory or signalling roles in relation to the soil biota.

Fundamentally, soil organic matter contains energy-rich bonds that represent an energy source for the soil biota. Hence soil organisms carry appropriate biochemistries and life strategies to assimilate such energy to enable their growth and reproduction; in doing so, the compounds are further transformed and cycled between compartments. The soil biota can hence be viewed as the 'biological engine of the earth', driving many key processes that underpin soil function, and delivering the ecosystem goods and services described above.

Soil biodiversity transcends that found in all other compartments of the biosphere: the variety of life belowground always exceeds that aboveground. For example, the total fresh weight mass of the biota below an old temperate grassland can exceed 45 tonnes per hectare, at least equal to the aboveground biomass, and equivalent to a stocking rate of about one thousand sheep per hectare. A handful of such soil typically contains tens of milliards of bacteria, hundreds of kilometres of fungal hyphae, tens of thousands of protozoa, thousands of nematodes, several hundred insects, arachnids and worms, and hundreds of metres of plant roots. These large numbers are matched by extreme levels of biodiversity, particularly at the microbial scale. The generic concept of 'biodiversity' is a somewhat plastic one (Huston, 1994; Gaston, 1996), and in soil systems can be applied at a genotypic, phenotypic, functional and trophic level (Ritz et al., 2003). The highest magnitude of soil biodiversity is apparent at the genetic level, and has only become apparent since the 1990s and the advent of appropriate techniques to analyse nucleic acids derived from the soil microbiota. It has been demonstrated that upward of 10 000 genetically distinct prokaryotic types (operational taxonomic units, OTUs) can prevail in 100 g of soil (Torsvik et al., 1990), and subsequently many high-resolution analyses of soil community DNA based on PCR amplification and sequencing of ribosomal DNA and RNA have confirmed such high levels of prokaryotic diversity. Such data has led to the general calculation using species-abundance curves that there may be upward of 4×10^6 prokaryotic taxa in soils at a global level (Curtis et al., 2002), and greater numbers still are postulated (Gans et al., 2005).

The inorganic and organic constituents of soil are arranged in space such that they form a porous network of remarkable properties. The origins of soil pore networks are that the fundamental sand-silt-clay components, mediated by the panoply of organic materials, aggregate via electrostatic, chemical and physical means to form larger units that are always heterogeneous in their morphology. In general, the forces binding such aggregates together are greater at smaller size scales, and hence there tends to be a greater stability of soil structure at smaller scales. These small units then aggregate further to create larger structures, with an incipient hierarchy of scale and structural stability (Tisdall and Oades, 1983; Figure 8.3). Since the units are non-uniform, their packing creates a porous matrix, and since the constituents carry such a wide size-range, the porous network is heterogeneous across a concomitantly wide range of scales, more so than in most other porous media of biological or geological origin. This exceptional heterogeneity of the soil pore network imparts some significant properties to the soil system. Firstly, it modulates the movement of gases, liquids and associated solutes, particulates and organisms through the matrix, by virtue of the connectivity and tortuosity of the network. Path lengths for movement will be increased by tortuosity factors, and size-exclusion mechanisms can operate whereby larger organisms may be prevented from accessing organic matter (potential energy-containing substrate) or prey since such material is located in pores that are smaller than their physical size. Secondly, it means that water can be held in the matrix in a particular manner under gravity; capillary forces in small pores will mean that water is retained under gravitational pull or suction pressure from plants, proportionate to the size of the pore. Consequently, the availability of such water varies and hence modulates processes associated with hydration, such as dissolution and transport of solutes, and biological activity.

8.3 Soils as complex systems

Complex systems are characterized by being comprised of a very large number and variety of constituents within the system and by a concomitantly large number of interactions between such components. Whilst there are no formal thresholds for what constitutes 'large' in these contexts, it is clear that by any account soils are exceptionally diverse, heterogeneous and complicated in chemical, physical and biological terms. There are also some typical properties characteristic of complex systems that arise from these fundamentals, including an inherent non-linearity in the spatio-temporal dynamics of system properties, indeterminacy, emergent behaviour and self-organization. This section will consider how some aspects of soils reflect these criteria and may thus serve as a paradigm of complex systems, with some selected examples of how such properties are apparently manifest in soil systems.

8.3.1 Nonlinearity

It is typically the case that in a system that involves inter connected components, and where there is potential for feedback between such components, system responses are then nonlinear in relation to the behaviour of both some individual subprocesses and at the larger integrative scale. Soil components are consummately interconnected and the potential for feedback mechanisms is very

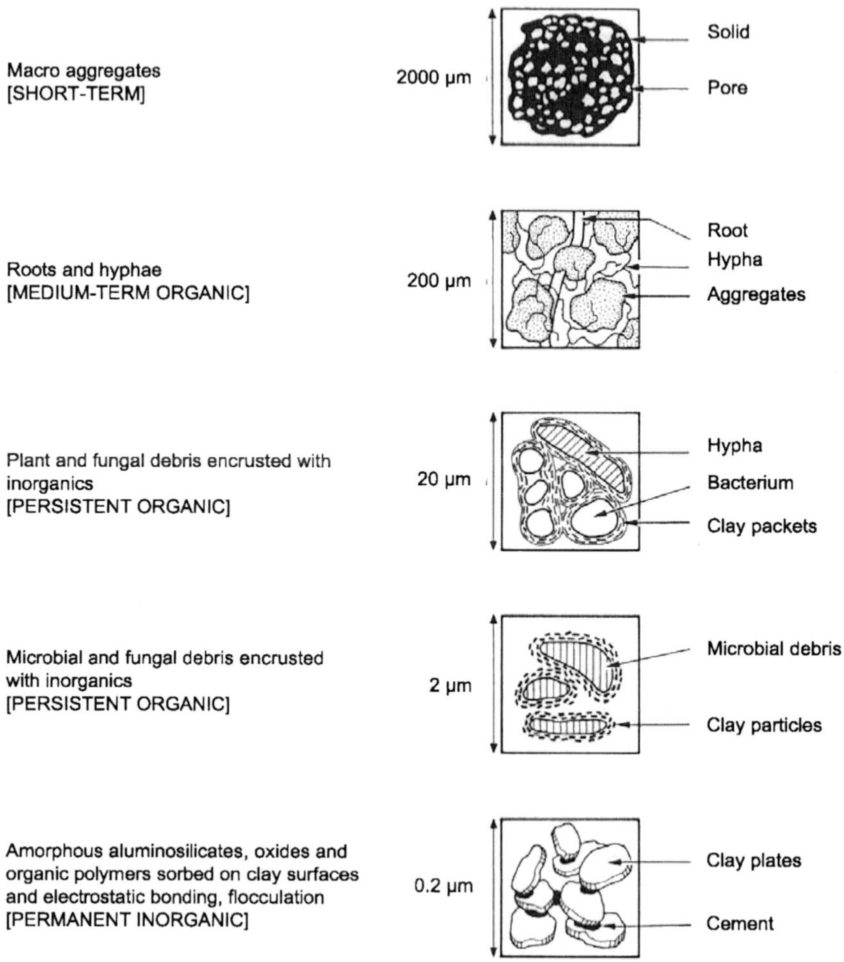

Figure 8.3: Conceptualization of aggregate formation in soils and associated hierarchy of size and stability of resultant aggregated units. Derived with permission from Tisdall and Oades (1983).

large. For example, in relation to the biota, trophic levels are interconnected in various ways (Figure 8.4), and the action of one type of organism can affect other organisms in a positive or negative manner (or remain neutral). In predator-prey relationships both synchronous and asynchronous cycling of populations occurs, and a variety of such patterns have been reported in the context of soil organisms (Moore et al., 2003; Zelenev et al., 2004; Hohberg and Traunspurger, 2005). However, soil organisms can interact in more subtle ways, for example interactions can mediated by volatile compounds, and such interactions can be highly diverse. Mackie and Wheatley (1999) studied such growth responses of some soil fungi to a range of randomly-selected soil bacteria. Growth of pure (single strain) cultures of four species of fungi were measured in the presence of single-isolate cultures of soil bacteria, where physical contact between the organisms was excluded, but since the cultures shared a common atmospheric headspace, volatile compounds could be exchanged. A wide spectrum of growth responses of the fungi were observed, from neutral through 60% inhibition to 35% stimulation compared to control cultures grown in the absence of bacteria. It was notable that growth of the fungal species that does not normally inhabit soil was always inhibited, demonstrating that where evolutionary contact between organisms has not been made, tolerance to bioactive compounds may not be manifest. For the soil-inhabiting species, there was a greater range of interactive effects. This example is also informative since it demonstrates that interactions between soil-inhabiting organisms can be strongly modulated via volatile compounds, which will be an effective mechanism in a spatially structured environment such as soil. Diffusion path lengths, and hence effective distances between organisms, will generally be much further if bioactive molecules diffuse in an aqueous rather than an atmospheric phase, and volatile movement will therefore tend to be faster. Many models of soil processes necessarily make simplified assumptions—the complexity of the system is such that a full consideration of all components is untenable and likely unnecessary—and show 'non-linearity' in their behaviour in that curvilinear behaviour of properties are often predicted. However, few studies explicitly consider soil systems and explore dynamical behaviour from the perspective of complex systems. A notable exception is the work of Manzoni et al. (2004), who modelled soil carbon and nitrogen cycling and suggest that under deterministic conditions, whilst 'traditional' linear models behave like exponential decay functions, explicitly nonlinear models may also show fluctuating behaviour. In their model, they show dynamic bifurcations between stable-node and stable-focus equilibria as a function of climatic parameters such as soil moisture and temperature. They suggest that both data-model comparison and linear stability analysis support the conclusion that linear models are less suited to describe fluctuations in dynamics that arise under certain conditions, and that strong nonlinearity appears when nitrogen-limitation feedback on decomposition is incorporated (Manzoni and Porporato, 2007). Other studies that explicitly consider soils as complex dynamical systems cover organismal population dynamics (e.g. Zelenev et al., 2005; Zelenev et al., 2006; van Bruggen et al., 2007).

8.3.2 Indeterminacy

Indeterminacy is not a prerequisite in the definition of a complex system but it certainly complicates the development of models of how soils function, and chal-

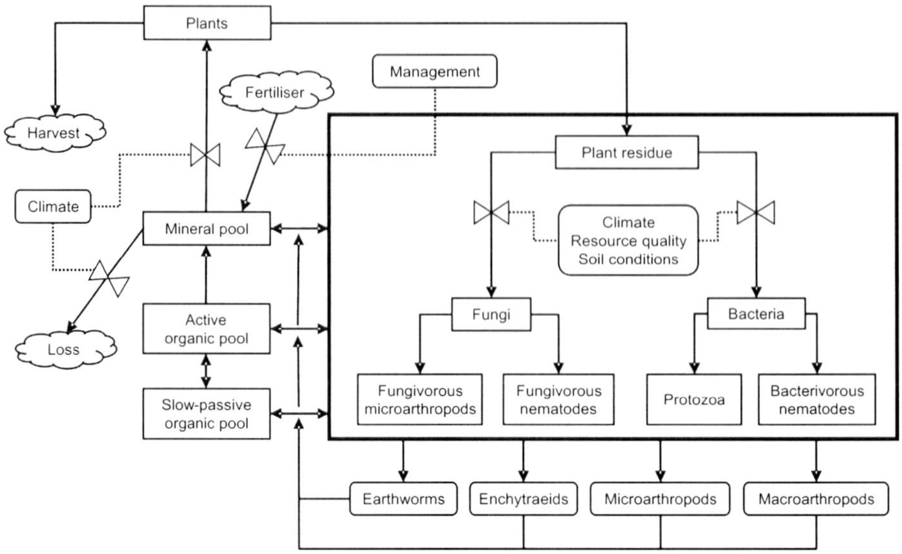

Figure 8.4: Example of major trophic relationships and feedback loops in the soil biological community of an agricultural soil under zero tillage. Derived with permission from Hendrix et al. (1986).

lenges experimentation. Soils are indeterminate in their constitution—which is inevitable given the variety of their constituents—and particularly in their structure. They also support some indeterminately-structured organisms. The pore network is manifest across many orders of magnitude of spatial scale and as such is effectively an indeterminate continuum. Within some limits, all parts of a soil are connected to all other parts at some scale, and as such are potentially or actually influenced by their immediate or remote neighbourhood. The often-rehearsed concept of the 'aggregate' as the fundamental unit of soil is rather flawed, since in the natural situation, notional aggregates are rarely present as such and are only manifest upon disintegration of the system (Young and Ritz, 2005). Aggregates can be considered as discrete volumes of soil out of their normal context, and in general most soils are not, as is often regarded, a bed of aggregates. The often articulated property of 'aggregate size distribution' for soils is also of limited utility since it is in fact contingent on the amount of energy used to create the distribution. Approaches that consider the soil as a spatial continuum, such as dimensional measures including fractal and spectral dimensions (Pachepsky et al., 2000), are generally more appropriate. In most respects, the property of indeterminacy challenges how to deal effectively with scaling issues in soil systems science, i.e. how small-scale processes influence system behaviour at the larger scale, and how to extrapolate experimental or sampling contexts upscale or downscale.

A biological example of indeterminacy in soil systems is demonstrated by the filamentous fungi. These micro-organisms can constitute a substantial proportion of the total biomass in soil, particularly where organic matter inputs are relatively high such as forests and grasslands. The fundamental growth unit of the so-called eucarpic fungi is the hypha, which is a tube-like filament that grows by apical extension and typically branches periodically, to create a branched network termed a mycelium. This structure is well-adapted for growth in porous systems, being able to explore space for food resources by relatively sparse branching, and exploit such material when located via a proliferation of branching. Fungi thus show foraging strategies, that often relate to the characteristic spatial distribution of the substrata they are adapted to utilize (Dowson et al., 1989). Materials can be translocated within hyphae between different zones of mycelia with profound implications for both the fungal organism and soil functioning (Ritz, 2006). Whilst the hypha is typically only a few micrometres in diameter, it can extend virtually without limit if the resources are available and the environment suitable, and mycelia can be colossal in extent. Indeed, the largest organisms on the planet are supposed to be fungal clones (genetically-identical material), which have been mapped in North American and some European forests and shown to extend over several hectares and are estimated to have masses of several hundred tonnes (Smith et al., 1992). Understanding the spatio-temporal community dynamics of microbes such as fungi is particularly challenging due to this indeterminacy. Many fungi show aspects of territoriality, considered to be related to competition for resources—in soil this can be energy-containing substrates or habitable space. When two individual mycelia encounter one another, they will interact with a number of potential outcomes: if they are fully compatible, the hyphae may fuse (anastomose) and effectively create a larger mycelium; they may intermingle and co-exist; they may lay down physical or biochemical barriers and thus exclude each other; or one of the pair may kill and displace the other, possibly assimilating part or all of it in the process. The outcome of such interactions is in itself not determinate and may depend on other factors such as the prevailing and relative sizes of the mycelia, the nutritional status of the organisms and the environmental context, for example the spatial distribution of resources.

8.3.3 Emergent behaviour

In complexity science, emergent behaviour is defined as a property of a system where local or overall behaviour appears when a number of constituent simple entities demonstrate more complex behaviours as a collective than individually. The spatio-temporal dynamics of soil fungal communities provide an example in this subcomponent of the soil system. White et al. (1998) studied this phenomenon where they established the probability of interaction outcomes between pairwise combinations of three species of soil fungi in an experimental system based upon pairs of adjacent square tiles (domains) of agar gel. Larger arrays of such tiles were then inoculated with prescribed spatial patterns of the different species and the resultant dynamics followed. The systems showed strong nonlinearity, and a marked sensitivity to the starting configuration. In some cases, there was a consistent pattern to the development of the spatial arrangement of the communities, but in other configurations a wide variety of patterns developed (Figure 8.5). The system was modelled using the two-tile interaction data but it was not possible to predict the spatio-temporal dynamics using this information alone (Bown et al., 1999). This was postulated to occur since in the larger arrays the context of each fungal domain changed, such that other local and distal factors were important, including the potential for compatible mycelia to anastomose, create larger entities and then out-compete other species in a way not apparent in the two-tile systems.

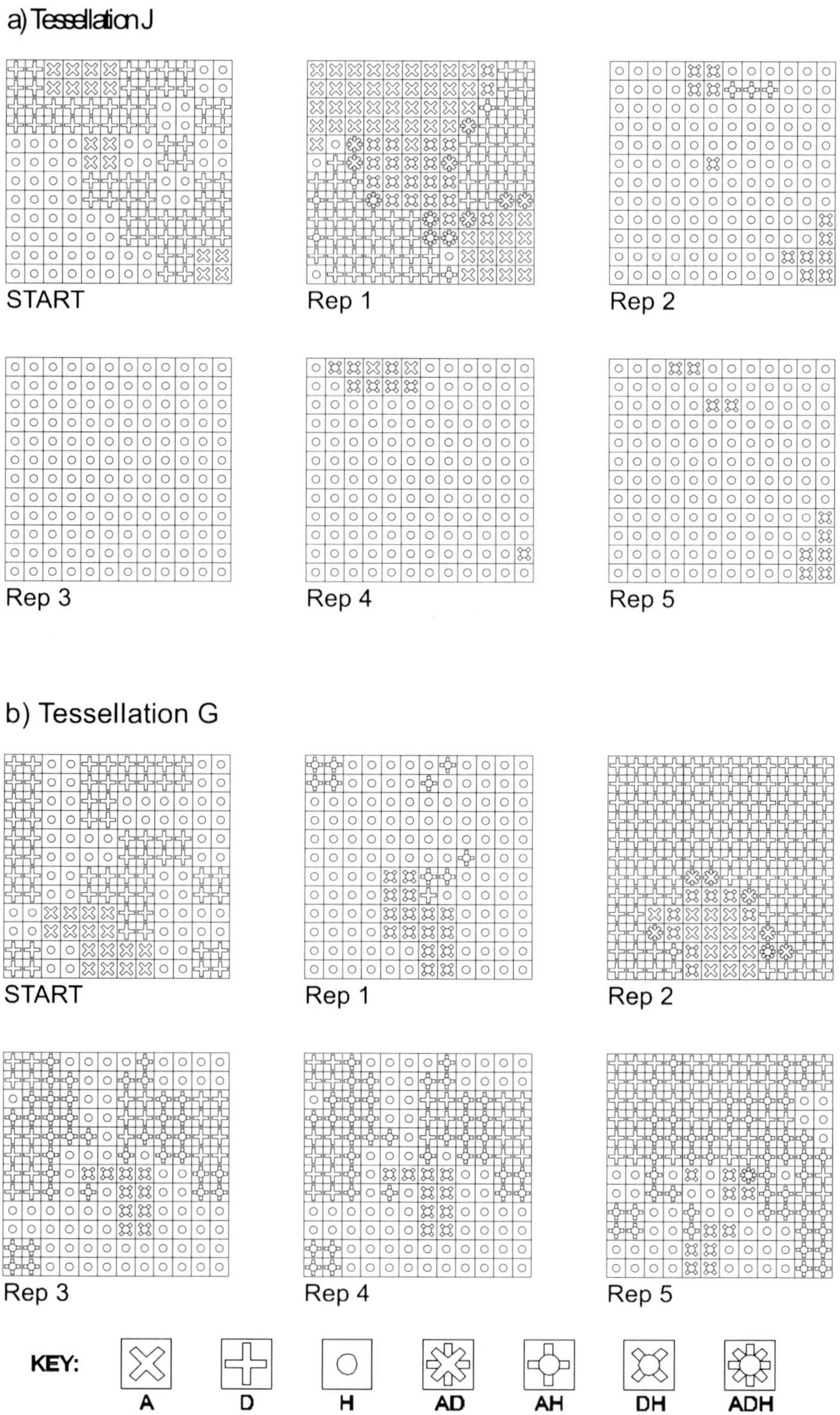

Figure 8.5: Maps showing spatial patterns in fungal communities in model systems based on tessellated arrays of agar tiles. For explanation see next page.

Explanation for Figure 8.5

The system involved inoculating sterile tiles with two prescribed patterns (a) and (b) of three fungal species (A, *Poria placenta*; D, *Coniophora marmorata*; H, *Paecilomyces variotii*) as shown in the "START" map. Each array comprised a matrix of 6×6 tiles each separated by an air-gap of 3 mm; the symbols denote which species were present within each quadrant of each domain. The associated five maps are examples of the emergent spatial configuration in precise replicates of the communities 7 weeks after incubation.. Note how the tessellation in (a) resulted in a more consistent spatial pattern in the emergent community that the tessellation in (b). Derived with permission from White et al. (1998).

8.3.4 Self-organization

Soils function by virtue of their spatial organization—the pore network provides the physical framework which imparts many of the key properties upon the system and modulates the resultant dynamics as described above. If soils are destructured by physical means, for example in laboratory circumstances, or at the field scale via climatic cycles, animal digging or trampling, or following tillage operations, they typically show a subsequent restructuring in that connected pore networks are established. The mechanisms that lead to this are briefly outlined above, and crucially, there needs to be an energy input to the system since it has to involve work, i.e. the physical displacement of solid materials. These processes are typically enhanced by biotic activity. For example, Feeney et al. (2006) grew seedlings of perennial ryegrass in pots of soil that had been destructured by passage through a 2 mm sieve. The plants were constrained to a central cylinder of a fine mesh such that roots were not able to pass into an outer region of soil, but microbes would have had such access. Soil structural properties were measured using X-ray tomography which enabled porosity (at 4.4 μm) and correlation length to be calculated. Compared to non-planted control soils, a significant increase in porosity and correlation length was recorded where plants were present after 30 days. Furthermore, both these properties were significantly greater in the volume of soil containing both roots and microbes compared to microbes alone. These sorts of observations have led to the notion of soils as self-organizing systems, driven by cycles of feed-forward and feed-back processes (Young and Crawford, 2004). It can thus be proposed that soil systems are organizationally impelled to create and modify the internal structuring of the system such that process dynamics ensue. This also represents a form of emergent behaviour, and the concept is certainly linked to that of soils as complex systems. These ideas need further development, in that a coherent framework (or 'theory of soil') remains elusive (Crawford et al., 2005). Complexity scientists should look at soils as a paradigm for the study of the discipline—and as argued in this chapter, they represent a potentially useful paradigm of a complex system with the incentive of an urgent need to better understand their behaviour in order to manage them sustainably.

8.4 References

Bown, J.L., Sturrock, C.J., Samson, W.B., Staines, H.J., Palfreyman, J.W., White, N.A., Ritz, K. and Crawford, J.W. Evidence for emergent behaviour in the community-scale dynamics of a fungal microcosm. Phil. Trans. R. Soc. (Lond.) B 266 (1999) 1947–1952.

Brady, N.C. and Weil, R.R. The Nature and Properties of Soils. Upper Saddle River, New Jersey: Prentice Hall (2002).

Commission of the European Communities. Thematic Strategy for Soil Protection. [COM(2006) 231 final]. Brussels (2006).

Crawford, J.W., Harris, J.A., Ritz, K. and Young, I.M. Towards an evolutionary ecology of life in soil. Trends Ecol. Evol. 20 (2005) 81–87.

Crawford, J.W., Ritz, K. and Young, I.M. Quantification of fungal morphology, gaseous transport and microbial dynamics in soil: an integrated framework utilising fractal geometry. Geoderma 56 (1993) 157–172.

Curtis, T.P., Sloan, W.T. and Scannell, J.W. Estimating prokaryotic diversity and its limits. Proc. Natl Acad. Sci. USA 99 (2002) 10494–10499.

Defra. The First Soil Action Plan for England. London Department of Environment Food and Rural Affairs Publications. (2004).

Dowson, C.G., Springham, P., Rayner, A.D.M. and Boddy, L. Resource relationships of foraging mycelial systems of Phanerochaete velutina and Hypholoma fasciculare in soil. New Phytol. 111 (1989) 501–509.

FAO World Reference Base for Soil Resources, 2nd edn. Rome: Food and Agriculture Organisation of the United Nations (2007).

Feeney, D., Crawford, J.W., Daniell, T.J., Hallett, P.D., Nunan, N., Ritz, K., Rivers, M. and Young, I.M. Three-dimensional micro-organisation of the soil-root-microbe system. Microb. Ecol. 52 (2006) 151–158.

Gans, J., Wolinsky, M. and Dunbar, J. Computational improvements reveal great bacterial diversity and high metal toxicity in soil. Science 309 (2005) 1387–1390.

Gaston, K.J. Biodiversity: A Biology of Numbers and Difference. Oxford: Blackwell (1996).

Hendrix, P.F., Parmelee, R.W., Crossley, D.A., Coleman, D.C., Odum, E.P. and Groffman, P.M. Detritus food webs in conventional and no-tillage agroecosystems. BioSci. 36 (1986) 374–380.

Hohberg, K. and Traunspurger, W. Predator-prey interaction in soil food web: functional response, size–dependent foraging efficiency, and the influence of soil texture. Biol. Fertil. Soils 41 (2005) 419–427.

Huston, M.A. Biological Diversity. Cambridge: University Press (1994).

Krupenikov, I.A. History of Soil Science: from its Inception to the Present. New Dehli, Amerind (1992).

Mackie, A.E. and Wheatley, R.E. Effects and incidence of volatile organic compound interactions between soil bacterial and fungal isolates. Soil Biol. Biochem. 31 (1999) 375–385.

Manzoni, S. and Porporato, A. Theoretical analysis of nonlinearities and feedbacks in soil carbon and nitrogen cycles. Soil Biol. Biochem. 39 (2007) 1542–1556.

Manzoni, S., Porporato, A., D'Odorico, P., Laio, F. and Rodriguez-Iturbe, I. Soil nutrient cycles as a nonlinear dynamical system. Nonlin. Process. Geophys. 11 (2004) 589–598.

Millennium Ecosystem Assessment . Ecosystems and Human Well-being: Synthesis. Washington, D.C.: Island Press, (2005).

Moore, J.C., McCann, K., Setala, H. and de Ruiter, P.C. Top-down is bottom-up: Does predation in the rhizosphere regulate aboveground dynamics? Ecology 84 (2003) 846–857.

Pachepsky, Y., Crawford, J.W. and Rawls, W.J. Fractals in Soil Science. Amsterdam: Elsevier (2000).

Ritz, K. Fungal roles in transport processes in soils. In: Fungi in Biogeochemical Cycles (ed. G.M. Gadd), pp. 51–73. Cambridge: University Press (2006).

Ritz, K. Underview: origins and consequences of belowground biodiversity. In: Biological Diversity and Function in Soils (eds R. D. Bardgett, M. B. Usher and D. W. Hopkins), pp. 381–401. Cambridge: University Press (2005).

Ritz, K., McHugh, M. and Harris, J. A. Biological diversity and function in soils: contemporary perspectives and implications in relation to the formulation of effective indicators. In Agricultural Soil Erosion and Soil Biodiversity: Developing Indicators for Policy Analyses. (ed. R. Francaviglia), pp. 563-572. Paris: OECD (2003).

Smith, M.L., Bruhn, J.N. and Anderson, J.B. The fungus Armillaria bulbosa is among the largest and oldest living organisms. Nature (Lond.) 356 (1992) 428–431.

Tisdall, J.M. and Oades, J.M. Organic matter and water—stable aggregates in soils. J. Soil Sci. 33 (1983) 141–163.

Torsvik, V.L., Goksoyr, J. and Daae, F.L. High diversity in DNA of soil bacteria. Appl. Environ. Microbiol. 56 (1990) 782–787.

van Bruggen, A.H.C., Blok, W.J., Kaku, E., Terniorshuizen, A.J., Berkelmans, R., Zelenev, V.V. and Semenov, A.M. Soil health, oscillations in bacterial populations, and suppression of pathogenic fungi and nematodes. Phytopathol. 97 (2007) S154–??.

White, N.A., Sturrock, C.J., Ritz, K., Samson, W.B., Bown, J.L., Staines, H.J., Palfreyman, J.W. and Crawford, J.W. Interspecific fungal interactions in spatially heterogeneous systems. FEMS Microb. Ecol. 27 (1998) 21–32.

White, R.E. Principles and Practice of Soil Science: Soil as a Natural Resource. Oxford: Blackwell (2005).

Wild, A. Russell's Soil Conditions and Plant Growth (ed.). London: Longmans (1988).

Young, I.M. and Crawford, J.W. Interactions and self-organization in the soil-microbe complex. Science 304 (2004) 1634–1637.

Young, I.M. and Ritz, K. The habitat of soil microbes. In: Biological Diversity and Function in Soils. (eds R.D. Bardgett, M.B. Usher and D.W. Hopkins), pp. 31–43. Cambridge: University Press (2005).

Zelenev, V.V., Berkelmans, R., van Bruggen, A.H.C., Bongers, T. and Semenov, A.M. Daily changes in bacterial-feeding nematode populations oscillate with similar periods as bacterial populations after a nutrient impulse in soil. Appl. Soil Ecol. 26 (2004) 93–106.

Zelenev, V.V., van Bruggen, A.H.C., Leffelaar, P.A., Bloem, J. and Semenov, A.M. Oscillating dynamics of bacterial populations and their predators in response to fresh organic matter added to soil: The simulation model 'BACWAVE-WEB'. Soil Biol. Biochem. 38 (2006) 1690–1711.

Zelenev, V.V., van Bruggen, A.H.C. and Semenov, A.M. Modeling wave-like dynamics of oligotrophic and copiotrophic bacteria along wheat roots in response to nutrient input from a growing root tip. Ecol. Modelling 188 (2005) 404–417.

Complexity and Security
J.J. Ramsden and P.J. Kervalishvili (Eds.)
IOS Press, 2008

Chapter 9

Complexity in materials science and semiconductor physics

Paata J. Kervalishvili

Georgian Technical University, 0175 Tbilisi

Abstract. Today, materials science and technology development is fundamentally complex. Various compounds based on chemical elements, diluted and containing doping materials, isotopically modified substances, biopolymers, etc. demonstrat novel properties different from their elementary and chemically pure materials. The preparation and regulation of these new material properties requires multi-disciplinary science and technology and constitutes a complex system. The most ubiquitous of these new materials are those being used and developed in the field of semiconductors. Complexity in semiconductor physics is manifested by several new phenomena related to quantization effects and has given rise to new solid state electronics - nano-electronics and spintronics.

9.1 Controlled disorders, nanoscience, nanotechnology and spintronics

Nanostructures constructed from inorganic solids like semiconductors have new electronic and optical properties considerably different from that of the common crystalline state due to their size and consequent quantization effects (Ramsden, 2006; Kervalishvili, 2003). The quantization effects reflect the fundamental characteristics of structures as soon as their size falls below a certain limit. An example of the simplest nanostructure is the quantum dot formed from the energy well of certain semiconductor materials with 5–10 nm thickness sandwiched between other semiconductors with normal properties. Quantum dots, for example, have lead to important novel lasers, optical sensors and other electronic devices. The application of nanolayers to data storage, switching, lighting and other devices can lead to substantially new products, for example energy cells, and eventually to the quantum-based internet. Quantum dots are playing a key role in overcoming the problem of an expensively increasing variety of semiconductor materials. A much wider spectral range of optical devices is now available from a restricted number of less expensive material families. For instance InP based lasers are being replaced by InAs quantum dot lasers on GaAs wafers. Nanoscience and nanotechnology encompass the development of nano-electronics and spintronics, spintronic material production, nano spintronic measuring devices and technologies. Advanced nano-electronics is mainly based on self-assembly coupled with novel switching elements and a defect-tolerant architecture.

As the feature size of electronic devices and their integrated elements decreases (while in accordance with Moore's law, their functionality doubles every two years), the complexity of their preparation process rapidly increases. For instance, when the number of electrons decreases, the statistical fluctuation in their number can become an appreciable fraction of the total number of electrons present in the material. However, the number of electrons in a device will approach one in about fifteen years, if present scaling trends continue. Already, single electron transistors exist. In their turn, such devices promise to revolutionize computing systems by integrating tremendous numbers of devices at low cost. These trends will provide new computing opportunities and have a profound impact on the architectures of computing systems.

To move towards this goal it will be necessary to develop the tools able to exploit efficiently the huge computing capabilities promised by nanotechnology in the domain of simulation of complex systems composed of huge numbers of relatively simple elements. A high level modelling and simulation (HLMS) tool (Figure 9.1) enables the convenient generation of a software description (model) of the complex target system and its exploration through simulation on a conventional computer. This tool allows an easy description of the target system by means of an interactive menu. Then, it generates a software description (model) of the system. Based on this description, the tool can perform preliminary simulation on conventional general-purpose computers. Thus, before engaging in the complex task of implementing the system in the nano-network, the user can simulate on conventional computers a reduced version of the system to validate various choices concerning the system parameters and the evolution laws of the system entities (Elliot, 1999).

The nano-network architecture fit (NAF) tool transforms the description of the complex system generated by the HLMS tool into a description that fits the constraints of the nano-network architecture (Figure 9.2). The generation of the latter has to take into account the communication resources of the nano-network and the strategies that could best solve the communication constraints and constrictions, which is often the most challenging task in highly parallel systems. Thus, determining relevant nano-network architectures and communication strategies is essential for implementing an efficient nano spintronics, based on the use of magnetic semiconductors, represents a new area of science and engineering. Principally new materials and devices (Figure 9.3) for information technologies operate via charge and spin degrees of freedom carriers, and are free from the limitations inherent in metal spintronic devices. Storing data

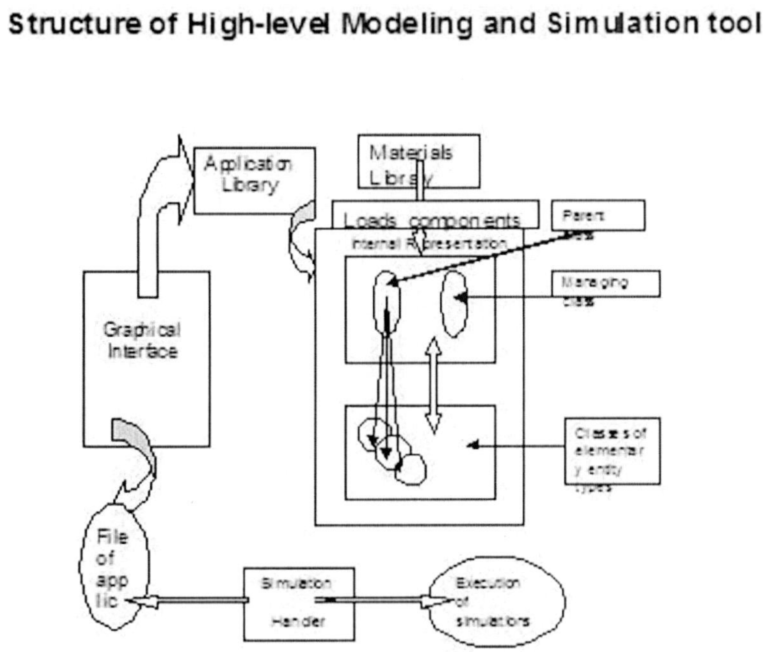

Figure 9.1: Sketch of the structure of the high level modelling and simulation (HLMS) tool.

as fluorine and hydrogen atoms (Bauschlicheret al., 1997) on a diamond surface enables about $10^{1}5$ bytes per cubic cm to be held. The vision of molecular nanotechnology (Drexler, 1992) is based on the development of programmable molecular assemblers, producing atomically precise machines similar to living cells.

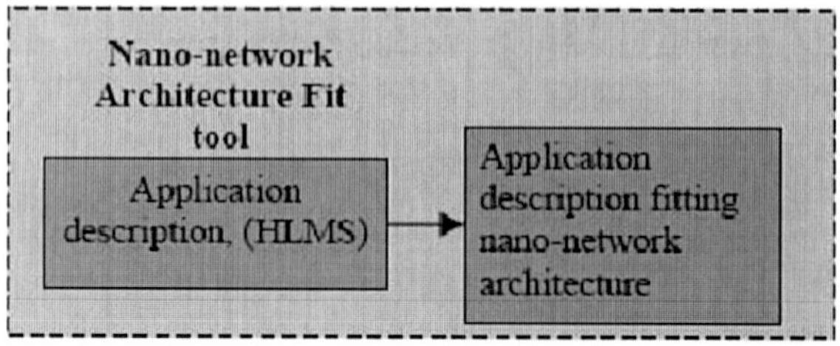

Figure 9.2: The architectural exploration (NAF) tool.

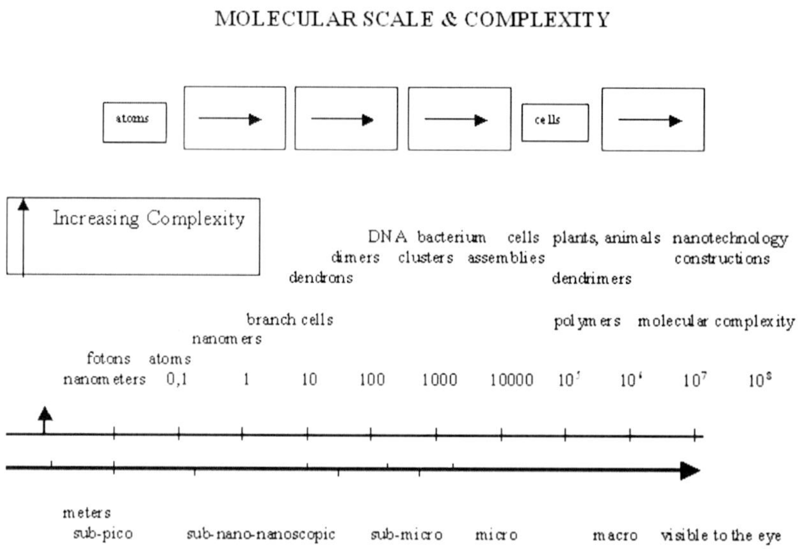

Figure 9.3: Complexity of Micro Nano size particles and structural elements.

The structural complexity of carbon nanotubes (CNT) (Kolmogorov and Crespi, 2000) is now well known. Ever since the landmark discovery of carbon nanotubes, they have been considered as ideal materials for many kinds of application due to their outstanding properties (e.g. mechanical strength, thermal conductivity, electrical conductivity, ultrastability, etc.). Various techniques, including the laser furnace technique, the arc discharge technique, and recently the catalytic chemical vapour deposition technique, have been developed for their high-quality generation on a large scale. Controllable manipulation of the shells of multi-walled carbon nanotubes (MWNT) has evoked interest because of the possibility of their application in nano-electromechanical systems (NEMS) such as ultralow friction bearings, GHz nanoöscillators, nanometer-scale actuators, switches, variable resistors, and tunable resonators (Figure 9.4). In situ manipulation of the nanotube core allows a reversible telescoping motion and, furthermore, allows the associated forces to be quantified. Acid etching is effective for opening nanotube caps but does not expose the inner layers in a controlled way.

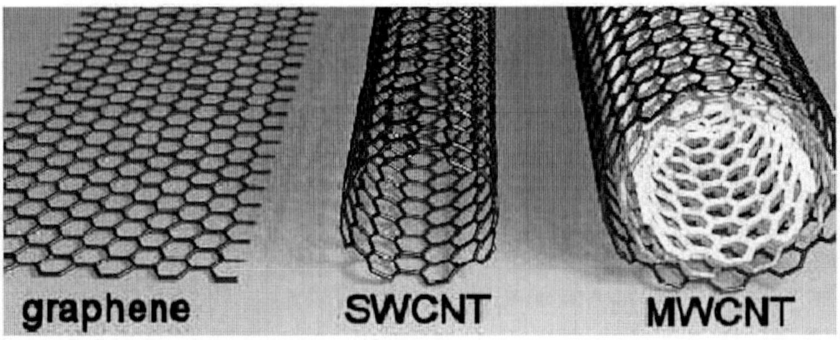

Figure 9.4: Single- and multi-walled carbon nanotubes.

Controllably locating and aligning MWNTs is important for both realizing electric breakdown on a large scale and understanding their behaviour. Most investigations of devices involving CNTs have employed random dispersions of nanotubes onto silicon chips, imaging the entire chip to locate the CNTs, and then fabricating the device at these locations. Though this method can be used to build single devices and for proving concepts in prototype studies, a deeper understanding of device performance and eventual commercialization of NEMS will require processes for handling, locating and aligning nanomaterials in a massively parallel fashion.

The interconnects in an integrated circuit distribute clock and other signals as well as providing power or a ground to various circuits on a chip. The International Technology Roadmap for Semiconductors emphasizes that high speed transmission will be the driver for future interconnect development. In general, the challenges in interconnect technology arise from both material requirements and the complexity of their processing. The susceptibility of common interconnect metals to electromigration at high current densities ($> 10^6$ A/cm^2) is a problem. On the processing side, current technology relies on three steps:

- dry etching to create the trenches/vias;
- deposition to fill metal plugs;
- planarization.

Innovative material and processing solutions are crucial to sustain the growth curve. Due to their high current carrying capability, high thermal conductivity, and reliability, (MWNT) have recently been proposed (Li et al., 2003) as a possible replacement for metal interconnects. Consequently, CNT via diode devices were fabricated for current (I)-voltage (V) electrical measurements. Based on the linear attributes of the $I - V$ curves, the resistance of the CNT diodes was calculated by Ohm's law.

There has been a great interest in the use of biomolecules specifically to actuate and assemble micro and nano-sized systems. In nano-electronics, DNA

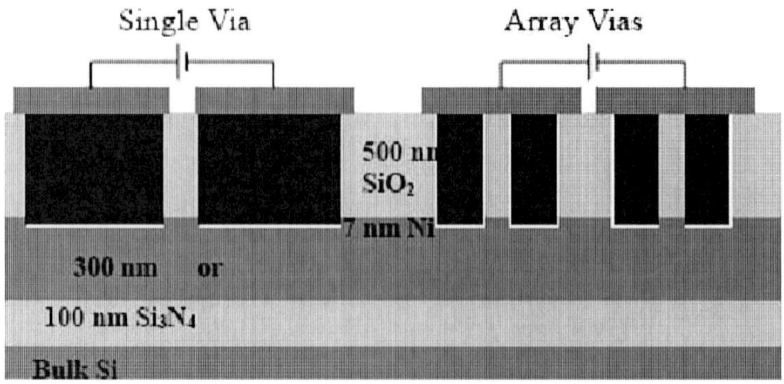

Figure 9.5: Cross section of the CNT via diode structure.

may be used as molecular switches for molecular memories or electronic circuitry to assemble future electronic transistors. In nano-robotics (Figure 9.6), structural elements may be carbon nanotubes, while the passive/active joints may be formed by appropriately designed DNA elements (Hamdi et al., 2005).

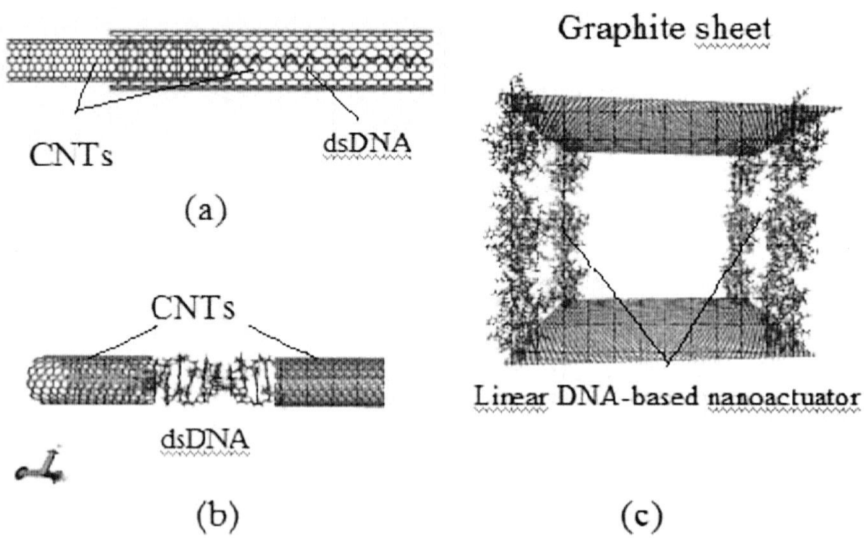

Figure 9.6: Basic DNA-based molecular components: (a) and (b) passive joints; (b) multidegrees of freedom elastic spring; (c) active joints (single degree of freedom parallel platform).

In bionanodevices, nature assembles components using molecular recognition. In the case of DNA, hydrogen bonding provides the specificity behind the matching of complementary pairs of single-stranded DNA to hybridize into a double strand of helical DNA. While these tasks have been performed by nature efficiently and perfectly, robotic engineers will need prototyping tools for the design process of future bio-nano-robotic systems.

Nanomaterials exhibit different properties, as compared to their bulk materials, due to quantum size effects. The quantum size effects are related to alterations in the electronic band structure of the metal atom and the presence of a high fraction of electronically unsaturated atoms at the surface of the metal nanoparticles. These peculiar properties of nanomaterials have important implications in material science such as surface enhanced spectroscopy, catalysis, optics, magnetics, microelectronics, information storage, sensors, photoelectrochemically activated electrodes, magnetic fluids, photonic crystals, biological sensors, medical diagnostics, ceramics, pigments in paints and cosmetics, etc. (Bachmann, 2001). The market for the nanotechnology sector is difficult to quantify, because there is no generally accepted definition and it is a very broad area comprising multiple technological fields and branches (Figure 9.7).

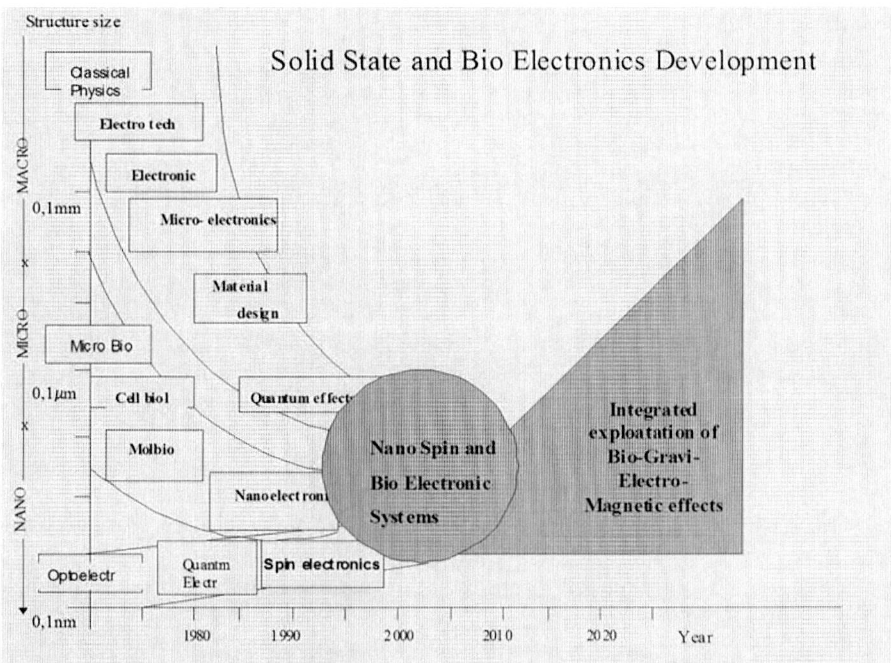

Figure 9.7: Solid state and bioelectronics development.

In addition, many of the nanotechnology areas are at a very early stage of development, which makes an assessment of future market potentials very difficult. Nonetheless there are some market studies available and Table 9.1 summarizes some nanotechnology products that seem to be the most relevant with regard to their market impact in the near future.

Nano-imperfections—high concentration of different defects connected with impurities of metal or non-metal chemical elements—occur for all solid states, even for so-called high temperature superconductors, where grain boundary imperfections and Josephson junctions showed the first evidence of macroscopic quantum behaviour, incoherence and dissipation (Figure 9.8) (Tafuri et al., 2006).

The following analysis has given additional insights into the understanding

Table 9.1: World markets of nanotechnology products (values in milliards of U.S. dollars per year).

Products (forecast)	2002	2007
Nanomaterials		
Nanosized particles	0.5	1.1
Carbon nanotubes	0.01	1.2
Polymer nanocomposites	0.01	1.1
Dyes and pigments	12.0	15.0
Nanotools		
Mask-making lithography	0.5	0.9
Steppers	5.3	7.9
Scanning electron microscopy	0.5	0.8
Chemical vapour deposition	3.6	5.7
Nanodevices		
GMR hard disks	21.8	27.8
Laser diodes	4.7	8.3
OLEDs	0.1	2.5
Field emission displays	0.01	0.05
Nanobiotechnology		
DNA chips	1.0	2.1
Protein chips	0.1	0.6
Drug delivery	0.01	0.05

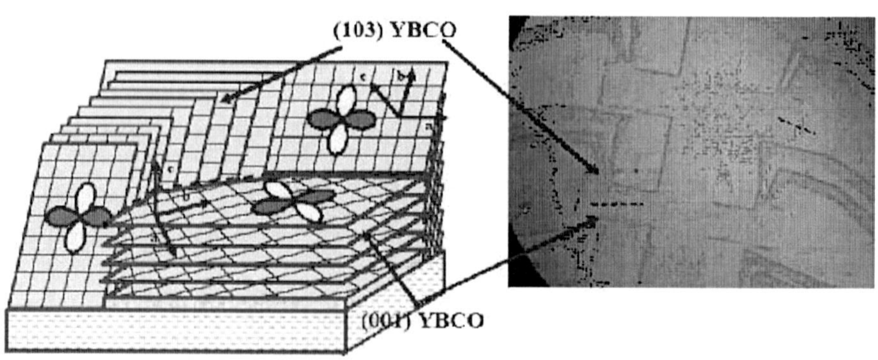

Figure 9.8: (a) Sketch of the biepitaxial grain boundary structure for three different interface orientations indicated the two limiting cases of 0° and 90°, and an intermediate situation defined by a generic angle q. (b) Optical image of various grain boundary junctions with different interface orientations.

of the relaxation processes of low energy quasi-particles. Through a 'direct' measurement of the Thouless energy, the energy scale we found sets a lower bound to the low-temperature relaxation time of the order of picoseconds. This conclusion is consistent with the successful observation of macroscopic quantum effects in Josephson junctions. Novel and very exiting properties of semiconductors are determined by impurities and their disordered distribution. At the same time, the most recent theoretical and experimental achievements have shown that disorders in semiconductors can be controlled and regulated. A contemporary paradigm of complexity and multitask development typical of modern electronics is shown in Figure 9.9.

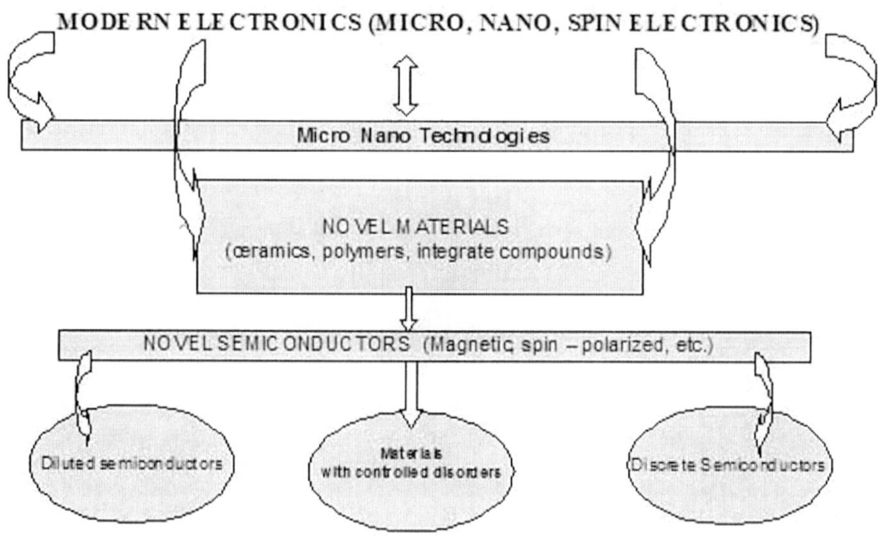

Figure 9.9: Principles of task distribution in novel electronics materials science and engineering.

9.2 Travelling electrical domains on localized states—disorder of semiconductor electronic structures

Current oscillations observed (Shklovsky and Efros, 1984) in weakly compensated semiconductors in the hopping conductivity region are associated with a periodic hopping domain formation and electrons travelling in the bulk of the sample; the domain motion probably occurs along dead end sections of the hopping network, responsible for the drooping region in the current-voltage characteristic. It has been shown that the hopping domains are of a triangular shape with equal leading and back edges, as shown in weakly compensated p-Si depending on the supply voltage applied. With increasing bias voltage, the current pulse shape changes to form a hopping domain. The change in the shape is ascribed to voltage redistribution in corresponding regions in the bulk of the crystal. The very interesting phenomenon is the current instability in the hopping conductivity region, which becomes apparent as a periodic current increase and decrease observed in the external electric circuit. The current oscillations were observed in samples with a sublinear current-voltage characteristic in the portion with a negative differential resistance (Aladashvili and Adamia, 2006). Figure 9.10 shows the dependence of the current density j of p-Si with boron impurity concentration $N_a = 5.9 \times 10^{16}$ cm^{-3} and compensation degree $K = 4 \times 10^{-5}$ with electric field E at $T = 10$ K. As soon as the $j - E$ characteristic begins to flatten, one observes an onset of current oscillation which suggests a region of negative differential conductivity.

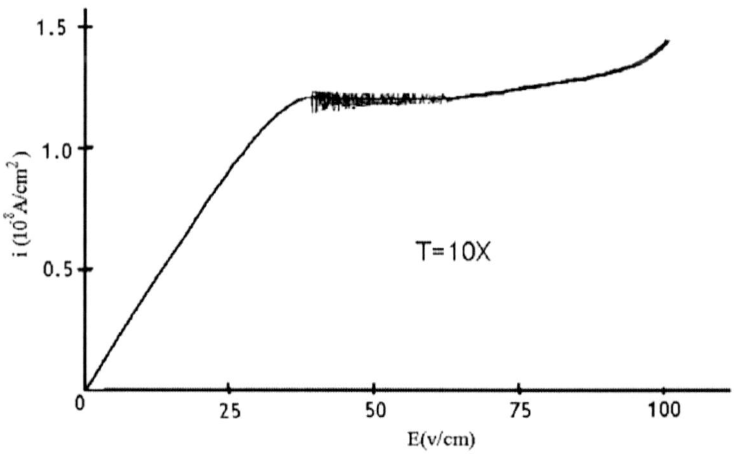

Figure 9.10: Behaviour of the current density as E increases slowly with time (the total duration of the sweep is 10 min).

In general terms the phenomenon is in accord with notion of electrons trapped in a network of acceptors, which governs the ohmic hopping transport. It is governed by an infinite cluster of acceptors, whose separation does not exceed $r_c + a/2$, where $r_c = 0.87 N_a^{-1/3}$ is the percolation radius. The extraction of the infinite cluster is shown in Figure 9.11. The characteristic length of the infinite cluster network is $L = (1/3)(r_c/a)^{\nu} N_a^{-1/3}$, where $\nu = 0.88$ is a critical index. Typical dead ends of the infinite cluster are of the same order. In a weak electric field when $eEL \ll kT$, an electron can easily hop in and out of such dead ends. But in a strong electric field when $eEL \gg kT$, the probability of escape from a dead end oriented against the field is reduced by a factor

$\exp(-eEL/kT)$.

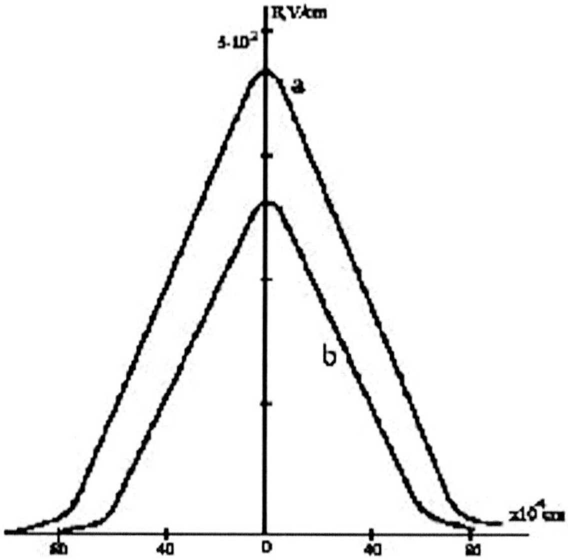

Figure 9.11: The $E(x)$ profiles of uniformly travelling stable domains formed after the application of different voltages (curve a corresponds to the field strength $E_{\text{th}} = 86$ V cm^{-1}; curve b was obtained in the trigger regime with $E_{\text{tr}} = 78$ V.

If the escape out of such dead ends along the electric field can occur only by a hop with the length greatly exceeding r_e then that dead end is a trap for an electron. Thus with the growing electric field, a great fraction of electrons are found in traps. As a result the conductivity decreases exponentially with the increase of E:

$$s(E) = s_0 \exp[-eEL/(2kT)] \quad (9.1)$$

Equation (9.1) leads to a negative differential resistance when $E > 2kT/(eL)$. The nature of domain instability, the potential varying distribution along the sample surface, is revealed using a capacitor probe. The time evolution was observed directly on the oscilloscope connected to the capacitor probe through a high impedance was electrometric amplifier (no lateral coordinate dependence was detected). Samples investigated have resistance of the order of 10^9 Ω and very small operating currents (10^{-9} A), and thus a protective electrode screen was used in the cable connector. The protective screen was connected to the follower with small output impedance, which effectively excluded capacitive and resistive leakage currents since the potential difference between signal wire and its environment was zero. Therefore, at the sample voltage close to the threshold of formation of current oscillations there exists a situation when the first domain is formed. In this case conditions are created for main diode transit régimes—the quenching régime and the lagging régime similar to the transit régime of a Gunn diode. With sufficiently high bias voltage, the conditions of voltage redistribution between the domain moving towards the anode and the next domain being formed on the cathode are realized. The formation of adjustable inhomogeneities in the sample makes it possible to control the current with time, enabling the creation of analogue devices working at low temperatures.

9.3 Diluted magnetic semiconductors

Diluted magnetic semiconductors (DMSs) synthesized by the introduction of magnetic ions into semiconductors are now forming a new horizon for spintronics, especially III-V-based DMSs (In,Mn)As and (Ga,Mn)As. These materials undergo ferromagnetic transitions at comparatively high temperatures and have high connectivity to sophisticated III-V artificial superstructures. A remarkable feature of the ferromagnetism is that it is mediated by the doped holes. It was demonstrated by illumination-driven ferromagnetic transition (Meilkhov and Farzetdinova, 2002) and by the disappearance of the ferromagnetism by counter-doping.

Ferromagnetism in compounds such as $Ga_{1-x}Mn_xAs$ is based on the indirect exchange interaction, which leads to the correct estimation of the Curie temperature within the framework of the traditional mean field theory (MFT). Mn atoms (with concentration N_μ) substituting for Ga atoms introduce in the system their own magnetic moments and, in addition, as acceptors deliver free holes (with concentration n). It is precisely those holes that become carriers responsible for the interaction. However, equality of the concentrations $n = N_\mu$, holds only at low Mn concentrations ($x \approx 0.05$), so that the carrier concentration is usually less than the concentration of magnetic impurities: $n = \gamma N_\mu$, where the coefficient of the impurity 'efficiency' $\gamma < 1$. Nevertheless, the concentration of magnetic impurities, delivering carriers, in actual systems is usually so high that an impurity band is formed which at $x \geq 0.01$ merges into the valence band. Furthermore, it is important that the carrier concentration is almost independent of the temperature: $n = \gamma N_\mu \approx \text{Const}$. The Mn concentration determines the Curie temperature in as-grown samples, and was similar to that reported previously, when the growth conditions were optimized. For example, a 10 K difference in the growth temperature makes the conduction change from metallic to insulating even though the growth mode observed by refractive high energy electron diffraction does not change. This high sensitivity arises from the fact that the growth is very far from the equilibrium; it will be an obstacle for the application of these materials.

Magnetic features of two-dimensional semiconductor systems with magnetic impurities interacting via carriers of arbitrary degeneracy have shown that reducing the system dimension (from 3D to 2D) results in significant lowering of the Curie temperature (Meilkhov and Farzetdinova, 2006). Properties are described using the generalized mean field theory for systems with the indirect interaction of magnetic impurities, and taking into account the randomness of their spatial arrangement by an Ising approximation, and supposing that the indirect coupling between magnetic moments of the impurity atoms is described with the help of the distribution function of local values of the field arising as a result of magnetic ions' coupling with their own surroundings. In real systems, the scattering of those fields proves to be very substantial. The following expression has been derived for the energy $w(r)$ of indirect interaction for two parallel spins S_1, S_2 of magnetic ions spaced at the distance r in the two-dimensional system with degenerated carriers:

$$w(r) = -\frac{m}{4\pi\hbar^2}\left(\frac{J_{\text{ex}}}{N}\right)^2 F(r)S_1S_2, F(r) = -\int_0^{k_F} kN_0(kr)J_0(kr)\mathrm{d}k \qquad (9.2)$$

where J_{ex} is the exchange energy for interaction of a spin with a free charge carrier of mass m, N is the concentration of lattice atoms ($N = 1/a^2$ for the square lattice of the period a); J_0, N_0 are Bessel functions. The result generalizes to the case of the arbitrary degeneracy (with the Fermi energy ε_F of any value) reads

$$F(r,T) = -\frac{1}{r^2}\int_0^{\infty} \frac{yN_0(y)J_0(y)\mathrm{d}y}{1+\exp[(\hbar^2y^2/2mr^2 - \varepsilon_F)/k_BT]}\,. \qquad (9.3)$$

The behaviour of the function (9.3) is determined not only by the temperature as such but also by the temperature dependence of the Fermi energy. In the framework of the standard two-dimensional band and under an invariant carrier concentration, the ratio $\eta = \varepsilon_F/k_BT$ is defined by the relation

$$\exp(\eta(T)) = \exp(\pi\hbar^2 n/mk_BT) - 1 \qquad (9.4)$$

that predicts negative η values at $T = 100$ K if $n \le 10^{12}$ cm^{-2}. Taking into account (9.4), the expression (9.3) could be written in the form $w(\rho) = -J_{\text{eff}}\phi(\rho,\tau)S_1S_2$, where $J_{\text{eff}} = (ma^2/4\pi\hbar^2)J_{\text{ex}}^2$;

$$\phi(\rho,\tau) = -\frac{1}{\rho^2}[e^{2\pi^2\gamma x/\tau} - 1]\int_0^{\infty} \frac{yN_0(y)J_0(y)\mathrm{d}y}{e^{2\pi^2\gamma x/\tau} + e^{y^2/\rho^2\tau} - 1}\,. \qquad (9.5)$$

and $\rho = r/a$ is the reduced separation between interacting impurities, $\tau = 2\pi ma^2k_BT/\hbar^2$ is the reduced temperature (for GaMnAs, $\tau \approx 10^{-3}T$ [K]), and $x = N_\mu/N$ is the relative concentration of magnetic impurities.

Let the system consisting of randomly arranged and oriented Ising spins be in the state characterized by the average reduced magnetization $0 \le j \le 1$. The total interaction energy $W = \sum_i w_i$ of a given spin S_1 with other spins $S_i (i = 2, 3, \ldots)$ is a random value that we shall define by the effective local magnetic field $H = -W/\mu(\mu = g\mu_B[S(S+1)]^{1/2})$ and described by the distribution function $F(j; H)$ depending on the average concentration N_μ of effective magnetic ions and the reduced system magnetization $j = 2\xi - 1$, where ξ is the average fraction of spins of "magneto-active" ions directed 'up'. This equation has the solution corresponding to the ferromagnetic state ($j > 0$) under the condition

$$H_j/\sigma > \sqrt{\pi/2}\,. \qquad (9.6)$$

The upper boundary τ_{Cmax} of the temperature range where the cited condition is satisfied determines the maximum attainable temperature of the ferromagnetic ordering at infinite interaction energy ($I \to \infty$). The Curie temperature at the finite interaction energy could be determined by solving the equation. Relevant non-monotonic dependencies $\tau_C(\gamma)$ are displayed in Figure 9.12. To compare, the dashed line in Figure 9.12 reproduces the dependence $\tau_C(n)$ obtained in the framework of the standard MFT. The optimal carrier concentration is on the order of 10^{12} cm -2.

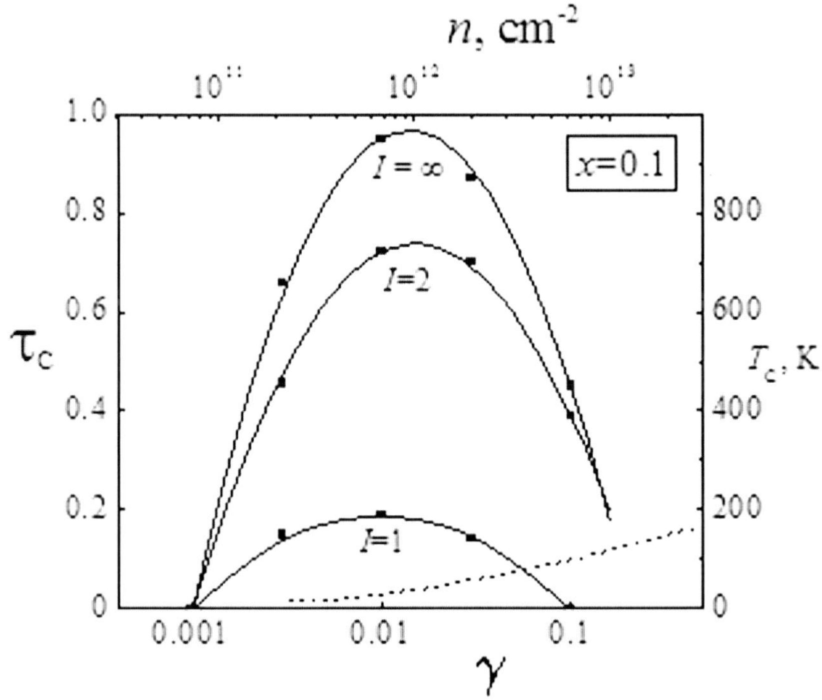

Figure 9.12: Dependencies of Curie temperature τ_C on the carrier concentration for the system with the concentration of magnetic impurities $x = 0.1$ for various interaction strengths I. The dashed line corresponds to $I \approx 1$.

In addition, Figure 9.13 demonstrates a threshold value of the interaction strength I to drive the system into the ferromagnetic state. This is to be contrasted with the result of the standard mean-field theory that predicts no such threshold.

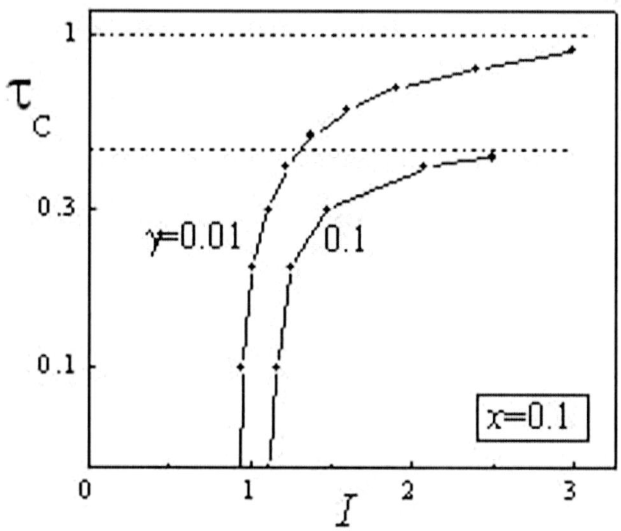

Figure 9.13: Curie temperature $\tau_C(I)$ dependencies for a two-dimensional system with concentration of magnetic impurities $x = 0.1$ at various carrier concentrations determined by the parameter γ.

9.4 Novel polymer nanocomposites for microsensors

One more example of complexity in semiconductor material science is the development of novel polymer nanocomposite materials possessing a wide spectrum of giant tensometric and magnetic properties (Wu et al., 2002). The so called "bottom-up approach" is used to obtain materials containing nanoparticles in a polymeric matrix (e.g. self-assembled monolayers, forming clusters, organic lattices, supermolecular structures and synthesized macromolecules, using chemical synthesis and different deposition methods). The average size of the conductive nanoparticles should be 4–5 nm; the average distance between nanoparticles in a nanomaterial is about 5–6 nm and the conductor content is 10–20% w/w (Kovacevic et al., 2002). Such a high concentration of metal nanoparticles in a nanocomposite film promotes the occurrence of new unique properties having important applications such as ultra-sensitive micro–sensors of pressure, temperature, tension. Also, these materials could be used for detecting various pollutants in the atmosphere, and also could serve as specific and highly effective catalysts. The main problem in the synthesis of nanocomposites with tunnel conductivity and the giant tensile effect is to make an array of conducting nanoparticles embedded into an elastic polymeric matrix.

Practical applications of these materials are connected with their giant sensitivity to mechanical pressure and deformations because the conductivity of such materials is determined by the quantum process of the electrons tunnelling between the conducting nanoparticles (Shevchenko et al., 1995).

The conducting polymeric composites based on compositions of caoutchouc possess a high sensitivity to mechanical effects when the density of the conducting components (metal, carbon particles) is close to the percolation threshold. This property is connected with the change in their electrical resistance under mechanical deformation. It is important to point out that similar compositions are obtained with conducting particles of rather larger size (0.1–1 μm). Utilizing nanometer-size conducting particles leads to the quantum tunnelling character of conductivity sharply increasing the sensitivity to deformations. Calculations show that near the percolation threshold the tensoresistivity of such a nanocomposite is enormous, and the conductivity may change by a few orders of magnitude at deformations of 1–10%, when the volume content of conducting nanoparticles in the matrix is about 20%. To compare, the coefficient of tensoresistivity K (the ratio of corresponding conductivity and deformation changes) is about 100 for existing semiconductor materials. The novel nanocomposites have a much higher sensitivity ($K \sim 10^3$–10^6), and will be capable of measuring deformations with submicron accuracy.

Practical interest in these novel materials (smart polymers) is due to the possibility of creating new electronic devices and resistive-strain sensors for the constant monitoring of the deformation of buildings, for the control of temperature and different mechanical parameters (for example, measuring thickness, distances, etc.), to measure (submicron) deformations in different areas of science and engineering, in particular, in audio devices, for the development of the high-sensitivity quick-response pressure and temperature sensors etc, and for producing flexible keyboards and switches (Tchmutin et al., 2003).

The magnetically-controlled elastic composite materials (magneto-elastics) produced by depositing ultrafine magnetic particles in a polymer matrix gives rise to the giant magnetostriction effect, which was first observed in composites with magnetic particles of the size of 3 microns. The synthesis of magneto-elastic nanocomposites heralds the novel and unique possibility to produce nanocomposites with giant tenso-resistivity and giant magnetostriction effects in one material.

Other possibilities for the application of magnetic nanocomposites are connected with the giant negative magnetoresistance effect. Significant changes of the negative magnetoresistance (up to 12%) have been observed in these nanocomposites at room temperature. This spintronic effect may enable the creation of microsensors of magnetic fields.

The interphase layers that extend some distance away from the bound surface give rise to properties dramatically different from that of the bulk polymer , because of the differences in structure (Varfolomeev et al., 1999). The interphase is important mechanically because its distinct properties control the load transfer between matrix and filler. The concept of interphase is not unique for nanocomposites, but due to the large surface area of nanoparticles, the interphase can easily dominate the properties of the material. The 1 nm thick interphase on 1–10 mm particles in a composite represents as little as 0.3% of the total composite volume. However, the 1 nm thick interphase on 4–8 nm particles can reach 30% of their total volume.

Another critical parameter of nanocomposites is the dispersion of the nanoparticles in the polymer matrix (David et al., 2004). The dispersion of inorganic nanoparticle filler in thermoplastics is not easily achieved because nanoparticles have the strong tendency to agglomerate in order to reduce their surface energy. This may be controllable by the use of magnetic fields to arrange the magnetic and structural disorder in the polymers.

9.5 Spin-polarized transport in semiconductors

The study of spin-polarized transport in nanosize multilayer structures consisting of alternating layers of ferromagnetic metals and non-magnetic semiconductors is currently popular. Operation of a spintronic device requires efficient spin injection into a semiconductor, spin manipulation, control and transport, and spin detection. The investigation of new ferromagnetic materials that are reliable and good spin injectors has lead to the use of discrete magnetic alloys. These are multilayer systems composed of submonolayers of a ferromagnetic material in the matrix of a non-magnetic semiconductor, for example, Mn/GaAs or Mn/GaSb. These alloys have high Curie temperatures and sufficiently high spin polarizations. However, it is important to control and manage the ferromagnetic metal-semiconductor boundary surface during the synthesis of these materials (Meilikhov et al., 2000). Only the methods of MOS hydride epitaxy and laser epitaxy using pulsed annealing of the epitaxial layers are effective. These technologies allow doping of the layers in an oversaturated condition (Figure 9.14). Spintronics has perspectives for the development and creation of new types of a nonvolatile random access memory (MRAM), quantum single-electron logical structures, and ultradense information storage media. Thus, the elementary information storage unit will be represented by an electron spin. This case will probably represent the limits of magnetic information recording The realization of the spin-polarized current transfer enables new possibilities for solid-state electronics. For instance, there are observations of spin-polarized luminescence, and the creation of high frequency diodes whose output characteristics may be changed by an external magnetic field. Another example is the possibility for a new generation of narrow-band devices in solid-state electronics at millimetre and submillimetre wavelength ranges like generators, amplifiers, receivers and filters, modulated and frequency-tuned by a magnetic field.

The discovery of the giant magnetoresistance effect (GMR) by Fert and colleagues in 1988 can be considered as the beginning of spintronics. This phenomenon is observed during the study of thin films with alternating layers of ferromagnetic and non-magnetic metals. It is found that, depending on the width of a non-magnetic spacer, there can be a ferromagnetic or antiferromagnetic interaction between magnetic layers. The antiferromagnetic state of the magnetic layer can be transformed into a ferromagnetic state by an external magnetic field. The spin-dependent scattering of conduction electrons is minimal, resulting in a low resistance, when the magnetic moments are aligned in parallel, whereas for an antiparallel orientation of the magnetic moments, the

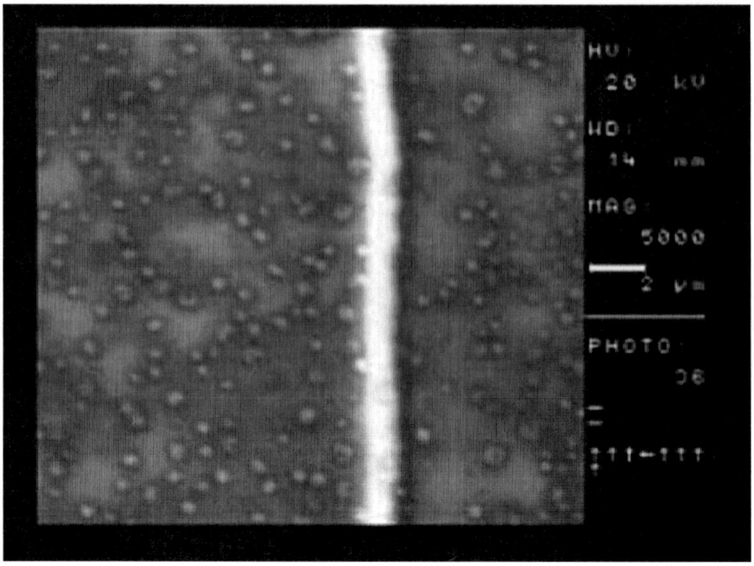

Figure 9.14: SEM images of Mn/GaAs samples prepared by laser epitaxy.

situation is reversed. The GMR effect has demonstrated that spin-polarized electrons can carry a magnetic moment through non-magnetic materials while preserving spin coherence (spin transport).

Sensors operating with magnetic tunnel junctions (MTJ) fall into the second class of spintronics devices. Here, the ferromagnetic electrodes are divided by a very thin dielectric layer, and electrons tunnel through a nonconducting barrier under the influence of an applied voltage. The tunnel conductivity depends on the relative orientation of the electrode magnetization, and tunnel magnetoresistance (TMR) is small for parallel alignment of magnetizations and large in the opposite case. In contrast with the GMR devices, the electrodes are magnetically independent and have different critical fields for changing the magnetic moment orientation. The first laboratory samples of ($NiFe/Al_2O_3/Co$) MTJ structures were demonstrated by Modera and colleagues in 1995, where the TMR effect reached 12% at room temperature (Ziese, 2001). Some of the largest manufacturers of electronics, including IBM, have declared recently the development of new memory devices: the so-called MRAM. They include storage units based on MTJ structures that provide increased storage density and access speed, and non-volatile data saving when the power supply is removed. The first industrial designs of such memory devices appeared in 2003. A disadvantage of these devices is the small scale of integration, and the necessity of additional controlling transistors. These limitations may be overcome with the development of semiconducting spin electronics, and in particular with the creation of spin transistors. In this case, spin electronic devices will not only switch or detect electrical and optical signals, but also to enhance them.

The third class of spin electronic devices is based on the development of multilayer nanostructures of ferromagnetic semiconductors, which demonstrate properties not available to their metal analogues. These devices are controlled by an electric field and rely on the giant planar Hall effect, which exceeds by several orders of magnitude the Hall effect in metal ferromagnets. The super-giant TMR effect with Hall effect control was observed for the first time in epitaxial (Ga,Mn)As/GaAs/(Ga,Mn)As structures. There are no effective ways to inject the spin-polarized current into nonmagnetic semiconductors at the present moment. Spin injection from magnetic semiconductors into nonmagnetic ones gives good results in a number of cases, but only operates at low temperatures, far from room temperature.

Interest in so called diluted magnetic semiconductors was given an impetus by the recent demonstration of the ferromagnetic critical temperature $T_c = 110$ K in GaMnAs. To date, most theoretical models proposed assume that the holes occupy a Fermi sea in the valence band. Theoretical models based on the virtual crystal approximation have been used to study the influence of disorder on transport and the magnetic properties of magnetic semiconductors. The Boltzmann equation with Born-approximation scattering rates has provides estimates of the anisotropic magnetoresistance effect of order up to 12%. The key property of the kinetic and magnetic anisotropy effects is a strong spin-orbit coupling in the basic semiconductor valence band (Kervalishvili, 2005). The most striking feature in off-diagonal conductivity coefficients, for example in (GaMn)As and other arsenides and antimonides, of diluted magnetic semiconductors is the large anomalous Hall effect, due to spin-orbit interactions. In metals, the standard assumption is that the anomalous Hall effect arises because of the spin-orbit coupling component in the interaction between band quasiparticles and crystal defects, which can lead to skew scattering with a Hall resistivity contribution proportional to diagonal resistivity. For diluted magnetic semiconductors, the anomalous Hall effect is based on spin-orbit coupling in the Hamiltonian of the ideal crystal and implies a final Hall conductivity even without disorder. The effects of the As_{Ga}-As_i-V_{Ga} transition (where As_{Ga} is an As atom on a Ga site, As_i is an interstitial As, and V_{Ga} is a Ga vacancy) on the ferromagnetism of (GaMn)As can be explained by the Mulliken orbital populations of the d-shell for both the majority and minority spins and the corresponding spin polarization for the ferromagnetic configuration. In this case, the ferromagnetic coupling is strengthened considerably by the distortion, which together with the energy splitting and Mulliken orbital population of As_iV_{Ga} gives rise to effects very similar to those of defect-free (GaMn)As. These suggest that the ferromagnetic order in (GaMn)As is unaffected by the presence of As_iV_{Ga} pairs. This result is in agreement with the hole-mediated picture of ferromagnetism, and can be understood by noting that As_iV_{Ga} defect energy levels show minimal splitting in (GaMn)As. More detailed studies of disorders will combine the Kondo description of the spin interactions as well as relevant Monte Carlo techniques applied to both metallic and insulating conditions.

9.6 Modelling of quantum systems—the way of quantum device design

The development of novel nanotechnology necessitates instruments that are able to manipulate at the level of Ångström (0.1 nm) units. Usually the building of such-sized particles is based upon self-assembly. The simulation and modelling of self-assembling processes is thus essential. Quantum system simulation and modeling elaborate atom- and molecular-scale physical effects, such as superconductivity and superfluidity and other important phenomena.

Quantum simulation is very complex. For the system of n simple quantum particles (quantum bits or qubits), the number of degrees of freedom increases as 2^n. For example, to model the hydrogen molecule, where four particles are participating (eight qubits) it is necessary to model 256 degrees of freedom. The current route for solving this problem for the case of systems of hundreds of electrons relies upon suitable approximations in the model-building process.

There are several possibilities for the construction of a quantum computer: nuclear magnetic resonance for liquid substances may yield a structure of up to 10 qubits, the Josephson transitions method for solids yields 2 qubits. At the same time so-called virtual experiments and programmes suitable for quantum computing are well developed (Thaller, 2004), and some use existing computers (Kervalishvili, 2007).

9.7 Conclusion

Modern electronics (micro-nano-spin-electronics) and its developments are based on complex systems that include novel materials (metals and non metals), their preparation technologies and new properties. Modification of material properties by different structural imperfections (structural defects: impurities, isotopes, etc.) is the instrument for control of the characteristics of these new systems.

Acknowledgment. Most of the research results included in this paper were obtained within the framework of the ISTC multinational integrated project N 1335.

9.8 References

1. Ramsden, J.J. (2005). What is nanotechnology? Nanotechnology Perceptions 1, 3–17.

2. Kervalishvili, P. (2003). Nanostructures as structural elements for semiconductor crystalline layers preparation and their testing. Proceedings of the International Conference on Materials Testing, AMA - Nuremberg, 13-15 May, pp. 107–112.

3. Elliott, J. (1999). Understanding Behavioral Synthesis: a Practical Guide to High-Level Design. Kluwer Academic Publishers.

4. Drexler, K. (1992). Nanosystems: Molecular Machinery, Manufacturing and Computation. Wiley.

5. Bauschlicher, C., Ricca, A., Merkle, R. (1997). Chemical storage of data. Nanotechnology 8, 1–5.

6. Kervalishvili, P., Kutelia, E., Petrov, V. (1985). Electron-microscopic investigation of the structure of amorphous boron. Am. Inst. Phys., N-0038 5654/85/05 0853/03, pp. 8–13.

7. Kolmogorov, A., Crespi, V. (2000). Smoothest bearings: interlayer sliding in multiwalled carbon nanotubes, Phy. Rev. Lett. 85, 4727–4730.

8. Li, J. et al. (2003). Bottom-up approach for carbon nanotube interconnects. Appl. Phys. Lett. 82, 2491–2493.

9. Hamdi, M., Sharma, G., Ferreira A., Mavroidis D. (2005). Molecular mechanics study of bionanorobotic components using force feedback. IEEE International Conference on Robotics and Biomimetics, 30 June–3 July, Hong Kong, pp. 105–110.

10. Bachmann, G. (2001). Market opportunities at the boundary from micro- to nanotechnology. MicroSystemTechnology News 3, 13–14.

11. Tafuri, L. et al. (2006). Coherent quasiparticle transport in grain boundary junctions employing high-Tc superconductors. Proc. Conf.European Nanosystems, 13–15 December Paris, pp. 140–144.

12. Shklovski, B., Efros, A. (1984). Electronic properties of doped semiconductors. Springer.

13. Aladashvili, D., Adamia, Z. (2006). Features of the traveling electrical domains on localized states. Novel Materials 1, 31–35.

14. Meilikhov, E., Farzetdinova, R. (2002). Ultrathin Co/Cu (110)-films as a lattice of ferromagnetic granules with dipole interaction. JETP Lett. 75, 142–146.

15. Meilikhov, E., Farzetdinova, R. (2006). Quasi-two dimensional diluted magnetic semiconductors with arbitrary carrier degeneracy. Novel Materials, 1, 60–63.

16. Wu C., Zhang M., Rong M., Friedrich K. (2002). Tensile performance improvement of nanoparticle-filled polypropylene composites. Composites Science and Technology 62, 1327–1331.

17. Kovacevic, V., Lucic, S., Leskovac, M. (2002). Morphology and Failure in Nanocomposites. Part I: structural and mechanical Properties. J. Adhesion Sci. Technol., 16, 1343–1347.

18. Shevchenko, V., Ponomarenko, A., Klason, C. (1995). Strain-Sensitive Polymer Composite Material, J. Smart Materials and Structures, 4, 31–36.

19. Tchmutin, I., Ponomarenko, A, Krinichnaya, E. (2003). Electrical properties of composites based on conjugated polymers and conductive fillers. Carbon, 41, 1391–1395.

20. Varfolomeev, A., Volkov, A., Cherepanov, V. (1999). Magnetic properties of polyvinyl alcohol-based composites, containing iron oxide nanoparticles. Advanced Materials for Optics and Electronics 9, 87–93.

21. David, K., Dan, N., Tannenbaum, R. (2004). Competitive adsorption of polymers on metallic nanoparticles. Report N 156, Georgia Institute of Technology.

22. Meilikhov, E., Aronzon, B., Gurovich, B., Kuleshova, E. (2000). On extreme density of data storage in patterned magnetic media. MRS Proc. Magnetic Materials, Structures and Processing for Information Storage, F1, 614, 5–11.

23. Ziese, M., Thornton M. (2001). Spin Electronics. Springer.

24. Kervalishvili, P., (2005). Semiconducting nanostructures—materials for spintronics. Nanotechnology Perceptions 1, 161–166.

25. Thaller, B. (2004). Visual Quantum Mechanics. Springer.

26. Kervalishvili, P. (2007). Quantum processes in semiconducting materials and spinelectronics. Rev. Adv. Mat. Sci. 14, 14–34.

Part III

Climate and Energy

Complexity and Security
J.J. Ramsden and P.J. Kervalishvili (Eds.)
IOS Press, 2008

Chapter 10

Introduction to global warming

Graham C. Holt[a] and Jeremy J. Ramsden[b]

[a] *Collegium Basilea (Institute of Advanced Study), Basel, Switzerland*
[b] *Cranfield University, Bedfordshire, UK*

Global warming is the ultimate security challenge. Even with our unprecedentedly sophisticated technologies, weather, much less climate, remain essentially beyond human control. As an example, a few hours of unusually heavy rainfall in Yorkshire on 25 June 2007 generated catastrophic floods that caused several tragic deaths and have given rise to insurance claims of around £1500 million—equal to the entire cost of home insurance claims in the UK in 2006, and roughly equal to the gross domestic product (GDP) of Liberia. As another example, an even shorter period of heavy rainfall on 20 July 2007 in Middlesex resulted in the cancellation of about 150 flights at Heathrow airport, and the temporary stranding of thousands of passengers.

Global warming is a complex issue indeed. There is first of all the problem of establishing it as a fact. Climate change, as a general feature of the past (and doubtless future) history of this planet is incontrovertible, as the enormous body of palaeontological and, more recently, ice core data shows. The basis of the evidence for climate change will be discussed below. Assuming then that it is occurring, the question arises whether anything should be done about it. One of the major points of discussion is whether anthropogenic influence is playing a role. Prima facie this seems to be perfectly possible—it is well-known, for example, that the mean temperature in London is several degrees higher than that of the surrounding countryside.

Climate change and the advent of global warming has become an issue that anyone aspiring to have political *savoir faire* must have a view on, whether believing that human activity is the cause, or that the evidence is merely compatible with the consequence of natural cycles. These polarized views have become expressed with an almost religious fervour by many of the proponents, and until recently the USA, the biggest industrialized polluter, appeared to be almost waging war on the advocates of the anthropogenic cause of world temperature rise. Obviously commercial interests are at stake in this debate; indeed any attempt to predict the future must take account of human activity in which pecuniary considerations have an impact on the final outcome. This makes the understanding of climate change a very complex system involving not only all the sciences, but the economics and politics of human behaviour as well.

In some ways it might be considered the height of arrogance and perhaps man's greatest achievement that after some 5 million years on Earth he has altered the weather! Certainly something has happened over the last hundred years or so wherever the hand of man shows a presence. The depletion of the ozone layer due to chlorofluorohydrocarbon ("aerosol") gases (mainly used in refrigeration and as propellants in aerosol cans) was observed in the early 1970s, and as a consequence of the Montréal protocol of 1987 and the subsequent regulation of aerosol gases, the ozone layer is now recovering. More recent observations have noted the global temperature rise and the increase in carbon dioxide in the atmosphere. Now, the almost infamous picture of average global air temperature (Figure 10.1) shows a steady rise from the beginning of the 20th century. Is it a coincidence then that this roughly corresponds to the start of the output of the second (mini)-Industrial Revolution (the industrialization of Germany and the USA), built upon the burning of fossil fuel, coal in the first instance, to power the machines of manufacturing and production? And if so, why was there no rise corresponding to the first Industrial Revolution in Britain, which began over a hundred years earlier? This epoch was also marked by the start of a rapid rise in population, see Figure 10.2, but the increase in world gross domestic product (GDP) lagged significantly behind. Indeed, the temperature rise, if anything, coincides with the rise of world GDP.

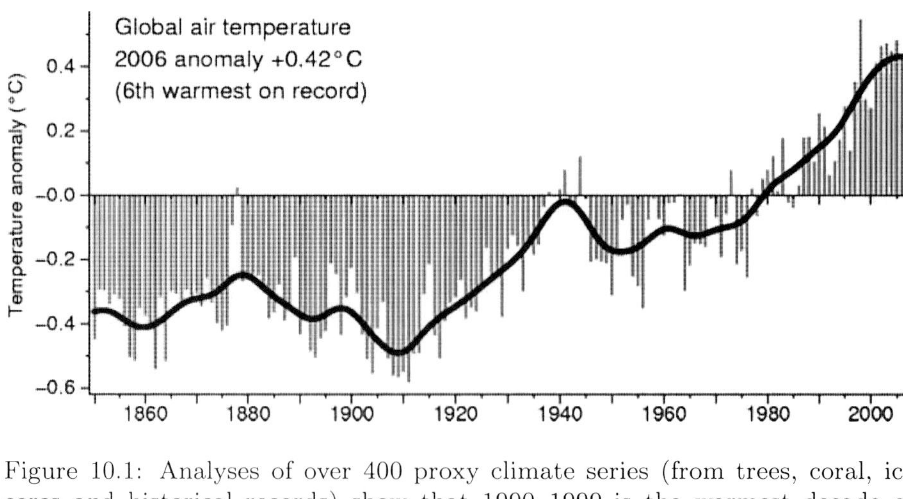

Figure 10.1: Analyses of over 400 proxy climate series (from trees, coral, ice cores and historical records) show that 1990–1999 is the warmest decade of the millennium and the 20th century the warmest century. (*After:* P. Brohan, J.J. Kennedy, I. Harris, S.F.B. Tett and P.D. Jones, Uncertainty estimates in regional and global observed temperature changes: a new dataset from 1850. J. Geophys. Res. 111 (2006) D12106.

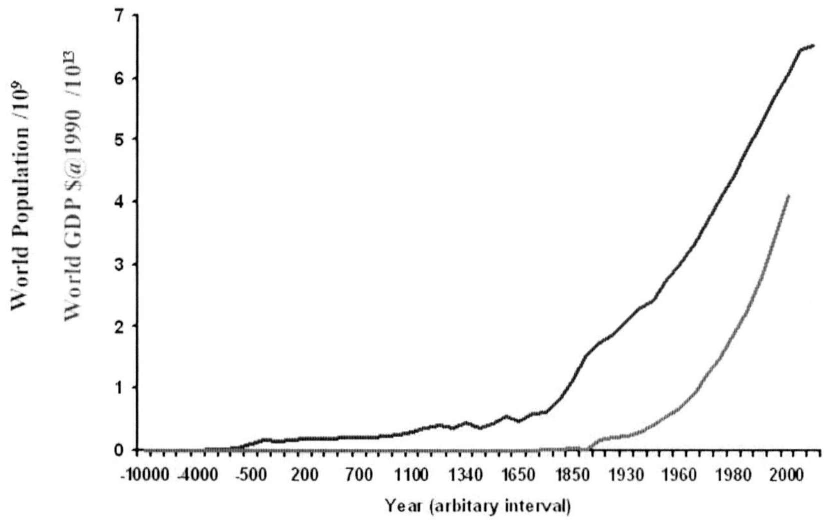

Figure 10.2: World population and world GDP normalized to the 1990 U.S. dollar.[a]

[a] A. Maddison (Monitoring the World Economy, 1820–1992. Paris: OECD, 1995) has constructed estimates of real GDP per capita for the world from 1820 to 1992. His estimates are best thought of as Laspeyres'[b] purchasing power parity estimates in 1990 international dollars.[c]
[b] E. Laspeyres, Die Berechnung einer mittleren Warenpreissteigerung. Jahrbücher f. Nationalökonomie u. Statistik 16 (1871) 296–314.
[c] That is, he: (i) compared income levels across countries not using current exchange rates, but instead changing one currency into another at rates that keep purchasing power constant ("purchasing power parity"); (ii) valued goods in relative terms using the prices found in a country in the middle of the world distribution of income ("international"); and (iii) calculated a value for 1990 GDP per capita in the United States equal to U.S. current-dollar GDP per capita in 1990 ("1990 dollars"). See: J.B. DeLong. Estimating world GDP, one million B.C.–present. Preprint, University of California at Berkeley (1998).

It is hardly surprising then that given the nature of man such coincidences should be the source of new investigations to understand the fundamental principles that govern our world. These first signs tend to become the rules of thumb to heuristically 'explain' the observed phenomena, rules that act as precursors to a full theory developed from reductionist principles;[1] Such a rule of thumb for climate change might be derivable through observing the factors contributing to the Earth's energy balance, namely the solar radiation warming the Earth, the Earth's own heat generation, and its reradiation into space. What has been known for some time is that the atmosphere is capable of operating as a one-way filter of this energy. Solar radiation is concentrated at the higher frequencies of the spectrum, while the Earth's reradiation is at a lower frequency, and gases such as carbon dioxide are translucent at higher frequencies but absorb energy at lower frequencies, engendering the so-called "greenhouse effect".[2]

10.1 The measurement of temperature and solar output

Until the invention of a convenient mercury thermometer by Fahrenheit in the early 18th century, reliable measurements were few and far between. The Royal Society of London published (in its monthly *Philosophical Transactions*) daily weather records, including temperature, in the 18th century, and there are a few other locations where reasonably reliable temperature data goes back a few centuries. Unfortunately the temperature of London air is heavily dependent on relatively local human activity, and these records are thus of little use in establishing global trends. The most reliable data now available comes from satellite-based monitoring—the spectrum of radiation emitted from a patch of the Earth is used to determine the temperature T from the wavelength at which the radiated energy has the greatest amplitude, according to Wien's displacement law:

$$\lambda_{\text{max}} = b/T \tag{10.1}$$

where b is Wien's displacement constant, with a value of 2.9×10^6 nm K. Satellite monitoring only goes back at best a few decades—a minuscule interval for establishing any kind of trend, and hence for extrapolation into the future, especially when it is remembered that the Earth's temperature is subject to numerous more or less periodic processes with frequencies ranging from hours (the rotation of the Earth about its axis), to months (the orbit of the earth around the sun), to years (oscillations in solar activity), and to much longer cycles lasting tens of thousands of years, whose origins are not particularly well understood. These potential sources of variation will be elaborated upon below (§10.3).

[1] A good example of the former is Fleming's right-hand rule in electromagnetism.

[2] The glass panes of a real greenhouse are typically symmetrical, allowing the same radiation through in both directions, but constitute a physical barrier to convection, that is, the loss of heat through the escape of gas molecules excited by adsorbed radiation. To some extent this also occurs on Earth as a whole: unless the excited molecules reach escape velocity (about 11 km/s) they and their energy remain in the atmosphere, trapped by the Earth's gravity.

Solar output

Detailed measurements have only been possible in the last few decades, with greater accuracy coming from recent satellite-based observations. Solar output would appear to have three superimposed cycles of maximum and minimum radiation. The shortest cycle time is about 11 years and corresponds to sunspots and solar flare activity, see Figure 10.3. A longer term view of the solar output is given by proxies such as the recorded history of sunspots,[3] and by looking at beryllium-10 (^{10}Be) created in the atmosphere by the action of solar radiation and captured in ice cores and fossils. Solar flares disturb the Earth's magnetic field and change the trajectory of cosmic rays, such that increased solar activity results in less beryllium-10. Such evidence, see Figure 10.4, shows that since the beginning of the 18th century, the Sun's output has increased by some 40%.

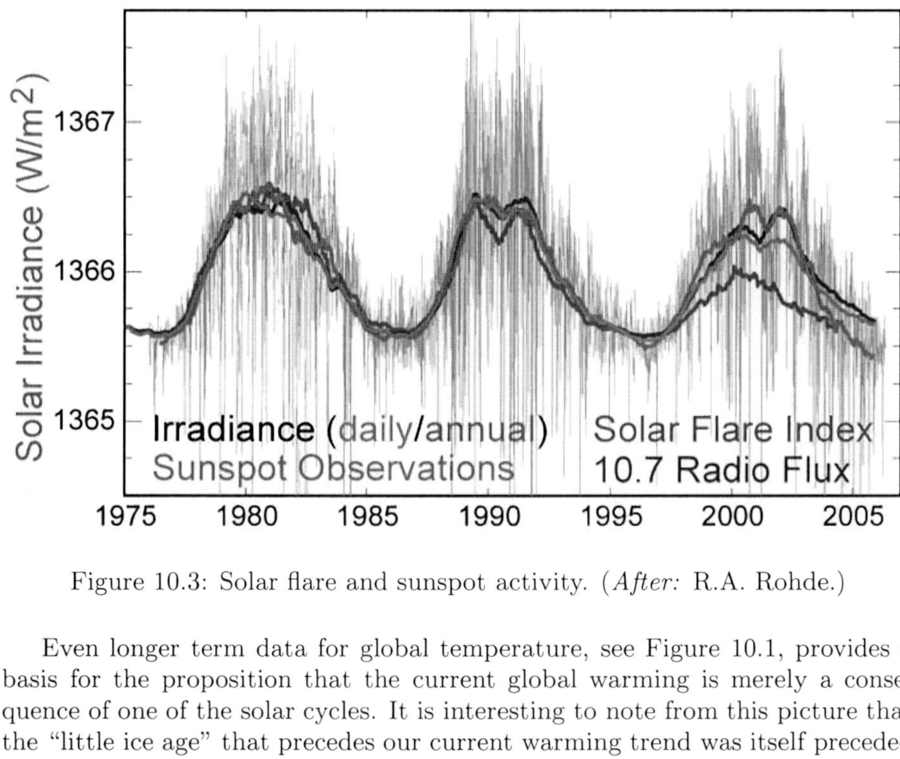

Figure 10.3: Solar flare and sunspot activity. (*After:* R.A. Rohde.)

Even longer term data for global temperature, see Figure 10.1, provides a basis for the proposition that the current global warming is merely a consequence of one of the solar cycles. It is interesting to note from this picture that the "little ice age" that precedes our current warming trend was itself preceded by a mediaeval warm period when vineyards were plentiful in England. Proxies established over geological time scales (Figures 10.1 and 10.1) show a correlation between global temperature and ice volume, and indicate a cycle time of approximately 100 000 years.

The principle of temperature recontruction from ice cores relies on the one hand on careful analyses of the isotopic composition of the ice (hydrogen oxide, H_2O), and on the other on assumptions about the rate of formation of the ice. Ice naturally contains a small proportion of heavy hydrogen, deuterium (^2H or D), which has a slightly higher boiling point; similarly oxygen-18 (^{18}O) is

[3]Sunspots are darker, cooler regions of the sun's surface associated with high magnetic flux. More sunspots indicate a more active sun with stronger (and more complicated) magnetic fields.

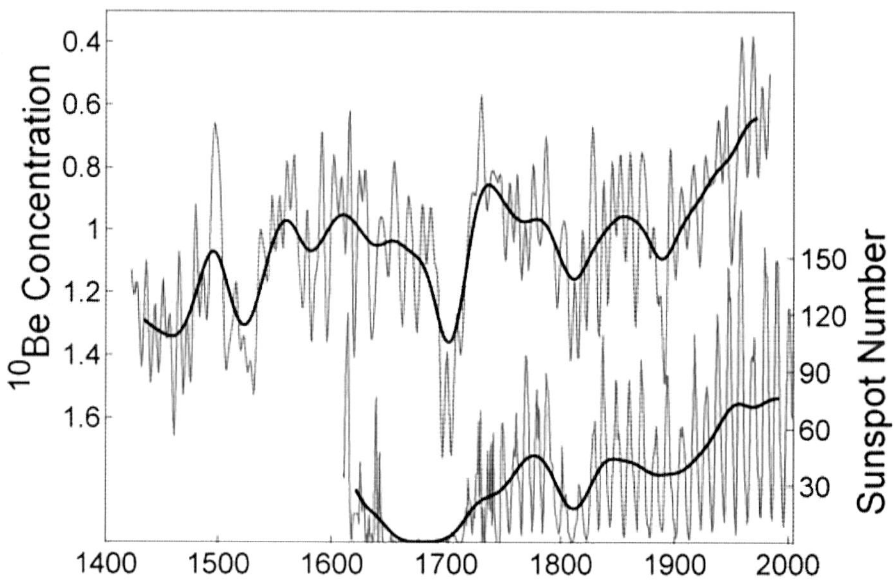

Figure 10.4: Proxy solar flare and sunspot activity. This figure shows two different proxies of solar activity during the last several hundred years. The lower trace is the so-called group sunspot number (GSN) as reconstructed from historical observations by D.V. Hoyt and K.H. Schatten, Group sunspot numbers: a new solar activity reconstruction. Solar Phys. 181 (1998) 491–512. The upper trace is the beryllium-10 concentration (in units of 104 atoms/(grams of ice)) as measured in an annually layered ice core from Dye-3, Greenland (J. Beer et al., An active sun throughout the Maunder minimum. Solar Phys. 181 (1998) 237–249). Both of these proxies are related to solar magnetic activity.

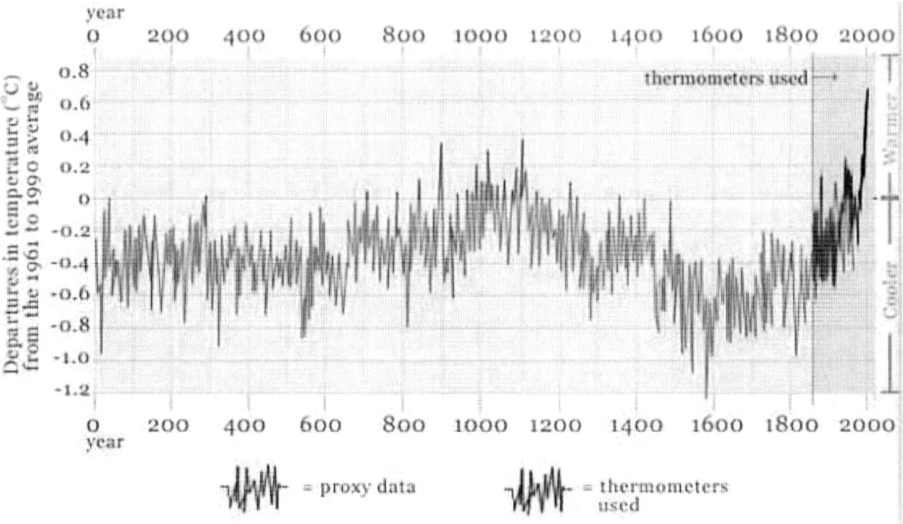

Figure 10.5: Northern hemisphere temperature.[a]

[a] Graph based on data reported by A. Moberg et al., Highly variable northern hemisphere temperatures reconstructed from low- and high-resolution proxy data. Nature (Lond.) 433 (2005) 613–617.

naturally present in the ocean along with the most common isotope, ^{16}O; as the least massive isotope of oxygen, ^{16}O evaporates slightly more readily than ^{17}O or ^{18}O, hence ^{16}O is preferentially evaporated from warm water and gets deposited preferentially on ice sheets, which are therefore enriched in H and ^{16}O and the ocean is depleted, but conversely enriched in D and ^{18}O. Hence the amount of deuterium and oxygen-18 in ancient ice can be used to reconstruct past temperature: for example, one part per million (ppm) enrichment in ^{18}O corresponds to an ocean temperature increase of approximately 1.5 °C. The age of a given depth of ice (from which a sample is taken) can be calibrated from layers of dust, which are presumed to originate from major volcanic eruptions, provided the dates of those eruptions are independently known, and from independent knowledge of the occurrence of insolation peaks, and so forth. There are of course many disturbing factors, such as strain-induced thinning of the ice.[4]

Ice cores also contain small bubbles of ancient air, which can be analysed to determine the concentrations of gases such as carbon dioxide and methane, hence allowing their correlation with temperature determined from the isotope ratios of the ice in which they were trapped. This relies, inter alia, on the gases being truly entrapped (i.e., on the ice being crack-free), and on the absence of disturbing chemistry sequestering CO_2 (e.g., carbonate formation).[5]

Furthermore, coral incorporates oxygen from the water in which it lives to produce the aragonite (a form of calcium carbonate) that constitutes its mineral

[4] J.R. Petit et al., Climate and atmospheric history of the past 420 000 years from the Vostok ice core, Antarctica. Nature (Lond.) 399 (1999) 429–436.

[5] H. Fischer et al., Ice core records of atmospheric CO_2 around the last three glacial terminations. Science 283 (1999) 1712–1714).

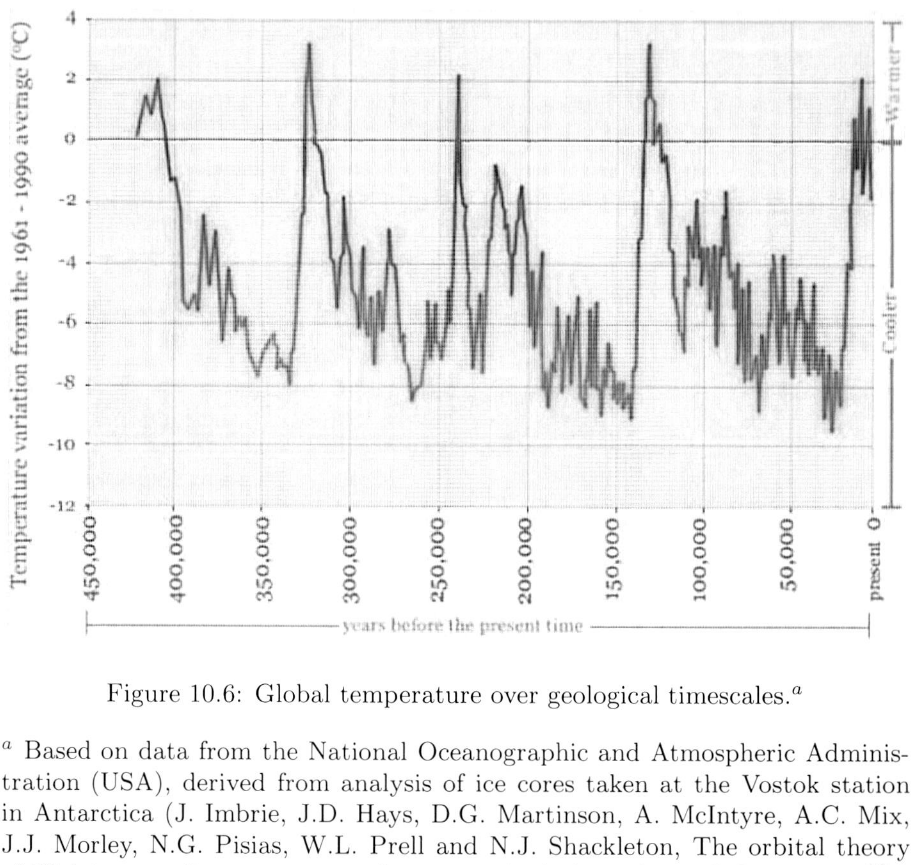

Figure 10.6: Global temperature over geological timescales.[a]

[a] Based on data from the National Oceanographic and Atmospheric Administration (USA), derived from analysis of ice cores taken at the Vostok station in Antarctica (J. Imbrie, J.D. Hays, D.G. Martinson, A. McIntyre, A.C. Mix, J.J. Morley, N.G. Pisias, W.L. Prell and N.J. Shackleton, The orbital theory of Pleistocene climate: support from a revised chronology of the marine ^{18}O record. In: A. Berger, J. Imbrie, J. Hays, G. Kukla, and B. Saltzman (eds), Milankovitch and Climate, Part 1, pp. 269–305. Dordrecht: Reidel, 1984.

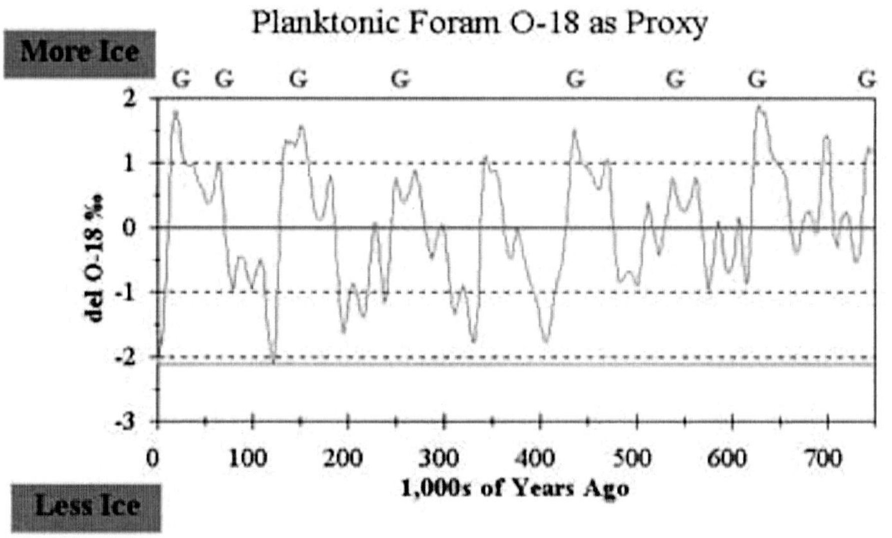

Figure 10.7: Ice volume deduced from ^{18}O enrichment. (Courtesy of Illinois State Museum, 2007.)

part, the oxygen isotope ratio of which therefore also depends on the ambient ocean temperature at the time of its formation. Moreover, annual bands can be distinguished in coral, enabling dating to be attempted.

Yet another proxy is provided by the foraminifera ("hole bearers"), or forams for short, a large group of amoeboid protists (zoöplankton) of the order *Foraminiferida*, with reticulating pseudopods (fine strands of cytoplasm that branch and merge to form a dynamic network) emerging from a perforated mineral shell. Plankton speciation changes in a definite way with changing temperature, which may therefore be deduced from observing the relative abundances of planktonic ("top-living") and benthic ("bottom-living") foraminifera in ocean sediment cores.[6] They too form their mineral parts using oxygen from the surrounding water, hence enrichment in ^{18}O in the mineral part of a foram indicates enrichment in the ocean, hence a higher temperature (from which a diminution of global ice volume is inferred).[7]

10.2 The Earth's energy balance

Although the deep interior of the Earth is indeed hot (probably several, at least five, thousand degrees)—a vestige of the gravitational compression that led to the initial formation of the planet—today it is the Sun that provides the overwhelming quantity of energy to the Earth; the flux of geothermal heat from the core to the surface—less than 0.1 W/m^2—is generally considered to

[6] J.D. Hays et al., Variations in the Earth's orbit: pacemaker of the ice ages. Science 194 (1976) 1121–1131.

[7] Boron isotope ratios of the shells can also be used to make inferences about the acidity of the oceans in the past.

be negligible compared with solar heating. The temperature of the planet is thus determined by the balance of heat received from outside (i.e. the Sun) and heat radiated into space. The Earth's surface (let us say the top 10–20 m of the crust) has a (fairly constant) mean temperature T of about 5 °C. This mean temperature is of course an extraordinarily difficult number to obtain experimentally. Satellite data is probably the most reliable today, and allows the virtually simultaneous observation of large areas.[8]

The Earth is constantly irradiated by the sun. The solar energy flux (i.e. the radiant power at the distance of the Earth's orbit round the sun) is approximately 1.37 kW/m^2 (this is called the solar constant, S_0), and therefore as a first estimate the Earth (modelled as a disk of radius r_E) must absorb, in total,

$$\hat{S} = S_0 \pi r_E^2 (1 - \mathfrak{A}) , \tag{10.2}$$

where r_E is the radius of the Earth (6400 km) and $\mathfrak{A}$ is the albedo (the fraction of incident radiation that is reflected, relative to a perfectly reflecting flat surface).[9] The Earth's average albedo is generally taken to be 0.31, to which clouds contribute 0.21–0.25, gases and particles in the atmosphere 0.05–0.08, and land and ocean surfaces 0.01–0.03. The contribution from the land (which is only 30% of the total surface area of the Earth) depends on the state of the Earth's surface—snow has a high albedo (possibly as much as 0.8–0.9, but it could be much lower depending on its detailed morphology[10]), whereas forests absorb a lot of radiation (albedo 0.05–0.10) (see Table 10.2 and the rest of §10.3.2). Therefore, the amount of solar radiation that the Earth absorbs depends on the amount of snow, desert, grassland, ocean, forest, etc. James Lovelock proposed with his well known "Daisyworld" model that the Earth's temperature might be regulated through its albedo—if it warms up (due to increased solar power, for example), the albedo increases through a proliferation of white, reflecting flowers, leading to cooling, and vice versa.[11] Higher temperature would also lead to increased evaporation from the oceans, hence leading to increased precipitation as snow, with a high albedo, on high ground; but at lower altitudes plant growth in deserts, or more luxuriant forest growth elsewhere, might be encouraged, lowering albedo and leading to further warming. Despite their necessary complexity, the models used to relate albedo to temperature are only useful for very coarse predictions, not least because many of the model parameters are not known with sufficient precision. Noteworthy is the fact that cloud is the dominant contributor to the Earth's albedo, and should be the first target of any attempt to model $(T, \mathfrak{A})$ relations.

The radius of the earth used to calculate the absorbed energy should properly include part of the atmosphere, since in particular clouds within the atmosphere

[8]See for example C. Ulivieri and G. Cannizzaro, Land surface temperature retrievals from satellite measurements. Acta Astronautica 12 (1985) 977–985; M. Moriyama et al., Comparison of the sea surface temperature estimation methods. Adv. Space. Res. 16 (1995) (10)123–(10)126.

[9]The albedo does not seem to have a universally agreed definition. Those current differ in several significant details (such as the spectral range), and hence it must be considered as an approximate quantity.

[10]See e.g. W.J. Wiscome and S.J. Warren, A model for the spectral albedo of snow. J. Atmos. Sci. 37 (1981) 2712–2733.

[11]The "Gaia" concept asserts that the planet has evolved such that light-coloured, rather than albedo-lowering dark-coloured flowers are favoured at higher temperatures. It is, however, doubtful whether there is sufficient empirical support for this proposition.

contribute to the albedo (see Table 10.2). The troposphere, which contains about three quarters of the mass of the atmosphere and nearly all the water vapour (and hence clouds), is about 10 km thick. This is negligible in comparison with r_E and will therefore be neglected.[12]

The energy that is not reflected (at the same wavelength as it arrives) is absorbed by the Earth's atmosphere, crust and oceans, which reradiate energy, approximately as a black body, the peak wavelength of which, corresponding to its mean temperature T of about 5 °C, is about 10 μm (equation 10.1), i.e. in the infrared part of the spectrum of radiation.

The total radiated power $\hat{E}$ is[13]

$$\hat{E} = \sigma T^4 4\pi r_E^2 . \tag{10.3}$$

Assuming that the planet is approximately in thermal equilibrium, by equating incoming (from equation 10.2) and outgoing energy, i.e.[14]

$$\hat{S} = \hat{E}(T) \tag{10.4}$$

and solving for T, we find $T = 254$ K, i.e. about -19 °C. It is the discrepancy between this predicted temperature and the actual one that provides the first alert to the possible existence of a "greenhouse effect", i.e. the Earth is warmer than it ought to be.[15]

Fixing T as 278 K, and using equation (10.4) with the appropriate substitutions to solve for $\mathfrak{A}$, we obtain the mildly surprising result that an albedo of zero is required to agree with observation. Undoubtedly the Earth *does* reflect some light, however, suggesting that some of the emitted energy (equation 10.3) becomes trapped in the atmosphere on its way back into space. Conveniently, the long wavelength (10 μm) emitted radiation can be absorbed by some of the more complex molecules that are present in the atmosphere, especially water, carbon dioxide and methane, that are transparent to the incoming solar radiation.[16] The three most abundant gases are nitrogen (78% v/v), oxygen (21%) and argon (1%). Carbon dioxide is typically present at 0.03–0.04%, neon at 0.002% and methane at 0.0001–0.0002%. Water vapour has a much more variable presence and can range from zero to a few percent. These molecules are all practically transparent to most of the incoming solar radiation (ultraviolet light is efficiently absorbed by the thin ozone layer near the top of the

[12]The stratosphere extends to about 50 km from the Earth's surface and accounts for almost all the remainder of the atmosphere. The Kármán line, 100 km above the Earth's surface, is considered to form the boundary between the atmosphere and outer space (beyond which aeronautics is impracticable).

[13]The power E_0 radiated by a black body at an (absolute) temperature T is given by the Stefan-Boltzmann law, $E_0 = \sigma T^4$, where σ is the Stefan-Boltzmann coefficient, equal to 5.67×10^{-8} W m^{-2} K^{-4}. Admittedly the Earth's crust is far from being a perfect black body, but the Stefan-Boltzmann law remains valid over a wide range of greyness.

[14]Priority for introducing this idea should be accorded to J. Fourier (cf. his *Remarques générales* and *Mémoire*, respectively, *sur les températures du globe terrestre et des espaces planétaires*: Ann. Chim. Phys. 27 (1824) 136–167 and Mém. Acad. R. Sci. 7 (1827) 569–604).

[15]One possible correction would be to first subtract the power $S^{(\mathrm{atm})}$ lost by atmospheric absorption (0.25 kW/m^2) from the solar constant S_0, and then use the much lower albedo of the actual surface (land and sea), about 0.054, to calculate $\hat{S}$ from equation (10.2). This yields $T = 261$ K, i.e. about -12 °C, still significantly lower than reality.

[16]In addition, some of the heat of the Earth's surface is carried into the atmosphere by conduction and convection of atmospheric gases and by evaporation of liquid water.

stratosphere), although, as noted above, water condensed into droplets forms clouds strongly scattering and reflecting light.[17] Infrared absorption arises from the chemical bonds between atoms: carbon dioxide has two C=O bonds and methane has four C–H bonds; the actual absorption spectra depend on in-phase and out-of-phase stretching and bending of these bonds, which occur at infrared frequencies. Carbon dioxide has important absorption bands peaking at wavelengths of around 2.7, 4.3 and 15 μm. Therefore, some of the reradiated infrared radiation will be absorbed by these molecules, warming them up: this "excess heat" will then be distributed by collisions with the abundant nitrogen and oxygen, so the atmosphere as a whole will become slightly warmer than otherwise. Some of the excess heat will be reëmitted, both out into space and back towards the Earth's surface, at yet longer wavelengths.

Although laboratory measurements of the absorption spectra of the gases concerned are fairly straightforward, the spectra, i.e. the peak shapes and especially their widths, of the gases in their actual atmospheric environment may be different. It is in fact still difficult to calculate the effects of the absorption of the long wavelength (peaking around 10 μm) radiation on atmospheric temperature.

It must also be borne in mind that the atmosphere is a highly dynamic system. Its basic composition is due to the ceaseless activity of life, mainly at the surface of the Earth. The most significant effect (oxygenation) is the result of the activity of photosynthetic cells (e.g. plants):

$$6CO_2 + 6H_2O + 36h\nu \xrightarrow{\text{enzymes}} C_6H_{12}O_6 + 6O_2 \,. \tag{10.5}$$

The enzymes (i.e. proteins) are the catalysts of the reaction. 6 is a typical figure for the number of photons (one photon is denoted by its energy, $h\nu$) required per CO_2 molecule consumed. This process has been taking place ever since the emergence of cyanobacteria in the oceans, probably more than 2000 million years ago, and which converted a reducing atmosphere to an oxidizing one. This process is autocatalytic in the sense that photosynthesis itself results in more photosynthetic material (enzymes). Animals on the other hand consume oxygen and exhale carbon dioxide (respire) as part of their metabolism (§10.4.1); anaerobic bacteria such as those present in the stomachs of cows generate methane.

Besides the above, the (slightly acidic) oceans, seas, lakes and indeed all water bodies contribute passively to atmospheric carbon dioxide regulation, not only by simple dissolution:

$$(CO_2)_g \leftrightharpoons (CO_2)_{aq} \,, \tag{10.6}$$

but also by virtue of the reaction:

$$CO_2 + H_2O + CO_3^{2-} \leftrightharpoons 2HCO_3^- \,, \tag{10.7}$$

which can in turn lead to the formation of solid mineral carbonates by reaction with dissolved metal ions.

[17] The spectrum of solar radiation peaks at a wavelength of about 500 nm, which is roughly what would be expected from a so-called black body with the temperature of the surface of the Sun, i.e. about 6000 K (equation 10.1). The larger the droplet, the more the scattering, according to the Rayleigh-Tyndall law $I_s/I_0 = [N2\pi^5 d^6/(3\lambda^4)][(n^2-1)/(n^2+2)]^2$, where I_s and I_0 are the scattered and incient radiation respectively, N is the number per unit volume of particles of diameter d, λ is the wavelength of the light, and n the refractive index of the particle, at that wavelength.

Within the atmosphere itself, many photochemical reactions are taking place, for example

$$H_2O + h\nu \rightarrow H^\bullet + HO^\bullet \ . \tag{10.8}$$

The radicals (denoted by $^\bullet$) generated by such reactions are highly reactive and will undergo further reactions, such as

$$H^\bullet + O_2 \rightarrow HO^\bullet + O^\bullet \ . \tag{10.9}$$

Photochemical reactions like this one are especially important for the fate of some of the anthropogenic emissions from industrial processes and emissions from volcanos, such as sulfur dioxide and the nitrogen oxides (see also §§10.6 and 10.7).

Winds and thermal convection constantly mix and homogenize the composition of the atmosphere, which is being modified by processes (10.5)–(10.9).

10.2.1 Industrial activity

Man contributes much more carbon dioxide due to industrial activity than through his metabolism.[18] The most important industrial activity is combustion. Burning wood, straw or autumn leaves generates carbon dioxide that is essentially used by other plants to grow (reaction (10.5)). On the other hand, a great deal of energy is now generated by burning carbonaceous fossil fuels: coal (C), natural gas (CH_4) and oil (e.g. C_8H_{18}). This can be summarized as

$$C + O_2 \rightarrow CO_2 + \text{heat} \ , \tag{10.10}$$

the heat of combustion (see Table 10.1) being used to generate electricity, propel motor vehicles, etc. Other important activities include the manufacture of cement, most typically by the reaction:

$$CaCO_3(\text{limestone}) \xrightarrow{\text{heat}} CaO + CO_2 \ , \tag{10.11}$$

which needs a great deal of heat to take place, the generation of which also produces carbon dioxide according to reaction (10.10); the manufacture of iron and steel:

$$Fe_2O_3(\text{haematite}) + \frac{1}{2}O_2 + C \xrightarrow{\text{heat}} 2Fe + 2CO_2 \ , \tag{10.12}$$

and of silica:

$$SiO_2 + C \xrightarrow{\text{heat}} Si + CO_2 \ ; \tag{10.13}$$

these reactions also require heat in order to proceed at a reasonable rate (mixing sand with coal at room temperature does not produce silicon), which is usually provided by the combustion of fuel. One may note that all the necessary reduction activity for extracting metals from ores (cf. equation 10.12) is due to our oxidizing atmosphere having been at work for thousands of millions of years.

[18]See §10.4.1.

Table 10.1: Heats (standard enthalpies) ΔH of combustion.

Compound	Formula	Molecular weight	ΔH/MJ/mol
Methane	CH_4	16 g/mole	−0.494
Butane	C_4H_{10}	58 g/mole	−2.88
Octane	C_8H_{18}	114 g/mole	−5.51
2,4,6-trinitrotoluene (TNT)	$C_7H_5N_3O_6$	227 g/mole	−0.95

10.3 Variations in contributors to the energy balance

Summarizing the above, the mean Earth temperature principally depends on:

1. Solar flux (insolation)
2. Albedo
3. Atmospheric carbon dioxide concentration, due to
   (a) Plant activity
   (b) Anthropogenic activity
4. Other factors such as plate tectonics (leading to volcanism).

The first two are direct, the others indirect. Note that these factors are not all independent. For example, insolation affects plant activity, which affects albedo. In this section, the magnitudes of the variations will be compared.

10.3.1 Solar flux

The Sun is a variable star. The so-called solar "constant" S_0 is not, but fluctuates quasi-periodically. The number of sunspots (magnetic storms on the surface of the Sun) follows an approximately 11 year cycle. The corresponding effect on irradiance only amounts to a variation of about 1 W/m^2 however, i.e. of the order of 0.1%. There is evidence for longer cycles (e.g. a 206 year cycle is sometimes mentioned), but obviously no direct measurements are available.

The average irradiance received by the Earth also depends on the distance from the earth to the Sun. As a dynamical system, the solar system goes through certain cycles, notably of the variation of eccentricity of the (elliptical) orbit (*ca* 10^5 years), and of the precession of the equinox and the advance of the perihelion (both *ca* 2×10^4 years). These are combined with a further cycle, that of the obliquity of the ecliptic (i.e. the axial tilt of the Earth), which alternates between 22 and 24.5° with a period of *ca* 4×10^4 years), to produce an estimate of the fluctuation in intensity of solar irradiation.[19] These fluctuations, amounting to several percent, are much more significant than the actual variation in the energy emitted by the Sun. It should also be noted that changes in obliquity do not of course change the overall amount of solar radiation received by the

[19] M. Milankovitch, Théorie mathématique des phénomènes produits par la radiation solaire. Paris: Gauthier-Villars (1920)

Earth, but the greater the obliquity, the greater the variations between summer and winter, which seems to be important for triggering ice ages.[20]

In more detail, over the 10^5 year Ice Age cycle the orbit changes from almost circular to more elliptical, i.e. the major axis of the ellipse grows (by a maximum of about 5%). This means that the Earth spends more time of the year further away from the Sun with the result that the received energy is reduced (by an amount proportional to the square of the Earth-Sun distance). At present the eccentricity is about 3%. Hence the maximum solar output equals $S_0^{(\text{present})}/(1.03)^2 = 1.45\ \text{kW/m}^2$, and similarly for the minimum, equal to 1.31 kW/m^2. Evidently the Earth is now close to the nadir of the cycle, thus we stand at the beginning of a period of warming. Using the thermal equilibrium equation (10.4) to calculate the temperature of the Earth as a function of albedo then gives the graph of Figure 10.8. This shows that the maximum temperature difference is about 7 °C, which compares reasonably well (in view of the crudeness of the calculation) with the 12 °C amplitude estimated for the Ice Age cycles.

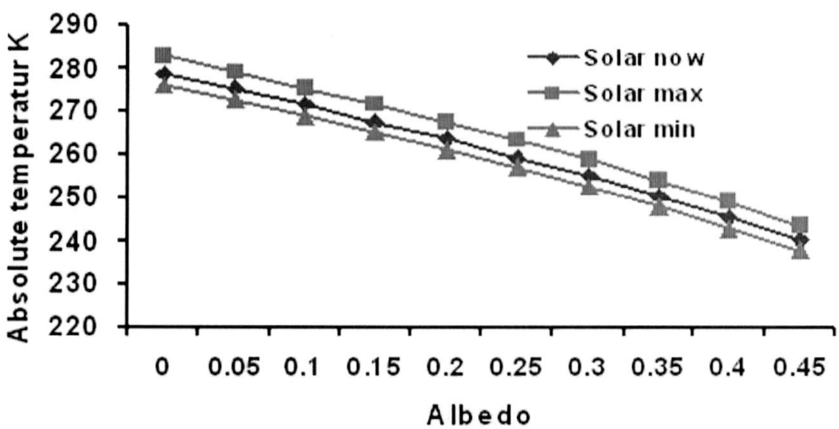

Figure 10.8: Thermal equilibria for the Earth's temperature, calculated according to equation (10.4) and based on the first Milankovitch cycle for the elliptical extremes.

Since Milankovitch published his book, there have been considerable advances in analysing ocean sediments[21] and Antarctic ice[22] in order to provide evidence for actual climate changes. Antarctic ice cores down to a depth of over 3.5 km have in particular now yielded climate and atmospheric history going

[20] Implicit in this statement is the proposition that there are several basins of attraction in the climatic system. For example, if a relatively small area of ice can be created where none was before, due to its high albedo more incoming radiant energy than before will be reflected, thus inducing cooling and the formation of more ice.

[21] E.g. J.D. Hays et al., Variations in the Earth's orbit: pacemaker of the ice ages. Science 194 (1976) 1121–1131.

[22] E.g. J.R. Petit et al., Climate and atmospheric history of the past 420 000 years from the Vostok ice core, Antarctica. Nature 399 (1999) 429–436.

back almost half a million years. We cannot in this chapter discuss in detail the actual methods used to make inferences of atmospheric composition and temperature. We only remark that the data shows a correlation between the cycles of irradiance and actual climate changes, such as the roughly 10^5 year cycle of ice ages. Palaeontology and geology have yielded information that goes back tens and even hundreds of millions of years. There is some evidence that the Earth was covered with snow about 700 million years ago; one should also bear in mind the possibility of complete reversals of the Earth's axis.[23] Extrapolating from previous fluctuations, we are now in the so-called interglacial period, the Holocene, and hence in about 30 000 years a new Ice Age should begin.[24] If the temperature peak in the Jurassic epoch around 160 million years ago is part of yet another cycle, the pattern of temperature fluctuations begins to have an almost fractal character, but only that of an approximate fractal, which makes predictions from historical data over the next few decades or even centuries a hopeless task, particularly as we have still very little idea about the actual mechanisms of the fluctuations, in particular how the relatively small fluctuations in insolation are so drastically amplified by some of the other factors listed at the beginning of this section. One surprising result, see Figure 10.9, emphasizing our current lack of knowledge, is that there is little historical evidence for increasing carbon dioxide levels *preceding* global warming, and sometimes indeed a CO_2 increase follows a temperature rise.[25]

10.3.2 Albedo

The Earth's albedo is dependent upon the type of reflecting surface, see Table 10.2. There are very significant differences depending upon which patch of the Earth one is looking at. While it can be determined reliably and meaningfully for a distant planet, microalbedos of patches of the surface pose difficult measurement problems. Furthermore, although insolation at the Earth's surface is generally considered as the yearly average amount of solar radiation that strikes the Earth's land and ocean surfaces, and is of course greater at low latitudes and less at high latitudes due to the curvature of the Earth and decreasing angle of the Sun at higher latitudes. However, there are significant other differences with location, above all due to the presence of clouds, which typically cover about half the planet. Clouds have a relatively high albedo (the figure depending upon the type of cloud, i.e. the size and density of its constituent water or ice droplets). Thus the equatorial belt receives less solar radiation at the surface than at around 20 degrees latitude. The reason is that the equatorial belt is where the major rain forests lie (Amazon, Congo basin, East Indies). There is a lot of rain to nourish the dark, light-absorbing forests—and lots of clouds to reflect sunlight. On the other hand the desert belts lie around 20° latitude (Sa-

[23] P. Warlow, Geomagnetic reversals? J. Phys. A 11 (1978) 2107–2130.

[24] Useful summaries of the fluctuations can be found in H. Blattmann, Neue Zürcher Zeitung, 9 May 2007, p. B1, and 11 July 2007, p. B1.

[25] It is relatively straightforward to determine the amounts of carbon dioxide, oxygen etc. and their isotopic compositions in a given piece of ice, and the assumption that the gases have remained entrapped since the ice was formed seems to be robust, in view of the minute diffusivities. What is much more difficult is to establish an independent chronology of the ice formation. This requires the introduction of many assumptions regarding the rate of ice formation and so forth, and while they might be reasonable, they can but rarely be independently tested.

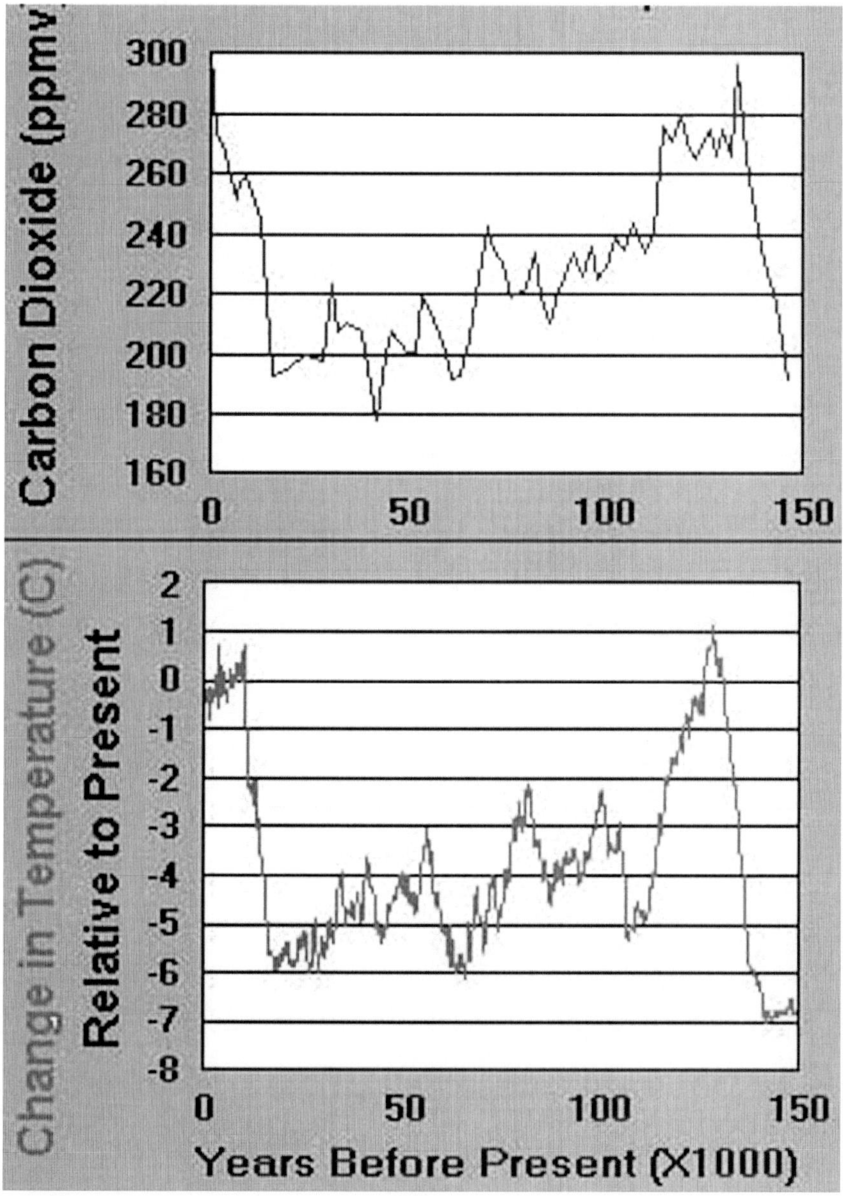

Figure 10.9: Ice core data and carbon dioxide concentration, based on the analysis of entrapped air from ice cores extracted from permanent glaciers from various regions around the globe. These show that global warming could be said to have begun 18 000 years ago, accompanied by a steady rise in atmospheric carbon dioxide (H. Fischer et al., Ice core records of atmospheric CO_2 around the last three glacial terminations. Science 283 (1999) 1712–1714).

hara, Arabian, Indian, Kalahari and other deserts). The air is dry and there are few clouds to reflect the sunlight, so it practically all reaches the surface. The remaining 69% (that is not immediately reflected) of incoming solar radiation is absorbed by the Earth; atmospheric gases, clouds, and dust are considered to absorb about 24%, and land and water surfaces absorb the rest (45%).

Table 10.2: Areas A of different zones and their corresponding albedos $\mathfrak{A}$.

Type of cover (zone)	$A/10^{13}$ m^2	$F^{(\text{zone})\ a}$	$\mathfrak{A}$
Cloud	25	0.49	0.5
Ocean[b]	36	0.70	0.07
Land	15	0.29	0.2
Permanent ice and snow[c]	3.6	0.07	0.7
Desert, mountain (without snow)	3.7	0.07	0.3
Forest, woodland	5.7	0.11	0.1
Grassland, cultivated land, marsh	4.0	0.08	0.15

[a] Fraction of the Earth's surface occupied by that zone. The Earth's total surface area is 5.15×10^{14} m^2.
[b] Total volume of the oceans is 1.35×10^{18} m^3. Ocean sediment is about 1 km thick on average (D.L. Divins, NGDC Total Sediment Thickness of the World's Oceans & Marginal Seas; available from http://www.ngdc.noaa.gov/mgg/sedthick/sedthick.html).
[c] About 1.6×10^{13} m^3 of this covers the land. As noted elsewhere, the albedo of snow and ice can vary considerably.

10.4 Variations in atmospheric carbon dioxide

The volume of the atmosphere, considered as a shell 10 km thick covering the Earth (r_E = 3.2 Mm), is 5.2×10^{18} m^3. Hence the total volume of carbon dioxide in the atmosphere (assuming a content of 0.0385% v/v) is 2.1×10^{15} m^3, which is roughly 9×10^{16} moles or 1000 Gt (1 Tt) of carbon, or almost 4 Tt of CO_2.[26] This figure is slightly higher than the one usually quoted because the calculation ignores the decreasing density of the atmosphere with height.

10.4.1 Biogenic factors

Photosynthesis

It has already been pointed out that the oxidizing atmosphere of our planet is essentially due to the activity of cyanobacteria several milliard years ago. Since then plants have evolved into some very sophisticated forms, and continue to make a substantial contribution to determining the composition of the Earth's atmosphere.

Plants sequester CO_2 from the atmosphere, incorporating it as carbon into their structure (equation 10.5), and reduce water to oxygen. The energy for these processes comes from absorbed sunlight, which is often assumed to be

[26] One of the difficulties of the field is the variety of units used to describe quantities. One gigatonne is 10^9 tonnes (metric tons) or one petagram (Pg); one teratonne is 10^{12} tonnes.

the limiting factor. The calculation in the following Table (10.3) uses the surface areas from the previous Table (10.2) and assumes that the photosynthesis equation (10.5) represents the average effect of carbon sequestration by trees, other plants, and photosynthetic ocean biomass. The (significant) fraction of the Earth's surface covered by reflecting cloud (assumed not to allow any photosynthetically active radiation to penetrate) also needs to be taken into account (Table 10.2). Furthermore, we are now interested in the flux of photosynthetically active radiation received at the surface, and therefore subtract the power lost by absorption in the atmosphere (250 W/m^2). It is also convenient to work with energy, rather than power, hence we multiply by the interval τ of interest:[27]

$$\tilde{S}_\phi^{(\mathrm{x})} = (S_0 - S^{(\mathrm{atm})})A^{(\mathrm{x})}(1 - \mathfrak{A}^{(\mathrm{x})})(1 - A^{(\mathrm{cld})}A^{(\mathrm{x})}/(A^{(\mathrm{E-th})})^2)\mathfrak{A}^{(\mathrm{cld})}\overline{PAR}\tau \tag{10.14}$$

(for zone x), where the subscript ϕ denotes photosynthetically active radiation, and $\overline{PAR}$ is the photosynthetically active radiation fraction of the solar spectrum, which we take as 0.45.[28] The mean energy of a photosynthetically active photon is

$$\mathcal{E} = hc/\lambda\,, \tag{10.15}$$

where h is Planck's constant (6.6×10^{-34} J s), c is the speed of light (3.0×10^8 m/s), and the wavelength will be taken as $\lambda = 500$ nm (roughly the median of the photosynthetically active radiation spectrum). Thus the number (in moles) of photons $\tilde{N}_\phi$ absorbed per year by the biomass is

$$\tilde{N}_\phi = \tilde{S}/(\mathcal{E}N_L)\,, \tag{10.16}$$

where N_L is the Loschmidt (Avogadro) number, 6×10^{23}. Then, using equation (10.5) the amount of carbon sequestered (in moles) is

$$\tilde{N}_\mathrm{C} = \tilde{N}_\phi\overline{BAF}/6 \tag{10.17}$$

where $\overline{BAF}$ is the biomass active fraction, taken as 1% of total biomass, corresponding to average annual growth rate.[29] The mass of carbon sequestered annually by photosynthesis is then

$$\tilde{M}_\mathrm{C} = \tilde{N}_\mathrm{C}M_r^{(\mathrm{C})}\,. \tag{10.18}$$

Taking the molar mass of carbon, $M_r^{(\mathrm{C})}$, as 12 g/mol, the number emerging from this calculation can be multiplied by 10^{-15} to obtain gigatonnes, Gt (or petagrams, Pg). Some results are given in Table 10.3.

As for the subsequent fate of the sequestered carbon, other than direct exploitation (logging for fuel, joinery or construction) some of it is lost through oxidation and bacterial activity, and in the oceans especially much is mineralized (as various metal carbonates).

---

[27] For one year, this will be $3600 \times 12 \times 365$ s, assuming a daily mean exposure of 12 h.

[28] M. Tsubo and S. Walker, Relationships between photosynthetically active radiation and clearness index at Bloemfontein. Theor. Appl. Climatol. 80 (2005) 17–25.

[29] It is assumed that there is no other limitation, e.g. due to lack of water, or reactive nitrogen, on plant growth. See also J. Barber and M.D. Archer, Photosynthesis and photoconversion. In: M.D. Archer and J. Barber (eds), Molecular and Global Photosynthesis. London: Imperial College Press (2004).

Table 10.3: Annual quantities of photosynthetically active incident radiant energy and carbon sequestered photosynthetically from the atmosphere.

Type of cover (zone)	$\tilde{S}_{\phi}^{(\mathrm{zone})}/\mathrm{J}$	$\tilde{N}_{\mathrm{C}}/\mathrm{mol}$	$\tilde{M}_{\mathrm{C}}/\mathrm{Gt}$
Ocean	1.7×10^{24}	1.2×10^{16}	147
Forest, woodland	3.9×10^{23}	2.7×10^{15}	33
Grassland, cultivated land, marsh	2.6×10^{23}	1.8×10^{15}	22
Totals		1.7×10^{16}	201

Plant respiration

All growing and living matter respires and releases carbon into the atmosphere. For plants (including phytoplankton), the rate of respiration is dependent on many factors including temperature and CO_2 and reactive nitrogen levels, but a simple empirical rule appears adequate for a level of approximation commensurate with the uncertainties surrounding the overall problem:[30] the respiration rate is one quarter of the photosynthesis rate plus a proportion of the active biomass of the plant, that is, for annual production,

$$\tilde{R}_{\mathrm{C}} = \tilde{M}_{\mathrm{C}}/4 + C\mathfrak{M} \tag{10.19}$$

where $\tilde{R}_{\mathrm{C}}$ is the annual amount of carbon respired, $\mathfrak{M}$ is the (dry) active (respiring) plant biomass (assumed to be all carbon, the dominant element in terms of mass), and C is a constant established by McCree, which has a value of 5.5.[31] The results of this calculation are given in Table 10.4).

Animal respiration

The growing human population is also a source of respired CO_2 but the amounts are still small by comparison. An average adult breathes 15 times per minute taking in 0.5 L of air and exhales 3.6% as CO_2. This equates to only 0.3 Gt carbon per year from the current population of over 6×10^9 living souls (about 1 kg emitted CO_2/day per person). Domestic animals have about three times the biomass of humans (Table 10.5) but even their contribution to carbon release by respiration would be small compared to plants (Table 10.4). The mass of other animal life forms is more difficult to estimate; ants (including termites) are believed to be the most numerous, with perhaps $\sim 10^{18}$ units, individually very small of course. Actually the majority of living biomass is probably procaryotic,[32] estimated at $\sim 5 \times 10^{30}$ cells, equivalent to $\sim 5 \times 10^{17}$ g

[30] K.J. McCree, An equation for the rate of respiration of white clover plants grown under controlled conditions. In: Prediction and Measurement of Photosynthetic Productivity. Proceedings of the IBP/PP Technical Meeting, 14–21 September 1969, Trebon. Wageningen: Centre for Agricultural Publishing and Documentation (1970).

[31] This is one of the weakest parts of the whole calculation. McCree's equation (including the values of the constant C (5.5) and the other one (1/4) in the first term on the right hand side) applies to one particular species (white clover) grown under one particular set of conditions. It is possible that, fortuitiously, these values correspond to averages over all species, weighted according to their abundances, and in turn averaged over the actual (average annual) conditions under which they are growing—this is essentially what we are assuming. When it comes to calculate the influence of changing conditions, in particular temperature and water supply, we shall need to know how these variables affect the 'constants'.

[32] W.B. Whitman et al., Prokaryotes: The unseen majority. Proc. Natl Acad. Sci. USA, 95 (1998) 6578–6583.

Table 10.4: Annual release of carbon through plant (equation 10.19) and human respiration.

Type of cover	Volume	Active biomass (dry)	Respired carbon	Notes
	V/m^3	$\mathfrak{M}/\mathrm{Gt}$	$\tilde{R}_\mathrm{C}/\mathrm{Gt}$	
Ocean	3.6×10^{16}	3.6	57	*a*
Forest, woodland	1.4×10^{14}	4.2	32	*b*
Grassland, cultivated land, marsh	4.0×10^{13}	1.2	12	*c*
Humans		0.15	0.3	*d*
Total			99	

[a] The active surface depth is considered to be 100 m, within which phytoplankton chloroplast abundance is 1.0 mg/m^3.The global distribution of chlorophyll has been averaged over the period 1 January 2002 to 28 February 2005 using data collected from MODIS on the Aqua satellite (NASA Goddard Space Flight Center). The chlorophyll is presumed to constitute 1% of the dry biomass $\mathfrak{M}$ of the phytoplankton.

[b] Forest leaf canopy height is assumed to be 2.5 m and the amount of mesophyll within the leafy volume is assumed to be 3 g/m^3. This is however presumed to be only 10% of leaf dry biomass that appears to correspond to $\mathfrak{M}$ in equation (10.19) (cf. S.A. James, W.K. Smith and T.C. Vogelmann, Ontogenetic differences in mesophyll structure and chlorophyll distribution in *Eucalyptus globulus* ssp. globulus. Am. J. Botany 86 (1999) 198–207).

[c] Mean plant height is assumed to be 1 m, mesophyll density is taken to be 3 g/m^3 (presumed to be only 10% of grass dry biomass, see note *b* above).

[d] Average mass 70 kg per individual.

carbon, with a production rate of perhaps 30% of this number, and an estimated distribution of $\sim 10^{29}$ cells in the open ocean, $\sim 2 \times 10^{29}$ in soil, $\sim 3.5 \times 10^{30}$ in ocean subsurfaces (sediments), and $\sim 1.5 \times 10^{30}$ in terrestrial subsurfaces. Turnover is fastest in the open ocean (tens of days), intermediate in soil (on the order of 1 year), and slowest in the subsurface zones (on the order of one thousand years). Their specific contribution to the carbon cycle is extremely difficult to assess at present. Table 10.5 compares the global masses of various life forms.

Table 10.5: Estimates of biomass.

Type	Dry biomass/kg [a]
Antarctic krill	5.0×10^{11}
Crops	2.0×10^{12}
Domestic animals	7.0×10^{11}
Forest leaf canopy	4.3×10^{12}
Grasslands	1.2×10^{12}
Humans	2.5×10^{11}
Ocean phytoplankton	3.6×10^{12}
Procaryotes	10^{15}

[a] "Ecological biomass" is regarded as dry weight (i.e. excluding water) and thus is approximately one third that of living biomass.

10.4.2 Volcanoes

Volcanic activity results in the emission of about 4×10^{12} moles of carbon (mainly as carbon dioxide) per year, which equates to only 0.05 Gt of released carbon. The eruption of Mount Pinatubo (in June 1991) emitted several cubic kilometres of rock (mostly as dust and smoke) and *ca* 1 teramole of sulfur. This dust and aerosols generated from sulfur oxides (cf. §10.7) can potentially have a tremendous cooling effect through directly preventing solar radiation from reaching the surface of the Earth (see §10.7), but an eruption of this magnitude typically only occurs once or twice every century.[33]

10.4.3 Anthropogenic factors

The annual release of carbon due to anthropogenic activity may be calculated from the data published by the oil industry for the annual mass of (fossil) fuel burnt $\tilde{\mathfrak{B}}$ according to equation (10.10), hence

$$\tilde{\mathfrak{R}}_{\mathrm{C}} = (\tilde{\mathfrak{B}}/M_r)n \tag{10.20}$$

where M_r is the relevant molar mass of the fuel (Table 10.1) and n is the number of carbon atoms per fuel molecule. $\tilde{\mathfrak{R}}_{\mathrm{C}}$ is the annual number of moles of carbon released. If mass is desired then it simply has to be multiplied by the molar mass of carbon (cf. equation 10.18). The results are given in Table 10.6.

[33] R.D. Cadle, A comparison of volcanic with other fluxes of atmospheric trace gas constituents. Rev. Geophys. Space Phys. 19 (1980) 746–752.

Table 10.6: Annual carbon release due to anthropogenic factors (industrial activity).

Material	$\tilde{\mathfrak{B}}$/kg	$\tilde{\mathfrak{R}}_C$/mol	$\tilde{\mathfrak{M}}_C$/Gt	Data source
Coal	4.9×10^{12}	2.07×10^{14}	2.5	World Coal Institute (2004)
Oil	3.9×10^{12}	2.71×10^{14}	3.3	BP Annual Report (2007)
Natural gas	2.57×10^{12}	1.61×10^{14}	1.9	BP Annual Report (2007)
Subtotals	1.14×10^{13}	6.39×10^{14}	7.7	
Cement	1.9×10^{12}	1.9×10^{13}	0.2	Mineral Commodities Summary (2004)
Iron and steel	1.3×10^{12}	1.29×10^{13}	0.2	International Iron and Steel Institute
Glass	1.3×10^{10}	1.77×10^{11}	0.0	Glass on Web (2006)
Silicon	3.1×10^{7}	4.34×10^{8}	0.0	Prometheus Institute (2005)
Subtotals	3.2×10^{13}	3.2×10^{13}	0.4	
Totals	1.5×10^{13}	6.7×10^{14}	8.1[a]	

[a] Cf. the 2004 global fossil fuel CO_2 emission estimate, 7910 million metric tons of carbon, which is an all-time high and a 5.4% increase compared with 2003. Globally, liquid and solid fuels accounted for 77.5% of the emissions from fossil fuel burning in 2004. Combustion of gas fuels (mostly natural gas) accounted for 18.1% (1434 million metric tons of carbon) of the total emissions from fossil fuels in 2004 and reflects a gradually increasing global utilization of natural gas. Emissions from cement production (298 million metric tons of carbon in 2004) have more than doubled since the mid 1970s and now represent 3.8% of global CO_2 releases from fossil-fuel burning and cement production. Gas flaring, which accounted for roughly 2 per cent of global emissions during the 1970s, now accounts for less than 1 per cent of global fossil fuel releases (Compendium of Data on Global Change. Oak Ridge National Laboratory: Carbon Dioxide Information Analysis Center (2005)).

The usage of fossil fuels by sector is shown in Table 10.7. There is considerable overlap in heating, electricity generation and industry since heating and industry use both fossil fuels directly and energy from electricity (and some heat is diverted to generate electricity). Hence the figures are only indicative. The global chemical industry is a large user of both raw materials and energy, but the former does not at present contribute greatly to carbon release, since only a small amount of the final products (such as polymers) is burnt (to provide district heating in cities using waste incinerators, for example). Exxon and Shell together supply about 5×10^{10} kg of oil annually as feedstock to the chemical industry; the global annual consumption of oil as chemical feedstock can be expected to be $\sim 10^{11}$ kg, an order of magnituide less than the mass of fuel burnt to produce cement.

Table 10.7: Usage of fossil fuel energy by sector.

Sector	Subsector	Fraction of total	Comments
Transport		0.28	
	Shipping	0.03	
	Rail	0.01	
	Road	0.20	
	Air	0.04	
Heating		0.19	
Air conditioning		0.03	
Industry		0.32	Annual Energy Review, Energy Information Administration (2006)
Electricity		0.40	Used subsequently in some of the other sectors

The heat produced from burning fossil fuels is

$$\tilde{\mathfrak{H}} = (\tilde{\mathfrak{B}}/M_r)\Delta H \; , \tag{10.21}$$

where $\tilde{\mathfrak{H}}$ is the annual quantity of energy released as heat. From the amounts burnt and the enthalpies of combustion, see equation 10.10, the energy released as heat per year from burning natural gas, oil and coal (Table 10.8) is about 3×10^{20} joules, which is less significant than geothermal heat (1.6×10^{21} joules), and is two orders of magnitude less than the solar radiation absorbed by the ocean (Table 10.3). Hence we may infer that it does not contribute to global warming directly.

Deforestation has proceeded at an accelerated rate in the 20th century as developing countries in particular have recklessly exploited their natural resources (Figure 10.10), using technologies (chainsaws and so forth) imported from the developed world. This is very much a greed-driven activity (cf. the Preface and Chapter 21), but in some countries, e.g. Madagascar, much of the population appears to take a wanton delight in forest destruction regardless of even short-term economic consequences, due to motivations that can perhaps be traced

Table 10.8: Annual energy release by burning fuel.

Fuel type	$\mathfrak{N}_{\mathrm{C}}$/mol	$\mathfrak{H}$/J
Coal	2.0×10^{14}	9.9×10^{19}
Oil	2.7×10^{14}	1.3×10^{20}
Natural gas	1.6×10^{14}	8.0×10^{19}
Subtotals	6.4×10^{14}	3.1×10^{20}
Timber[a]	5.7×10^{15}	2.9×10^{21}

[a] Due to extreme deforestation (clearing tropical rain forests, as in Figure 10.10). See text for the assumptions made.

back to fear (of a vast, gloomy domain).[34] Forest and woodland comprise a significant proportion of the carbon sequestration cycle and thus the impact of deforestation may well be more destabilizing than the effects of industrialization. Furthermore, making some simple assumptions that tree height is 10 m, tree density is 1 $\mathrm{Mg/m^3}$, half a tree is wood, that wood is 5% as effective as the same mass of coal in producing heat and that 1% of forests are burnt per year (Figure 10.11), then 2.9×10^{21} joules are liberated annually, which is comparable to the annual heating of the forests and woodland by the Sun. Deforestation therefore appears to be a significant direct contributor to global warming, apart from the loss of CO_2 sequestration ability.

10.5 The carbon cycle

The carbon cycle as a consequence of photosynthesis-driven sequestration, respiration and anthropogenic output is shown in the ubiquitously reproduced Figure 10.12. The simplistic figures derived in the previous sections agree only in orders of magnitude; they are simple (but transparent) approximations, whereas the paths to the figures given on the diagram are often obscure. The models used to derive this picture are very complicated, not least because even the detailed versions contain averages over different plant and phytoplankton types, each of which will have its own limiting factors for growth, photosynthesis and respiration.

It is apparent from this diagram that the atmospheric carbon balance depends upon small differences (equivalent to less than 1% of the carbon in the atmosphere, for example) between pairs of large numbers, and all that that implies in terms of the influence of uncertainties in those numbers on predictions made on that basis.

[34] There is insufficient space here to analyse the phenomenon of deforestation in detail. Europe, too, was previously largely covered with the forest, which was mostly destroyed (using local technology) in the course of development. Psychological factors (the fear of the forest—as can easily be discerned from numerous folk tales) must also have provided impetus for the destruction. It is a sobering thought that it has taken so long for the deleterious consequences of deforestation to be perceived. Perhaps we are now in possession of sufficient data to be able to undertake a proper risk assessment of it.

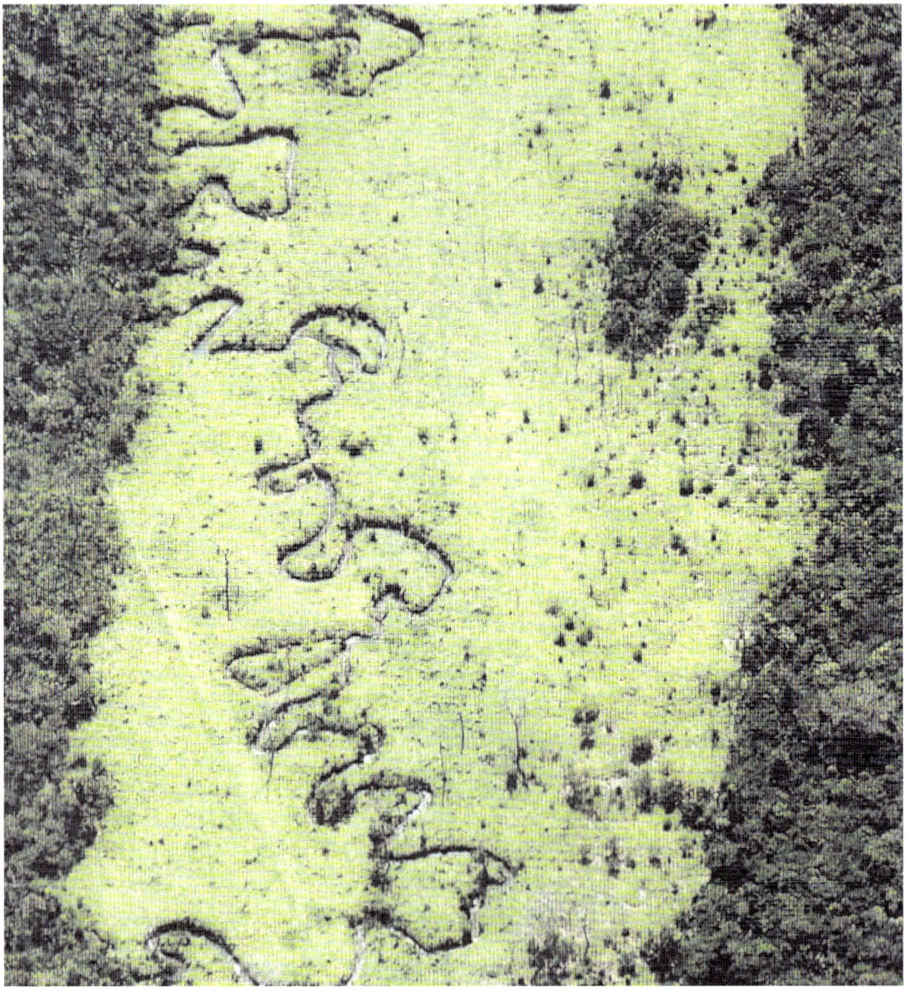

Figure 10.10: Amazon deforestation viewed from the air. The feature meandering through the centre is a water course not extant prior to the forest clearing. Ruderal vegetation (light green—high albedo) replaces the ancient forest (dark green—low albedo), with a great loss of active biomass.

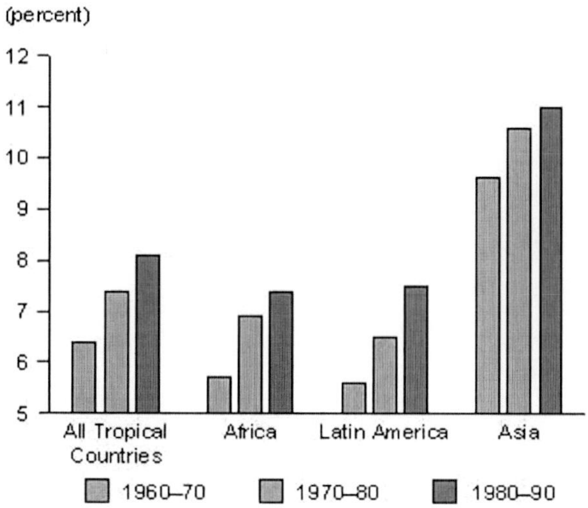

Figure 10.11: Estimated rate of tropical deforestation 1960–1990. (*Source:* The World Resources Institute.)

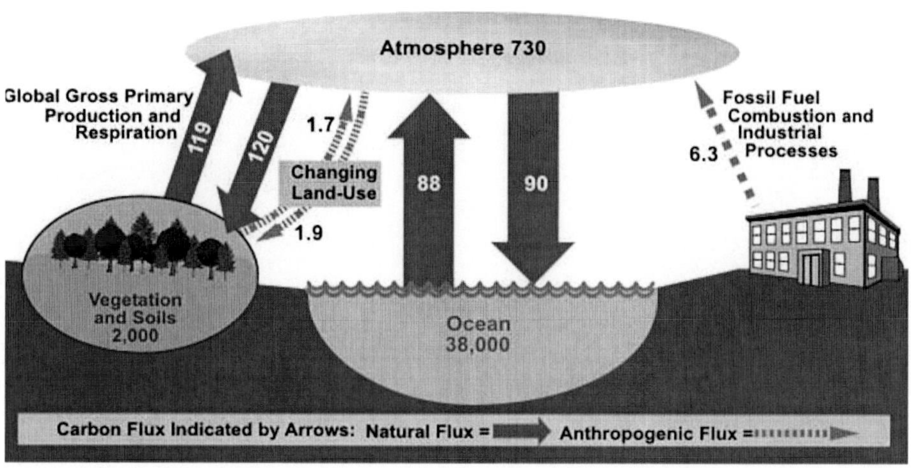

Figure 10.12: The carbon cycle in giga metric tonnes of ^{12}C. (*From:* Climate Change. Intergovernmental Panel on Climate Change (2001).)

10.6 The nitrogen cycle

Although nitrogen is the most abundant gas in the atmosphere, in the form of the N_2 molecule it is almost inert and hence unavailable to life, although it is an essential constituent of proteins. A small number of procaryote species, both bacteria and archaea, have developed the ability to "fix" nitrogen, i.e. convert it to a reactive, usable form. Agriculturalists know that certain vegetables, the legumes, live symbiotically with these prokaryotes; other plants depend on mineral reactive nitrogen (nitrates and other forms of oxidized nitrogen). The growth of agriculture meant that nitrogen deficiency became a major limiting factor, until the Haber-Bosch process was developed to synthetically produce ammonia from atmospheric nitrogen gas and hydrogen. The growth of the internal combustion engine has also led to the significant production of nitrogen oxides in combustion chambers, which are then dispersed into the atmosphere. Table 10.9 summarizes data on natural and anthropogenic reactive nitrogen production.

The nitrogen cycle is important to the present discussion because reactive nitrogen supply may limit tree growth (and possibly terrestrial plant biomass more generally),[35] thereby affecting net sequestration of carbon through photosynthesis (equation 10.5 and Table 10.3).

Table 10.9: Reactive nitrogen production.[a]

Source	Annual production/moles N
Natural biofixation	
Ocean	14×10^{12}
Land	17.2×10^{12}
Anthropogenic	
Haber-Bosch	24×10^{12}
Cultivation	7×10^{12}
Fossil fuel combustion	7×10^{12}

[a] Data from J.N. Galloway et al., Nitrogen cycles: past, present and future. Biogeochem. 70 (2004) 153–226.

10.7 The sulfur cycle

Sulfur is released into the atmosphere through volcanic activity and burning fossil fuels. Hydrogen sulfide and organic sulfides are burnt to sulfur dioxide, which is then

$$H_2S \xrightarrow{O_2} SO_2 \xrightarrow[H_2O]{O_2} H_2SO_4 \,. \tag{10.22}$$

Apart from the well known acid rain phenomenon, the sulfuric acid also exists as aerosols in the upper atmosphere, affecting albedo directly, and also acting as cloud condensation nuclei.

[35] See F. Magnani et al., The human footprint in the carbon cycle of temperate and boreal forests. Nature (Lond.) 447 (2007) 848–852, and R. Hyvönen et al., Impact of long term nitrogen addition on carbon stocks in trees and soils in northern Europe. Biogeochem. (2007) (DOI 10.1007/s10533-007-9121-3).

Most species of phytoplankton excrete dimethyl sulfoxide, which similarly can reside as an aerosol in the atmosphere, where it can also react with other molecules to create other aerosol-forming species.[36]

10.8 Consequences of global warming

This paper has deliberately focused on the actual scientific evidence, in the shape of both unconditional knowledge (data) and the interpretative framework, i.e. theory (conditional knowledge) used to draw inferences from the data. In this way it is very different from much of the other recently published literature, such as the Stern Review,[37] which gives cursory attention to the evidence, rather uncritically accepting the popular view that man is the culprit (see Chapter 13 and then devotes hundreds of pages to examining the deleterious economic consequences that follow from the assumption of an inevitable temperature rise with business as usual (BAU), but that can be alleviated by appropriate human intervention (moving to the low carbon economy). This is essentially a scenario-building exercise, but it must be remembered that the "evidence" for anthropogenic global warming is itself the result of scenario building,[38] hence we have a scenario built on a scenario, truly a house built upon sand. In this chapter, we have tried to show that what are essentially "back of the envelope calculations" are above all useful for indicating which factors are the most significant, according to the "order of magnitude" rule, and hence where human effort can most sensibly be directed towards assuring a measured response to threats to safety and, ultimately, survival.

If global warming is accepted as a fact (the current figure seems to be around 0.01 °C per annum, with an uncertainty that is extraordinarily difficult to establish), the next problem is to establish what are its (possible) consequences, before moving on to whether its cause is anthropogenic, and whether it can be influenced by human intervention.[39]

The heterogeneity of climate

It is commonly accepted that man is adapted to his (earthly) environment. As Sommerhoff has pointed out however,[40] adaptation does not merely mean existence: it means that man is able to survive, not only under the actual set of conditions, but also those likely to be encountered in the future. A crucial question is whether man's adaptedness to his environment is sufficient to cope with the consequences (which are not really known) of possible change. Ecosystems can be, as we know, extraordinarily resilient; at the same time they

[36] E.g., R.J. Charlson et al., Oceanic phytoplankton, atmospheric sulphur, cloud albedo and climate, Nature 326 (1987), 655–661; see also the critique of S.E. Schwartz, Are global cloud albedo and climate controlled by marine phytoplankton? Nature 336 (1988) 441–445.

[37] The Economics of Climate Change: the Stern Review. Cambridge: University Press (2007).

[38] It would be difficult to overestimate the role that models, with their temporary and shifting validities, have played in establishing this figure.

[39] Naïvely, it would appear to be obvious that if the cause is anthropegenic, then the solution is in man's hands. Were this to be the case, however, then suicide would be unknown. Collective blindness has occurred often enough in history, and possible reasons for collective suicide, such as despair in the face of a declining civilization, or greed, can easily be found.

[40] G. Sommerhoff, Analytical Biology. Clarendon: Oxford (1950).

can also be extraordinary fragile. A salutary reminder of the latter is the fate of the passenger pigeon in North America—apparently it was the most numerous bird on the continent in the 17th century, whose population (estimated at about 5000 million) equalled perhaps one third of all the birds there at the time. They were hunted with incredible zeal by the European settlers, and the last survivor died in Cincinnatti zoo in 1914.

The so-called global temperature anomaly (i.e. variation) for the regions of the Earth is shown in Figure 10.13. From these data it is worth exploring further possible connexions with urbanization, population density and recent flooding, see Figures 10.14 and 10.15.

Interestingly, the greatest temperature anomalies are predominantly in the cooler regions of the Earth and in the more industrialized Northern hemisphere. What becomes apparent is the difficulty of making correlations in such a complex system. Industrialized regions and climate zones are moreover often very localized even within a given country, and many large countries will have multiple subclimates so that the wrong (inapproriate) way of averaging (such as a global mean of the quantity in question) may often lead to a greatly disorted inference. From these figures it can be seen that man can live in a wide range of temperatures, and that presently some parts of the planet are too cold for any human being to live in comfort, and rather fewer parts are too hot. Therefore, at first sight it might seem that man—especially considered as a species—is well adapted to live in a wide range of temperatures; global warming would merely cause a shift in the places where it is convenient to live, and there are plenty of comparatively empty places that are currently too cold in which to live.

This very simplistic conclusion ignores a large number of secondary effects of increasing temperature. The plants on which man depends for food have become increasingly domesticated and hence less adaptable to changes in climate (yet at the same time man has developed increasingly sophisticated techniques for accelerating evolution (through genetic engineering) and could presumably readapt these plants to different conditions); increasing temperature would melt the polar ice caps and cause significant changes in sea level, hence flooding much land; on the other hand it will increase the evaporation from the oceans, resulting in more clouds (which will increase the Earth's albedo, hence less sunlight will be absorbed, countering the tendency for temperature to increase) and more precipitation, which may cause more water to be locked up in high mountain glaciers, but also flooding (Figure 10.15). Even these 'obvious' consequences of increasing temperature to extent counter each other making the net effect extraordinarily difficult to predict.

Furthermore, there are what we might call tertiary effects:[41] changes in the circulation patterns of water (i.e. ocean currents) and wind, possibly resulting in very significant changes of local climate and weather. There is already a suspicion that some of the weather extremes that many parts of the world have experienced in recent years are somehow due to the slight increase in mean global temperature. Minuscule though the temperature rise might appear to be, it must also be borne in mind that the dynamics of climate and weather are highly nonlinear, and riddled with delayed feedback, with all that this entails in terms of unpredictable outcomes.[42] Such dynamics are the very essence of com-

[41] The boundaries between secondary and tertiary are not well defined.

[42] A.B. Pippard, Response and Stability. Cambridge: University Press (1985).

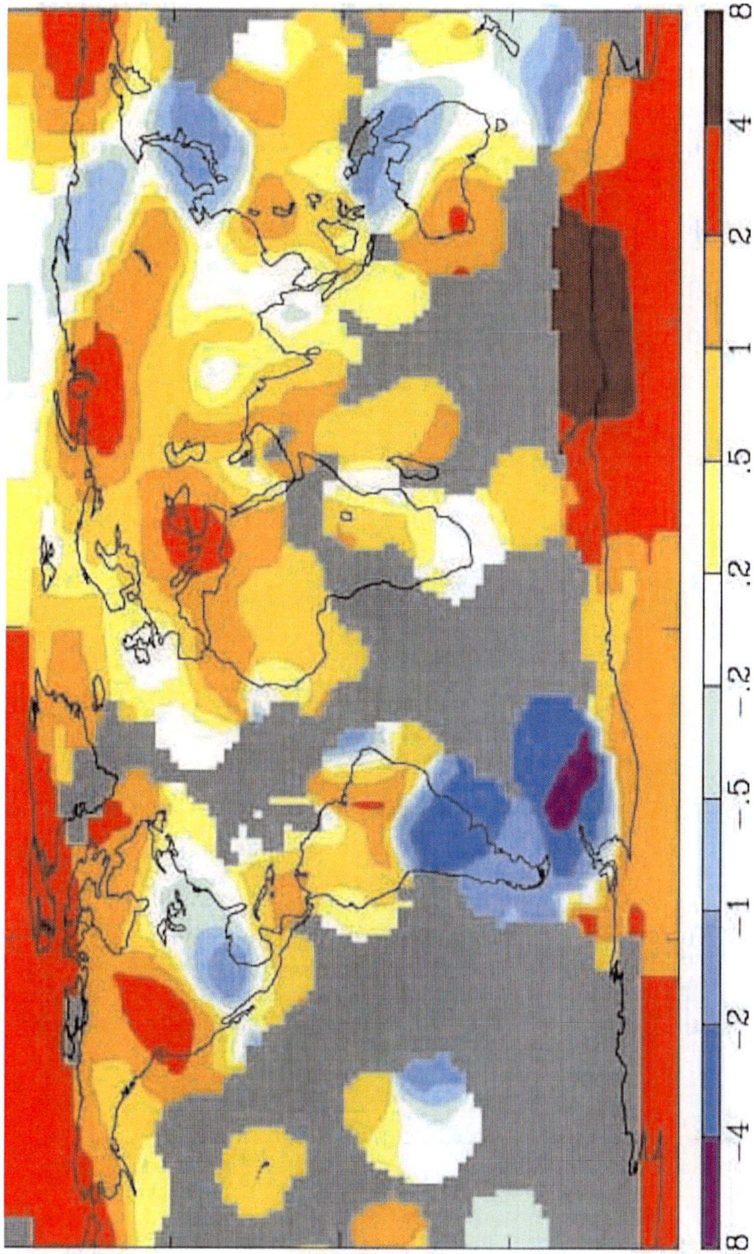

Figure 10.13: July 2007 Surface temperature anomaly in °C vs 1951–1980 mean. Grey areas indicate the lack of station data within a 1200 km radius. (*Source:* J. Hansen, R. Ruedy, M. Sato and K. Lo, NASA Goddard Institute for Space Studies and Columbia University Earth Institute, New York.)

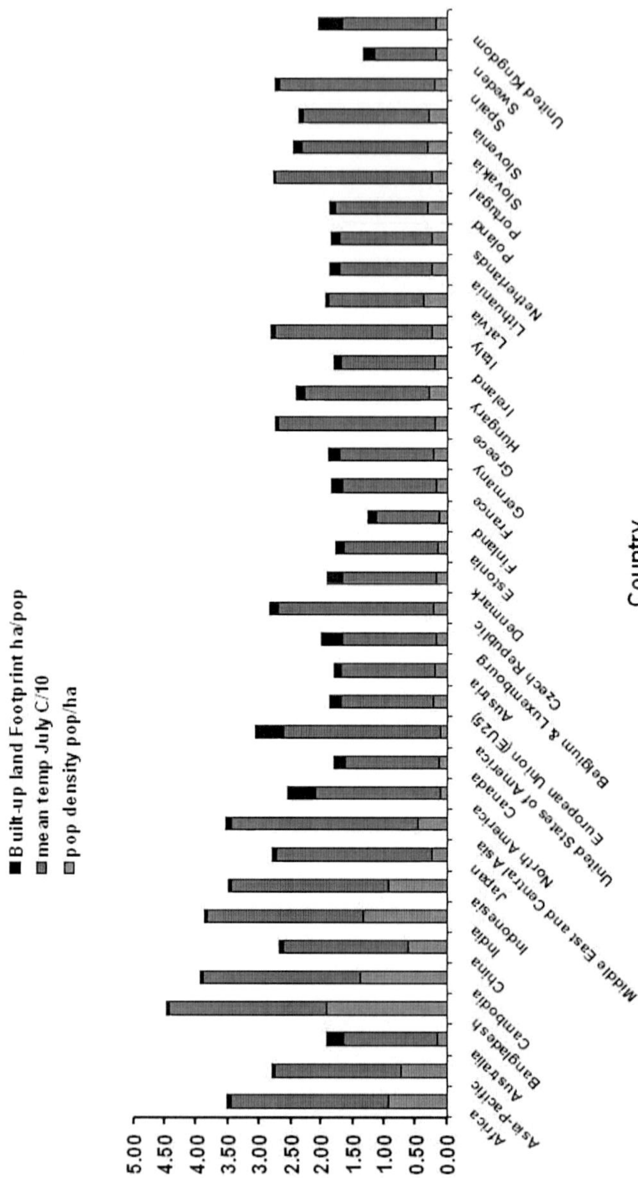

Figure 10.14: Countrywise population density (people per hectare); mean July temperature/10 °C; urbanization (hectares of urban land per person). (Density and urbanization data *from:* Global Footprint Network, Oakland, California, 2006 Edition.)

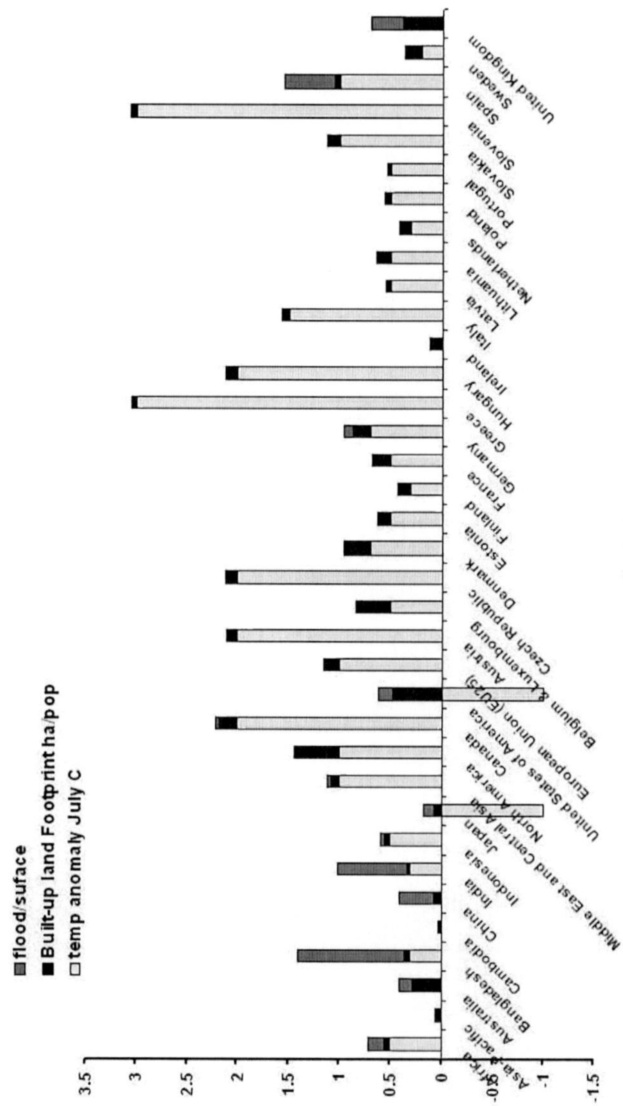

Figure 10.15: Temperature anomaly/°C; urbanization (hectares of urban land per person); ratio of flood-affected surface area (October 2006–August 2007) to regional surface area. (Flood data *from:* G.R. Brakenridge, E. Anderson and S. Caquard, 2007 Global Register of Major Flood Events, Dartmouth Flood Observatory, Hanover, New Hampshire.)

plexity. Invoking the so called "butterfly effect" implies that the Earth must be inherently unstable in order for the small contribution of carbon dioxide produced by man (amounting annually to less than one per cent of total atmospheric carbon dioxide) to unleash runaway global warming. But, evidence from the past would tend to indicate that the Earth's ecosphere is actually quite robust, surviving both extremes of global warming and carbon dioxide concentration, albeit perhaps not as high as present, but then the correlation between ice core data and the character of the atmosphere at the time of the core formation is yet another extrapolation.

Scope for political action

Deforestation, see Figures 10.11 and 10.10, has continued apace over the last four decades and the reduction alone in carbon sequestering capacity is almost a third of carbon emissions from the burning of fossil fuels.[43] Is this of more concern than the actual burning of fossil fuels? Perhaps not, not because scientifically it should be, but because the rich nations are unaffected directly by the price of wood so that their economies are safe from that particular impact, unlike the situation with diminishing fossil fuel reserves, which directly impacts the cost of living due to the inexorable rise in price as a consequence of market forces.[44]

Unpredictable, violent weather poses a potential threat to security since orderly communications—and hence the fabric of civilization—tend to be disrupted. The ramifications of unpredictable climate change, even over a restricted zone of the Earth, could be far-reaching, if not survival-threatening. For example, were the Gulf stream to cease to flow, drastic modifications to the built environment of the British Isles would doubtless be needed. Actually, the intensity of the threat strongly depends on the status of solidarity among the members of the communities affected. If the degree of solidarity is high (which one might identify with high ethical standards, see Part V) then people will band together and help each other, and hence minimize the effects of catastrophe. The system of insurance is of course a formal way of expressing this solidarity. At the same time it should be borne in mind that the global economy would grind to a halt if every manufactured artefact survived its engineered lifetime; destruction by force majeure is a very significant contributor to the global economy; essentially all the money paid out in meeting insurance claims is used to purchase freshly manufactured goods and is actually a boost to the economy.

Even minor rises in sea level resulting from the melting of polar ice may submerge some low lying idyllic (to some) islands. The impact is local rather than global. On a larger scale is the melting of permafrost in Siberia, that is already happening. In the past, the need to construct large buildings in zones of permafrost posed particular engineering and architectural problems, that were largely solved in Soviet times.[45] The hope now is that the softening of the ground on which many large structures rest, including those associated with

[43] Cf. Navigating the Numbers: Greenhouse Gas Data and International Climate Policy. World Resources Institute (2005).

[44] Or rather, the reallocation of resources: many citizens of the industrialized world find it unpalatable that much of their wealth is now being transferred to oil-producing nations whose contribution to human civilization is minuscule, although those nations may well invest their new-found wealth in the industrialized countries.

[45] See e.g. V. Conolly, Siberia Today and Tomorrow. New York: Taplinger (1976).

essential infrastructure, will follow a gradual and predictable course, enabling properly planned remedial measures to be implemented.

Difficulties of implementation

Construction is now in many countries one of the biggest sectors, and cement production now exceeds steel as the single largest anthropogenic industrial (non-fuel) carbon emitter. After the mass destruction of the built environment in World War 2, many countries built up large construction industries, a sector that has a low entry threshold for the entrepreneur, many of whom became veritable plutocrats. Although that there is no longer any real need for this huge industry, there is extreme reluctance to downscale it (and the entrepreneurs have become sufficiently influential to be able to block attempts to do so), hence the senseless pouring of concrete over the land—diminishing vital photosynthetic zones—continues unabated.[46] A better example of a vested interest deleterious to the general fabric of society, but providing a transitory material benefit for its factors, could scarcely be found.

"All the rivers run into the sea; yet the sea is not full; unto the place from whence the rivers come, thither they return again"—as has been known for thousands of years.[47] If the temperature increases, more water evaporates from the oceans, and hence there must inevitably be more rainfall. Recent severe floods in Britain, Switzerland and elsewhere have focused attention onto this problem. Evidence for the flooding is not simply anecdotal: there is actually data to support it.[48] On the other hand, before interest in the possibility of global warming emerged, interest in Switzerland for example focused on the tens of square metres of land that were daily being covered in concrete. This naturally prevents water from soaking into the soil, and increases flooding even if the level of rainfall remains constant. Indeed, plotting data on flooding incidents with data on the extent of built-up areas, yields some correlation for the UK, see Figure 10.15 (but not for all regions); in the UK the urbanization of flood plains may well have contributed to the scale of flooding on the river Avon in 2007. And one should not overlook a simple regulatory mechanism—an increase in evaporation from the surface will also increase clouds (from which the rain falls), which causes more sunlight to be reflected from the surface of the earth, and hence, all other factors being equal, should engender cooling and diminution in the evaporation that ultimately condenses and falls as rain.

10.9 Conclusions

Climate change has occurred throughout the Earth's history and is doubtless occurring now. Whether it is anthropogenic is much more difficult to establish. Not only are there numerous interfering factors, but these factors are inextricably interrelated with each other—forming a very complex system. Even if it is conceded that anthropogenic carbon dioxide, for example, is insignificant in

[46] In addition, there are many undesirable security and safety issues associated with this inflated industry. For example, manpower is provided by large numbers of migrant workers, whose status becomes more or less permanent, but who may have less of a stake in their chosen country of residence than the indigenous population.

[47] Ecclesiastes, 1, 7.

[48] For example, Figure 10.15.

quantity compared with natural sources, that does not allow one to conclude that it has no effect, the extreme nonlinearities known to be operating in the world's climate system provide abundant possibilities for amplifying even very slight effects.

The very fact that there has been a thousandfold increase in world population, and correspondingly in industrial output, since the beginning of the 18th century suggests that unprecedented anthropogenic influence may now be affecting global climate. The actual volumes of activity are certainly comparable with others known to have an effect. Therefore, at the very least, prudence would suggest that one should scrutinize man's activities rather more carefully than has traditionally been the case since the start of the industrial era of mass production, so characteristic of our age.

Even if a right action is formulated, there is no guarantee that it can be implemented. Politics is said to be the art of the possible,[49] and sometimes even a possibly wrong action is deemed to be better then inaction. Among politicians, the Intergovernmental Panel on Climate Change (IPCC) plays an influential role. When even erudite specialists in academic institutions cannot agree about the most fundamental factors believed to affect climate change, such as the infrared absorption of carbon dioxide in the upper atmosphere, one can sympathize with the difficulties faced by members of the IPCC: the issue (climate change) is so complex that it is practically impossible to grasp by any one individual, and unfortunately we have not yet perfected a way of sufficiently effectively coupling brains together to significantly enhance human reasoning ability to make it graspable. To be sure, the latest report of the IPCC[50] evokes scepticism, but the *acharnement* of its critics is as misplaced as the fervour of its supporters, for human resilience, if equal to the task of encompassing adaptation to the changed climate, is surely more than equal to the task of encompassing adaptation to policies based on the conclusions of the IPCC (and, it must be remembered, these policies are likely to be only very partially implemented). The real danger of a low carbon policy is that the economy will become so depressed that there will be a decline in technical capability (including scientific research activity, except perhaps into climate research!) that will diminish mankind's capacity to accommodate adverse environmental changes.

The situation might appear to be so complex that one might well be tempted to advocate essentially *laisser faire* (especially in view of what appear to be natural mechanisms of regulation). It is not, however, in the nature of man as a collective to let things alone: a majority of us want to proactively interfere. This in turn implies the need to investigate, both from the practical viewpoint, to provide a guide to action, and as part of the general duty of the scientist to deepen our understanding of the universe, regardless of the use that may be made of the knowledge—this work of deepening understanding is simply an epiphenomenon of human development [51]

Therefore, further investigation of climate change, accompanied by a vigorous and fully informed debate among all concerned—which actually means every denizen of the planet—is wholly justified, and is indeed a continuing matter of urgent priority.

---

[49] From Bismarck's definition of politics as "Die Kunst des Möglichen".

[50] See www.ipcc.ch/SPM2feb07.pdf.

[51] Cf. J.J. Ramsden, The New World Order. Moscow: Progress Publishers (1991).

The simplest solution

Finally, let us go back to the thermal equilibrium equation (10.4) and modify it slightly to include the (annual quantity of) heat energy $\tilde{\mathfrak{H}}$ produced anthropogenetically (Table 10.8 and a fraction $\mathfrak{F}$ of the reradiated energy retained by the atmosphere (due to absorption by greenhouse gases). The equation (with appropriate substitutions) then becomes

$$S_0 \pi r_{\mathrm{E}}^2 (1 - \mathfrak{A}) + \tilde{\mathfrak{H}}/\tau = \sigma T^4 4\pi r_{\mathrm{E}}^2 (1 - \mathfrak{F}) \,. \tag{10.23}$$

As pointed out before, solving this equation for T with $\mathfrak{A} = \tilde{\mathfrak{H}} = \mathfrak{F} = 0$ yields $T = 278$ K, about 5 °C; solving anew for $\mathfrak{F}$ with the albedo set to the known average value of 0.31 and with T set to 278 K, and keeping $\tilde{\mathfrak{H}} = 0$, yields $\mathfrak{F} = 0.3$, that is, about a third of the energy that would otherwise be radiated is trapped. The addition of the dominant anthropogenic source of energy, extreme deforestation by burning, only adds about 0.1 °C to the mean global temperature according to this calculation. All other anthropogenic heat sources (Table 10.10) are minor in comparison.

Table 10.10: Natural and anthropogenic energy sources.

Source	$\tilde{\mathfrak{H}}$/J	Notes
Total annual solar irradiation	3.8×10^{24}	
Extreme deforestation (1 year)	2.9×10^{21}	Emits CO_2
Geothermal energy (1 year)	1.6×10^{21}	
Burning of fossil fuels (1 year)	3.1×10^{20}	Emits CO_2
Conventional bombs dropped in WW2	10^{17}	Emits CO_2
Nuclear bombs dropped in WW2	5×10^{14}	
U.S. arsenal (50) of B-53 H-bombs	5×10^{18}	

These calculations leave quite open the relationship between carbon dioxide production, consumption, and the steady concentration in the atmosphere that is supposed to contribute to the greenhouse effect (here encapsulated in $\mathfrak{F}$).[52] This relationship is so exceedingly complex, and progress in understanding it as yet still so preliminary, such that even a relationship between $\mathfrak{F}$ and atmospheric carbon dioxide concentration cannot be said to have been established with any certainty,[53] that to use such a relationship as the basis of far-reaching policy would appear to be reckless in the extreme. Concerted effort applied to prevent what is indubitably a contributor to climate change, namely extreme deforestation, would appear to be the sole action that can be justified on the basis of incontrovertible facts at present.

The simplest, most direct, interpretation of "anthropogenic" suggests human population growth as the prime cause of temperature rise (cf. Figures 10.1 and 10.2). In fact, population growth clearly precedes temperature rise, which, however, follows the rise of world GDP rather closely. Phenomenologically, GDP

[52]Global knowledge of how photosynthesis (atmospheric carbon sequestration, equation (10.5), and plant respiration, equation (10.19), depend on environmental factors including temperature, humidity, and gas concentrations, is still very imperfect.

[53]Despite well over a century of investigation. See, for example, S. Arrhenius, On the influence of carbonic acid in the air upon the temperature of the ground. Phil. Mag. (Ser. 5) 41 (1896) 237–276, and references therein.

clearly follows, and is driven by, population growth. To reverse anthropogenic temperature rise, it is therefore almost self-evident that either world GDP must be reduced (back to the levels prevailing around 1925), or world population must be reduced (maintaining per capita GDP). Which is easier to achieve? We close the chapter with this open question.

10.10 List of the most common symbols

Symbol	Significance	Typical unit
$A^{(\mathrm{x})}$	area (of part x of the Earth)	m^2
$\mathfrak{A}$	albedo	–
$\tilde{\mathfrak{B}}$	annual mass of fuel burnt	Gt
$\overline{BAF}$	biomass active fraction	–
b	Wien's displacement constant	nm K
C	photosynthesis constant (equation (10.19))	–
c	speed of light	m/s
E	power emitted	$\mathrm{W/m}^2$
$\hat{E}$	total emitted power	W
$\mathcal{E}$	energy of a photosynthetically active photon	J
$\mathfrak{F}$	reradiated energy fraction retained atmosphere	–
h	Planck's constant	J s
ΔH	heat of combustion	J/mol
λ	wavelength of radiation	nm
M_r	molar mass of an atom or molecule	g/mol
$\tilde{M}_{\mathrm{C}}$	annual mass of carbon sequestered by plants	Gt
$\mathfrak{M}$	dry active biomass	Gt
$\mathfrak{M}_{\mathrm{C}}$	mass of carbon atoms released	Gt
N_L	Loschmidt number	mol^{-1}
n	number of carbon atoms in a fuel molecule	–
$\tilde{N}_{\mathrm{C}}$	annual no C atoms sequestered by plants	mol
$\tilde{N}_{\phi}$	annual no actinic photons absorbed by biomass	mol
ν	frequency of radiation	Hz
$\overline{PAR}$	photosynthetically active radiation fraction	–
r_{E}	Earth's radius	km
$\tilde{R}_{\mathrm{C}}$	annual no (or mass) of C atoms respired	mol (or Gt)
$\tilde{\mathfrak{R}}_{\mathrm{C}}$	annual no C atoms released from burning fuel	mol
$\tilde{\mathfrak{H}}$	annual heat energy produced anthropogenetically	J
S_0	solar constant (irradiance)	$\mathrm{W/m}^2$
$S^{(\mathrm{x})}$	power absorbed per unit area of x	$\mathrm{W/m}^2$
$\hat{S}$	total absorbed power	W
$\tilde{S}^{(\mathrm{x})}$	total annual energy absorbed by x	J
σ	Stefan-Boltzmann coefficient	$\mathrm{W\ m^{-2}\ K^{-4}}$
T	temperature	K
τ	duration (e.g. of daylight during 1 year)	s
V	volume	m^3

Complexity and Security
J.J. Ramsden and P.J. Kervalishvili (Eds.)
IOS Press, 2008

Chapter 11

Climate change and the complexity of the energy global security supply solutions: the global energy (r)evolution

Fulcieri Maltini

FM Consultants Associates[1]

Abstract. The impact of greenhouse emissions on climate change and the decrease in world fossil energy sources will have significant consequences for the future of the planet. Three recent major reports ("The Stern Review: the Economics of Climate Change", October 2006, "Where will the Energy for Hydrogen Production come from?—Status and Alternatives" by Ludwig-Bölkow-Systemtechnik GmbH—LBST/European Hydrogen Association, 2007, and "The Global Energy (r)evolution Scenario" by EREC-Greenpeace International, 2007) are analysed in this article. They reach the same conclusions about the complexity of the phase-out from the carbon society and the conversion to energy efficiency and renewable energy sources.

11.1 Climate change

Over the last century humankind has re-scripted its role in the natural world. Millions upon millions of people have been fed, many deadly diseases have been treated, technology has taken us into space, telecommunications run society. Much of nature has been bent to our will, but still it appears difficult to deal

[1]E-mail: fulcierimaltini@gmail.com

with the weather. For a number of years, it was believed that the changes in climate were caused by solar influence and cosmic radiation.[2] Now, in the face of disastrous flooding, the melting of glaciers and the threat of disappearance of entire islands or the considerable degradation of the ecosystem, a different reality has appeared.

A significant body of scientific evidence seems to indicate that the Earth's climate is rapidly changing, possibly as a result of increases in greenhouse gases caused by human activities. Since pre-industrial times (around 1800), atmospheric carbon dioxide, methane and nitrous oxide concentrations have increased, mainly as a result of burning fossil fuels, and deforestation and other changes in land-use. Figure 11.1 shows greenhouse gas emissions in 2000 by source.

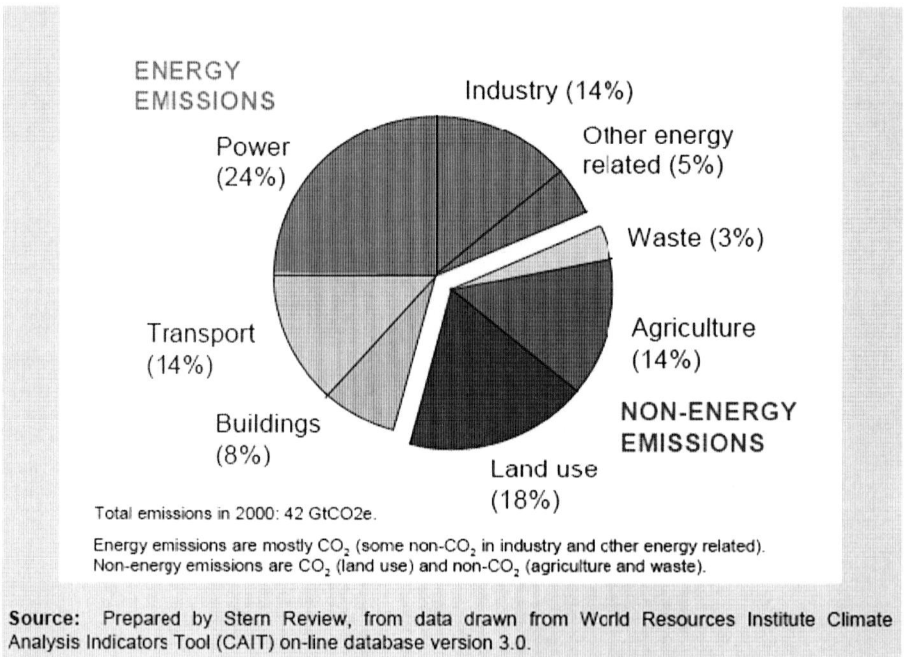

Figure 11.1: Greenhouse gas emissions in 2000 by source.

There is compelling evidence that the rising levels of greenhouse gases (GHG) will have a warming effect on the climate by increasing the amount of infrared radiation (heat energy) trapped by the atmosphere: this is the "greenhouse effect" (Figure 11.2, see also Chapter 10). The change in global average near-surface temperature between 1850 and 2005 has been on a rising trend as shown in Figure 11.3. Recent modelling by the Hadley Centre and other research institutes show that the observed trends in temperatures at the surface and in the oceans, as well as the spatial distribution of warming, cannot be replicated without the inclusion of both human and natural effects. Taking into account the rising levels of aerosols, which cool the atmosphere, and the observed heat uptake by the oceans, the calculated warming effect of greenhouse gases is more

[2] M. Lockwood and C. Froehlich, "Recent oppositely directed trends in solar climate forcings and the global mean surface air temperature", Proc. R. Soc. A doi:10.1098/rspa.2007.1880.

than enough to explain the observed temperature rise. Figure 11.4 summarizes the scientific evidence for the links between concentrations of greenhouse gases in the atmosphere, the probability of different levels of global average temperature change, and the physical impacts expected for each level.[3]

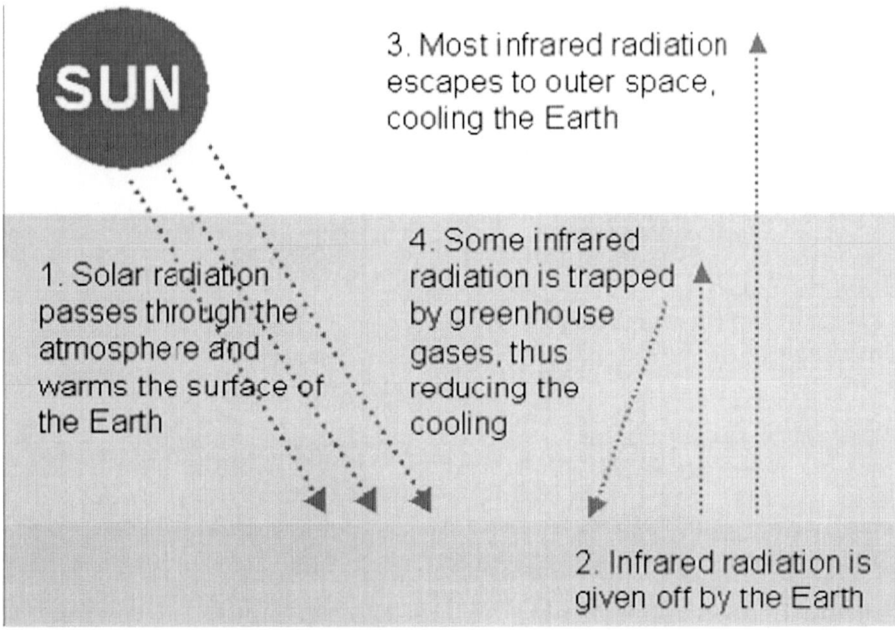

Figure 11.2: The greenhouse effect. (*Source:* based on data from DEFRA, UK, 2005).

Figure 11.4 illustrates the types of impact that could be experienced as the world comes into equilibrium with more greenhouse gases. The top panel shows the range of temperatures projected at stabilization levels between 400 ppm (parts per million) and 750 ppm CO2e (hereafter CO2e signifies CO_2 equivalent, including all GHGs) at equilibrium. Major GHGs are: CO_2 (carbon dioxide), CH_4 (methane), N_2O (nitrous oxide), PFCs (perfluorocarbons), HFCs (hydrofluorocarbons) and SF_6 (sulfur hexafluoride). The solid horizontal lines indicate the 5–95% ranges based on climate sensitivity estimates from the United Nations Intergovernmental Panel on Climate Change (IPCC) 2001 report and a recent Hadley Centre ensemble study. The vertical line on each range indicates the mean of the 50th percentile point. The dashes show the 5–95% ranges based on eleven recent studies. The bottom panel illustrates the range of impacts expected at different levels of warming. The relationship between global average temperature changes and regional climate changes is very uncertain, especially with regard to changes in precipitation.

The current level or stock of greenhouse gases in the atmosphere is today equivalent to around 380 ppm CO2e compared with only 280 ppm before the Industrial Revolution. These concentrations have already caused the world to warm by more than 0.5 °C and will lead to at least a further half degree warming

[3]The Economics of Climate Change: the Stern Review. Cambridge: University Press (2007).

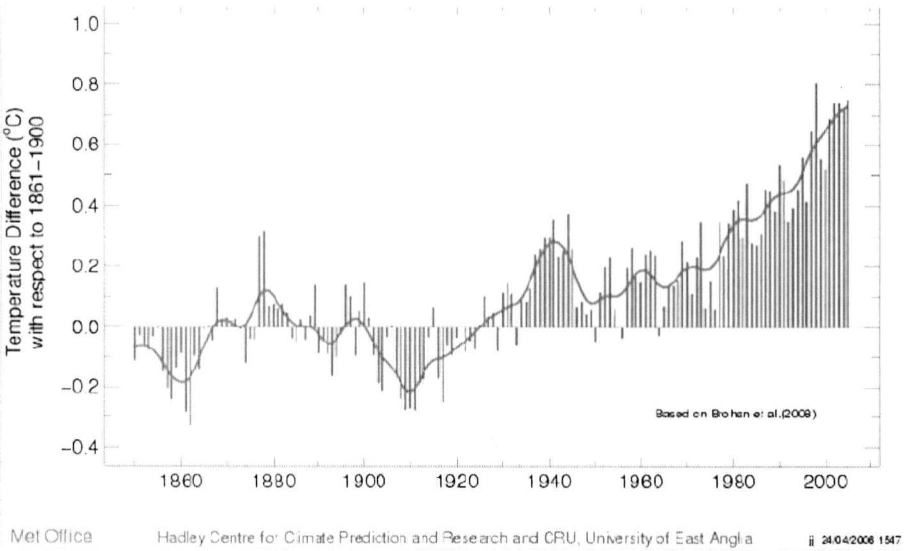

Figure 11.3: The Earth has warmed 0.7 °C since around 1850.

over the next few decades because of the inertia in the climate system. Stabilization at 450 ppm CO2e is already almost out of reach, given that we are likely to reach this level within ten years and that there are real difficulties of making the sharp reductions required with current and foreseeable technologies. But the annual flow of emissions is accelerating, as fast-growing economies invest in high carbon infrastructure and as demand for energy and transport increases around the world.

Figure 11.5 shows the link between greenhouse emissions and climate change. The Arctic has been predicted to be hit first by global warming, principally because warming at the northern pole is enhanced by positive feedback. Snow and ice reflect 80% to 90% of solar radiation back into space, but when these white surfaces disappear, more solar radiation is absorbed by the underlying land or sea as heat. This heat, in turn, melts more snow and ice. A major consequence is the considerable melting of Greenland ice sheet which in turn will affect the ocean circulation due to reduction of salt concentration in the sea. Warming will have many additional severe impacts, often mediated through water:[4]

- Widespread thawing of permafrost regions is likely to add to the extra warming caused by weakening of carbon sinks. Large quantities of methane (and carbon dioxide) could be released from the thawing of permafrost and frozen peat bogs. It is estimated, for example, that if all the carbon accumulated in peat alone since the last ice age were released into the atmosphere, this would raise greenhouse gas levels by 200 ppm CO2e.

- Melting glaciers will initially increase flood risk and then strongly reduce water supplies, eventually threatening one-sixth of the world's population,

[4]EREC—Greenpeace International "The Global Energy (R)evolution Scenario", 2007.

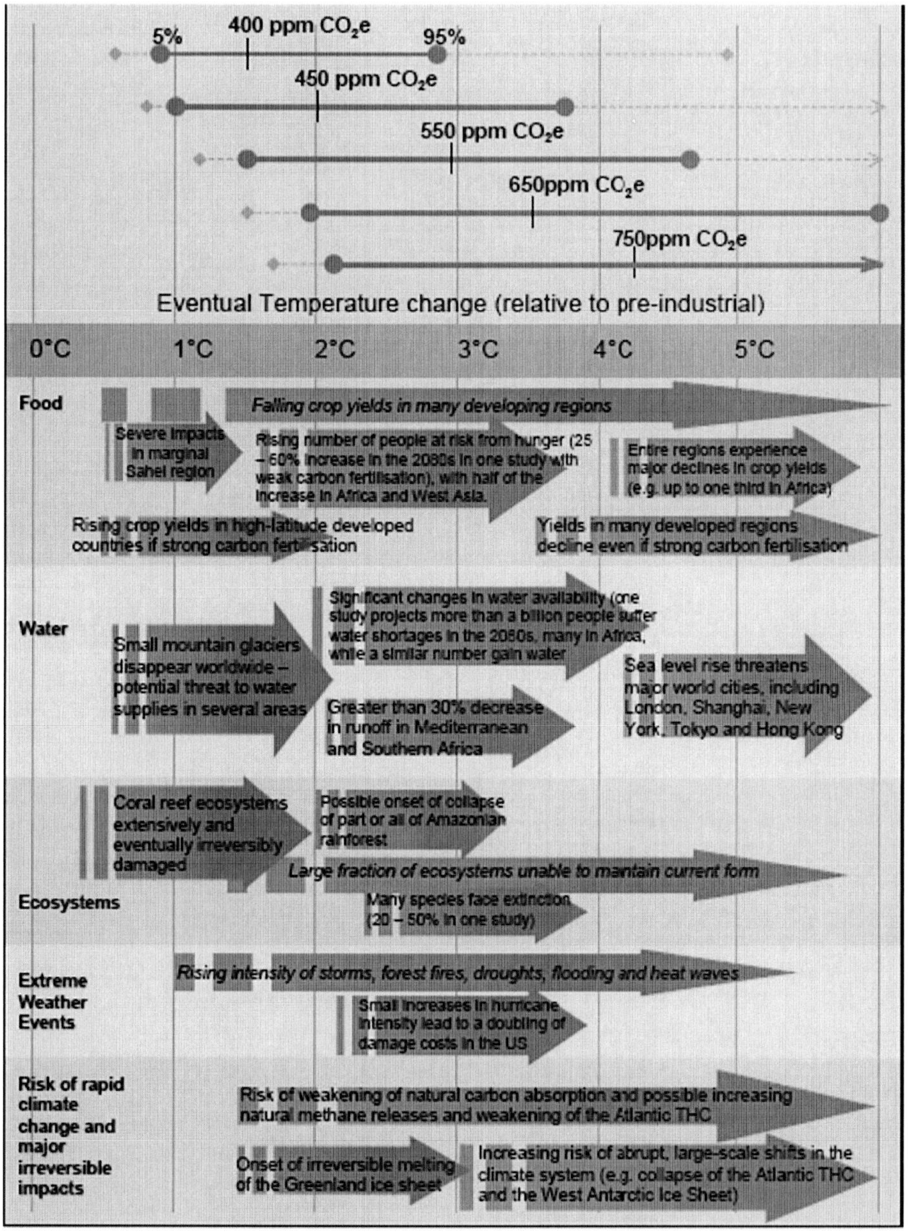

Figure 11.4: Stabilization levels and probability ranges for temperature increase. The figure illustrates the types of impact that could be experienced as the world comes into equilibrium with more greenhouse gases. *Source:* "The Economics of Climate Change: The Stern Review". Cambridge: University Press (2007).

predominantly in the Indian sub-continent, parts of China, and the Andes in South America.

- Declining crop yields, especially in Africa, could leave hundreds of millions without the ability to produce or purchase sufficient food. At mid to high latitudes, crop yields may increase for moderate temperature rises (2–3 °C), but then decline with greater amounts of warming. At 4 °C and above, global food production is likely to be seriously affected.

- In higher latitudes, cold-related deaths will decrease. But climate change will increase worldwide deaths from malnutrition and heat stress. Vector-borne diseases such as malaria and dengue fever could become more widespread if effective control measures are not in place.

- Rising sea levels will result in tens to hundreds of millions more people experiencing floods each year with warming of 3 or 4 °C. There will be serious risks and increasing pressures for coastal protection in South East Asia (Bangladesh and Vietnam), small islands in the Caribbean and the Pacific, and large coastal cities, such as Tokyo, New York, Cairo and London. According to one estimate, by the middle of the century, 200 million people may become permanently displaced due to rising sea levels, heavier floods, and more intense droughts.

- Ecosystems will be particularly vulnerable to climate change, with around 15–40% of species potentially facing extinction after only 2 °C of warming. And ocean acidification, a direct result of rising carbon dioxide levels, will have major effects on marine ecosystems, with possible adverse consequences on fish stocks. Higher temperatures will increase the chance of triggering abrupt and large-scale changes.

- Warming may induce sudden shifts in regional weather patterns such as the monsoon rains in South Asia or the El Niño phenomenon, changes that would have severe consequences for water availability and flooding in tropical regions and threaten the livelihoods of millions of people.

- A number of studies suggest that the Amazon rainforest could be vulnerable to climate change, with models projecting significant drying in this region. One model, for example, finds that the Amazon rainforest could be significantly, and possibly irrevocably, damaged by a warming of 2–3 °C.

An example of the effect of global warming is given in the photographs (Figure 11.6) of the Upsala glacier in Patagonia taken in 1928 and 2007.

The risks of serious, irreversible impacts from climate change increase strongly as concentrations of greenhouse gases in the atmosphere rise. Dr R.K. Pachauri, Chairman of the IPCC, stated in January 2007: "There is now growing awareness on the imperatives for a global energy future which marks a distinct departure from past trends and patterns of energy production and use. These imperatives emerge as much from the need to ensure energy security, as they do from the urgency of controlling local pollution from combustion of different fuels and, of course, the growing challenge of climate change, which requires reduction in emissions of greenhouse gases (GHSs), particularly carbon dioxide.

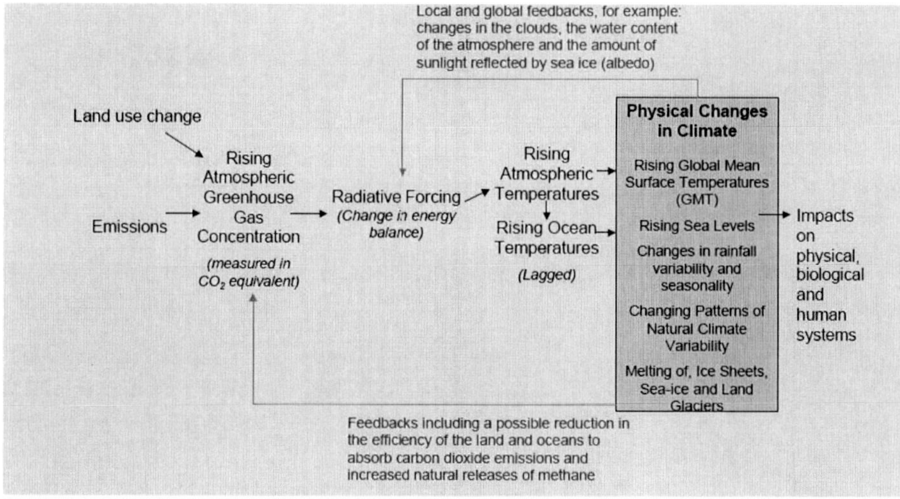

Figure 11.5: The link between greenhouse emissions and climate change.

The scientific evidence on the need for urgent action on the problem of climate change has now become stronger and convincing. Future solutions would lie in the use of existing renewable energy technologies, greater efforts at energy efficiency and the dissemination of decentralised energy technologies and options."

In response to the climate change threat, the Kyoto Protocol has committed its signatories to reducing their greenhouse gas emissions by 5.2% from their 1990 level by the target period of 2008–2012. This in turn has resulted in the adoption of a series of regional and national reduction targets. In the European Union, for instance, the commitment is to an overall reduction of 8%. In order to reach this target, the EU has also agreed to increase its proportion of renewable energy from 6% to 12% by 2010.

The Kyoto Protocol includes "flexible mechanisms" which allow economies to meet their greenhouse gas emission limit by purchasing GHG emission reductions from elsewhere. These can be bought either from financial exchanges, or from projects what reduce emissions in developing economies under the Clean Development Mechanism (CDM).

The Kyoto signatories are currently negotiating the second phase of the agreement, covering the period 2013–2017 and later. Within this time frame industrialized countries need to reduce their CO2e emissions by 18% from 1990 levels, and then by 30% between 2018 and 2022. Only with these cuts do we stand a reasonable chance of keeping the average increase in global temperatures to less than 2 °C, beyond which the effects of climate change will become catastrophic. Unfortunately among the 169 countries and other governmental entities that have ratified the agreement (representing over 60% of emissions from major countries), notable exceptions include the United States and Australia, both significant emitters

Costs rise significantly as mitigation efforts become more ambitious or sudden. Efforts to reduce emissions rapidly are likely to be very costly. An important corollary is that delay has a high price. Delay in taking action on climate change would make it necessary to accept both more climate change and, even-

Figure 11.6: The Upsala glacier in Patagonia, viewed in 1928 (top) and 2007 (bottom).

tually, higher mitigation costs. Weak action in the next 10–20 years would put stabilization even at 550 ppm CO2e beyond reach—and this level is already associated with significant risks. The level of 550 ppm CO2e could be reached as early as 2035.[11.4] At this level, there is at least a 77% chance, and perhaps up to a 99% chance, depending on the climate model used, of a global average temperature rise exceeding 2 °C. In economic terms, "The Economics of Climate Change: The Stern Review"[11.4] has evidenced that the cost of stabilizing CO2e levels at 550 ppm is 300 million euros/year, which is equivalent to 1% of global GDP, and opines that "this cost will be multiplied by 3–4 by 2050 if action is not taken today". Further, the Stern Review estimates that "if we don't act, the overall costs and risks of climate change will be equivalent to losing at least 5% of global GDP each year, now and forever. If a wider range of risks and impacts is taken into account, the estimates of damage could rise to 20% of GDP or more." The Review goes on to state: "In contrast, the costs of action—reducing greenhouse gas emissions to avoid the worst impacts of climate change—can be limited to around 1% of global GDP each year."

The investment that takes place in the next 10–20 years will have a profound effect on the climate in the second half of this century and in the next. "Our actions now and over the coming decades could create risks of major disruption to economic and social activity, on a scale similar to those associated with the great wars and the economic depression of the first half of the 20th century. And it will be difficult or impossible to reverse these changes."[11.4]

11.2 Primary energy resources

Alongside global warming, other challenges have become just as pressing. The global population on the planet will have increased by 2050 from 6.3 to 8.9 milliard individuals. Worldwide energy demand is growing at a staggering rate. Over-reliance on energy imports from a few, often politically unstable countries and volatile oil and gas prices have together pushed security of energy supply to the top of the political agenda, as well as threatening to inflict a massive drain on the global economy. But whilst there is a broad consensus that we need to change the way we produce and consume energy, there is still disagreement about how to do this. But a fundamental question has been asked repeatedly: "where will our energy come from in the coming decades?"

Today it mainly comes from finite fossil fuel; in the long term, it will have to come from renewable energies. The basic question of availability of raw energy, the impact on the environment which gravely affects the planet by the use of fossil fuels and the solution we can bring to preserve our future will be answered in this article. To do this, it is first necessary to clarify how long production rates can follow and meet the growing demand for crude oil, natural gas and coal. Furthermore, particularly for coal, we need to understand whether, to what extent and over what period of time, the separation and safe storage of carbon dioxide from burning fossil fuels is possible—a basic requirement for carbon-based energy production. In addition the contribution that nuclear energy can realistically make needs to be assessed.

The issue of security of supply is now at the top of the energy policy agenda. Concern is focused both on price security and the security of physical supply. At present around 80% of global energy demand is met by fossil fuels. The unrelenting increase in energy demand is matched by the finite nature of these sources. The regional distribution of oil and gas resources also does not match the distribution of demand. Some countries have to rely almost entirely on fossil fuel imports.

11.3 The fossil fuels[4,5]

11.3.1 Oil production

Oil is the life blood of the modern global economy, as the effects of the supply disruptions of the 1970s made clear. It is the number one source of energy, providing 36% of the world's needs and is the fuel employed almost exclusively for essential uses such as transportation. However, a debate has developed over the ability of supply to meet increasing consumption, a debate obscured by poor information and stirred by recent soaring prices. Figure 11.7 shows the historic trend in world oil production and its probable development in the future. The production is almost at a peak and will clearly decrease in the coming decades—the maximum crude oil production ("Peak Oil") represents a decisive turning point.

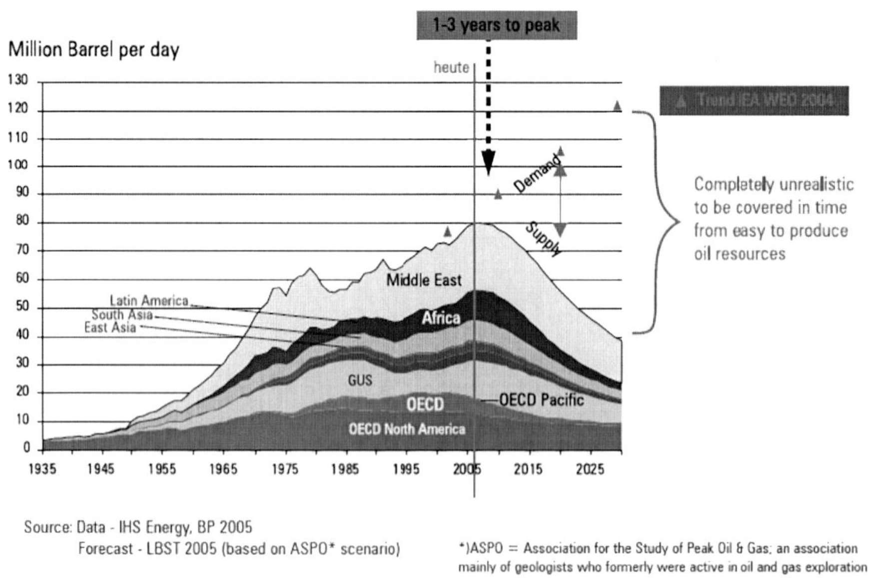

Figure 11.7: Conventional oil production.

A multitude of evidence supports this thesis: since 1980 we have been using more oil than we find each year and the gap is growing ever larger. More and more production regions have already exceeded their maximum production. This applies in particular to all the large old fields, which still make a significant contribution to world oil production. There are also clear signs that the oil-rich countries of the Middle East and the countries of the former Soviet Union cannot further extend their production. This is all in the face of the expectation of a further increase in worldwide demand, as highlighted in the International Energy Agency World Energy Outlook 2004 scenarios. The looming supply gaps will lead to serious distortions in the world economy. "Peak Oil" represents a structural interruption. The search for sustainable structures in energy supply can no longer be put off. There is a concern that there is not enough time remaining to organize a smooth transition to a post-fossil world.

For the last few years, *nonconventional sources of oil* production have been developed, namely the conversion of very heavy oils, such as Canadian tar sands or heavy oil in Venezuela, which, on a quantitative basis, come close to the Arabian oil reserves. Figure 11.8 shows the historical and predicted development of Canadian oil production. However, it cannot be concluded from this that oil from oil sands will replace the missing conventional crude oil. The following must be considered:[5]

1. This oil is only available in the soil in very small concentrations. Utilization requires significant strip mining activities. Within the best layers the concentration is around 20%. A considerable land surface is required, which in turn requires the destruction of large areas of forest.

2. The separation and purification of the oil uses a large amount of energy and water; the mining process is very slow and is more similar to the mining process for ores than conventional oil production. A large amount of hydrogen is required for the separation of sulfur and preparation of the oil. Natural gas is required in this process. However, only around half of the extracted bitumen is processed into synthetic crude oil in suitable refineries. In doing this, around 10% of the energy content of the bitumen is lost.

3. The lead times for projects are very long; the investments are high. For example, to develop a new mine with an extraction rate of 200,000 barrels/day, around 5–10 milliard USD must be invested.

4. The CO2e emissions from petrol from oil sands are comparable with those from coal.

5. The use of natural gas to process oil sands would be increasingly in competition with direct natural gas usage.

11.3.2 The chaos of the reserves[4]

Public data about oil and gas reserves is strikingly inconsistent, and potentially unreliable for legal, commercial, historical and sometimes political reasons. The most widely available and quoted figures, those from the industry journals "Oil & Gas Journal" and "World Oil", have limited value as they report the reserve figures provided by companies and governments without analysis or verification. Moreover, as there is no agreed definition of reserves or standard reporting practice, these figures usually stand for different physical and conceptual magnitudes. Confusing terminology ('proved', 'probable', 'possible', 'recoverable', 'reasonable certainty') only adds to the problem.

[5]Ludwig-Bölkow-Systemtechnik GmbH (LBST)/European Hydrogen Association, "Where will the Energy for Hydrogen Production come from?—Status and Alternatives", 2007 Data source: Oil, Gas, Coal- Nuclear Scenario, LBST Scenario 2005.

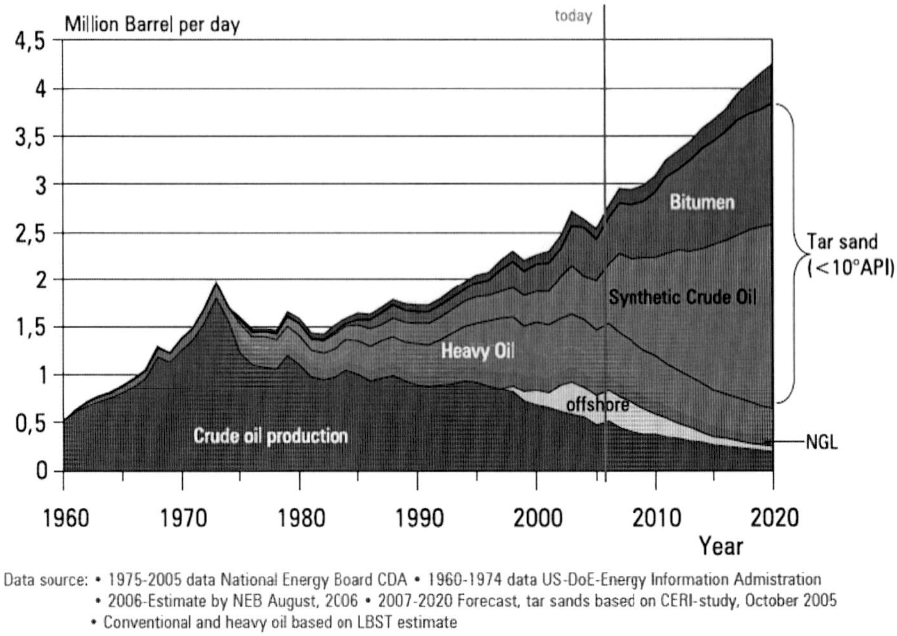

Figure 11.8: Nonconventional oil production. This diagram does not take some exotic, but potentially significant, sources into account, most notably deep sea gas (methane) hydrates.

Historically, private oil companies have consistently underestimated their reserves in order to comply with conservative stock exchange rules and through natural commercial caution. Whenever a discovery was made, only a portion of the geologist's estimate of recoverable resources was reported; subsequent revisions would then increase the reserves from that same oil field over time. National oil companies, almost fully represented by OPEC (Organisation of Petroleum Exporting Countries), are not subject to any sort of accountability so their reporting practices are even less clear. In the late 1980s, OPEC countries blatantly overstated their reserves while competing for production quotas, which were allocated as a proportion of the reserves. Although some revision was needed after the companies were nationalized, between 1985 and 1990 OPEC countries increased their declared joint reserves by 82%. Not only were these dubious revisions never corrected, but many of these countries have reported untouched reserves for years, even if no sizeable discoveries were made and production continued at the same pace. Additionally, the former Soviet Union's oil and gas reserves have been overestimated by about 30% because the original assessments were later misinterpreted.

Whilst private companies are now becoming more realistic about the extent of their resources, the OPEC countries hold by far the majority of the reported reserves, and information on their resources is as unsatisfactory as ever. In brief, these information sources should be treated with considerable caution. To fairly estimate the world's oil resources a regional assessment of the mean backdated (i.e. 'technical') discoveries would need to be performed. However the views of the International Energy Agency (World Energy Outlook 2004) are quite optimistic, as shown in Figure 11.9.

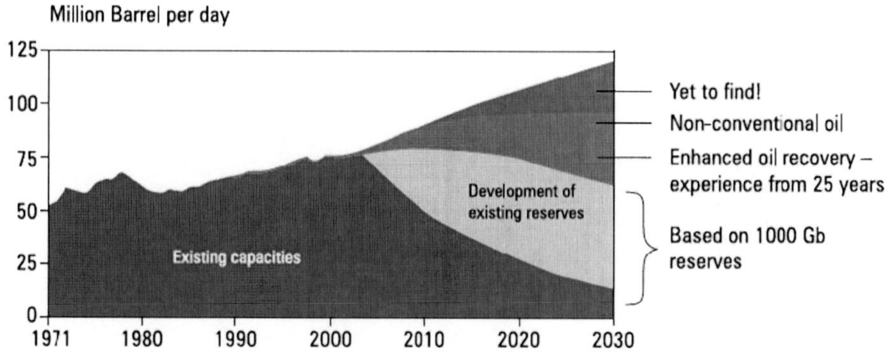

Figure 11.9: Future oil extraction as viewed by the International Energy Agency.

11.3.3 Natural gas

Natural gas has been the fastest growing fossil energy source in the last two decades, boosted by its increasing share in the electricity generation mix. Gas is generally regarded as a largely abundant resource and public concerns about depletion are limited to oil, even though few in-depth studies address the subject. Gas resources are more concentrated than oil so they were discovered faster because a few massive fields make up most of the reserves: the largest gas field in the world holds 15% of the"ultimate recoverable resources" (URR), compared to 6% in the case of the largest oil field .

Unfortunately, information about gas resources suffers from the same bad practices as oil data, because gas mostly comes from the same geological formations, and the same stakeholders are involved. Most reserves are initially understated and then gradually revised upwards, giving an optimistic impression of growth. By contrast, Russia's reserves, the largest in the world, are considered to have been overestimated by about 30%, as stated above. Owing to geological similarities, gas follows the same depletion dynamics as oil, and thus the same discovery and production cycles. In fact, existing data for gas is of worse quality than for oil and some ambiguities arise as to the amount of gas already produced because flared and vented gas is not always accounted for. As opposed to published reserves, the technical ones have been almost constant since 1980 because discoveries have roughly matched production.

The scenario shown in Figure 11.10 assumes that world gas production can still significantly increase and will only reach its maximum in the year 2020. This is based on the assumption that the production decrease in North America and Europe will be overcompensated for by an increase in production in Russia and the Middle East. This requires significant and timely investments in these regions. However, in spite of this optimistic picture, the future of gas production is rather overshadowed by risks. A further problem for production expansion in Russia and the Middle East is the requirement to significantly expand the infrastructure for the transport of liquefied natural gas. These investments require considerable resources, and time, and often also involve fighting local opposition to the construction of gasification terminals.

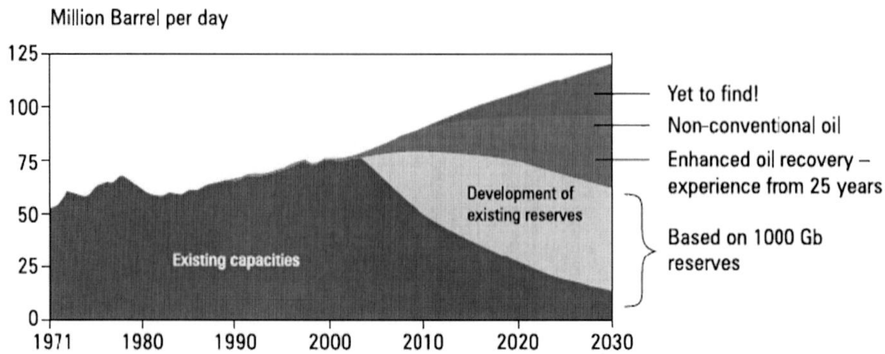

Figure 11.10: World natural gas production.

The scenario shows the possible development based on today's estimate of reserve situations and describes an upper limit. The actual development in the coming decades could of course be affected by regional bottlenecks.

11.3.4 Coal

Coal was the world's largest source of primary energy until it was overtaken by oil in the 1960s. Nevertheless, coal still supplies almost one quarter of the world's energy today. Despite being the most abundant of fossil fuels, coal's development is currently threatened by environmental concerns, hence its future will unfold in the context of both energy security and global warming.

Coal is abundant and more equally distributed throughout the world than oil and gas. Global recoverable reserves are the largest of all fossil fuels, and most countries have at least some. Moreover, existing and prospective big energy consumers like the US, China and India are self- sufficient in coal and will be for the foreseeable future. Coal has been exploited on a large scale for two centuries so both the product and the available resources are well known; no substantial new deposits are expected to be discovered. Based on the current data on worldwide coal reserves, a scenario of possible future production can be depicted. The aggregated production follows a logistic curve (adjusted to previous production and to reserves). The result is that annual worldwide coal production could be increased by 60% and would reach its maximum in around 2050. In theory, the decrease in crude oil and natural gas could, therefore, partly be offset by an increase in coal usage for primary energy. In the conversion to usable end energy, in particular to automotive fuel, significantly higher losses are generated with coal, so that replacement is of oil clearly more difficult.

Extrapolating the demand forecast, the world will consume 20% of its current reserves by 2030 and 40% by 2050. Hence, if current trends are maintained, coal would still last several 100 years. Figure 11.11 shows the historic development of the production and the estimate reserves of hard coal and lignite.[3]

It is important to note that the specific CO2e emissions of hard and lignite coal are significantly higher than with crude oil and natural gas. Average values are: hard coal: 346 g CO2e/kWh, lignite coal: 414 g CO2e/kWh, natural gas: 203 g CO2e/kWh, petrol/diesel: 264 g CO2e/kWh.

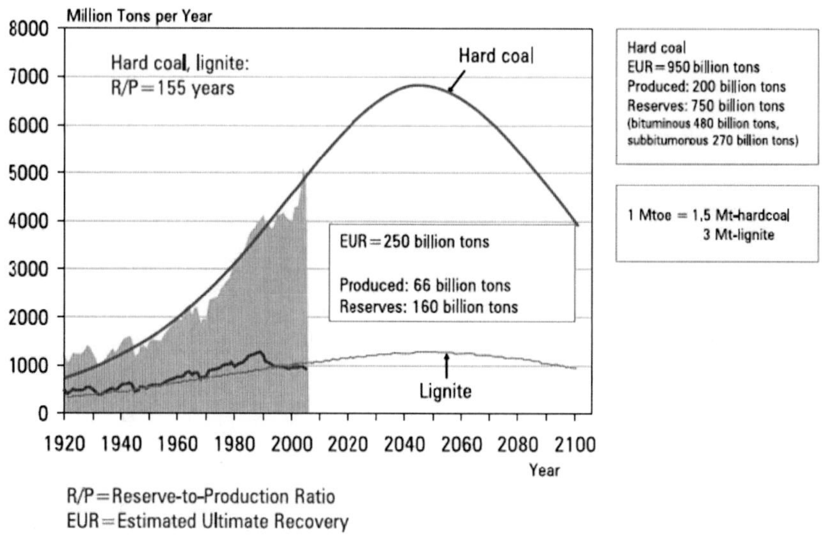

Figure 11.11: World coal production and reserves.

11.4 Carbon dioxide capture and storage and clean coal technologies[4,5]

Carbon dioxide capture and storage (CCS) technology offers the possibility for significantly reducing the amount of CO2e from the combustion of fossil fuels. Some technologies process the fossil fuel before it is burned, others treat the gas after combustion in order to improve the environmental performance of conventional coal combustion. It is a major challenge to attempt to collect the waste gases after the combustion process and to store them in geological formations.

Pre-combustion capture processes include the coal cleaning (to reduce the ash content) and various 'bolt-on' or 'end-of-pipe' technologies to reduce emissions of particulates, sulfur dioxide and nitrogen oxide, the main pollutants resulting from coal firing apart from carbon dioxide. Flue gas desulphurisation, for example, most commonly involves 'scrubbing' the flue gases using an alkaline sorbent slurry, which is predominantly lime or limestone based.

More fundamental changes have been made to the way coal is burned to both improve its efficiency and further reduce emissions of pollutants. They are included in the category of so-called clean coal technology (CCT):

Integrated gasification combined cycle (IGCC): coal is not burnt directly but reacted with oxygen and steam to form a 'syngas' (synthetic gas) composed mainly of hydrogen and carbon monoxide, which is cleaned and then burned in a gas turbine to generate electricity and produce steam to drive a steam turbine. IGCC improves the efficiency of coal combustion from 38–40% up to 50%.

Supercritical and ultrasupercritical: these power plants operate at higher temperatures than conventional combustion, again increasing efficiency towards 50%.

Fluidized bed combustion: coal is burned in a reactor comprised of a bed through which gas is fed to keep the fuel in a turbulent state. This improves combustion, heat transfer and recovery of waste products. By raising pressures within a bed, a high pressure gas stream can be used to drive a gas turbine, generating electricity. Emissions of both sulfur dioxide and nitrogen oxide can be reduced substantially.

Pressurized pulverized coal combustion: This is based on the combustion of a finely ground cloud of coal particles creating high pressure, high temperature steam for power generation. The hot flue gases are used to generate electricity in a similar way to the combined cycle system.

Other potential future technologies involve the increased use of coal gasification. Underground coal gasification, for example, involves converting deep underground unworked coal into a combustible gas which can be used for industrial heating, power generation or the manufacture of hydrogen, synthetic natural gas or other chemicals. The gas can be processed to remove CO2e before it is passed on to end users.

Storage of carbon dioxide in geological repositories such as depleted oil or gas reservoirs, aquifers and coal beds is today considered as the ultimate solution for the 'final' disposal of greenhouse gases. Some tests are presently taking place in Germany and in a depleted oil field on the North Sea but, like other potential geological reservoirs, these are located at a great distance from the power plants. Moreover, geological instabilities and leakage rates need to be explored and monitored during and after use. Storage of carbon dioxide in the deep ocean is also an option with potentially high environmental impacts like undersea acidification that could greatly affect the ecosystem.

As of today the technology has not provided an economic model to transform the CO2e produced. Employing CO2e capture and storage will increase the price of electricity from fossil fuels. Although the costs of storage depends on several factors, including the technology used for separation, transport and the kind of storage installation, experts from the UNIPCC calculate the additional costs at between 3.5 and 5.0 cents/kWh of power generated. This means the technology would more than double the cost of electricity today. As with nuclear waste, however, the question is whether this will just displace the problem elsewhere.

11.5 Uranium resources and nuclear energy[4,5]

Nuclear power makes use of the energy that is released by nuclear fission of the natural radionuclide ^{235}U. Uranium ores extracted in open pit or underground mines, typically containing about 20% uranium oxides, are refined to produce "yellow cake", urania, which contains on average 90% U_3O_8, containing a mixture of isotopes, but predominantly (99.275%) ^{238}U. The material is then enriched to a content of about 3% ^{235}U, the natural concentration of which is only a few parts per million. Currently, 435 nuclear power plants are in operation worldwide. They produce 6.5% of the world's energy and generate 15.7% of the world's electricity. As shown in Figure 11.12, the distribution of ore is almost as concentrated as oil and does not match regional consumption. Five countries—Canada, Australia, Kazakhstan, Russia and Niger—control three quarters of the world's supply. As a significant user of uranium, however, Russia's reserves will be exhausted within ten years. Secondary sources, such as old deposits, currently make up nearly half of worldwide uranium reserves. However, those sources will soon be used up. Mining capacities will have to be nearly doubled in the next few years to meet current needs.

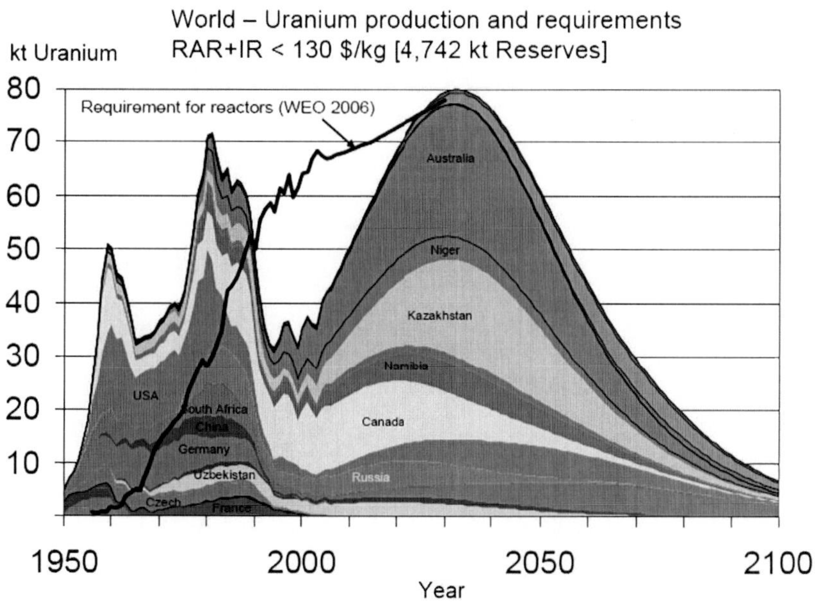

Figure 11.12: World uranium production and requirements.

The worldwide supply of nuclear fuel, which can be extracted for less than 130 USD per kilogram of uranium, is guaranteed for less than the next 70 years, assuming a yearly consumption of about 66 500 tons of uranium (Figure 11.13). In 2005, 41 870 tons of natural uranium were mined, which met 63% of the world supply. At the moment, and probably also for the next two decades, the additional supply of uranium is covered by the stocks of energy supplying companies, reprocessed nuclear waste and the decommissioning of highly enriched uranium from U.S. and Russian weapons.

The sustainable development of nuclear power is presently undermined by high plant capital cost, by government subsidies, the extremely long period of construction, local opposition and above all by the difficulties of converting and depositing nuclear waste. The amount of energy required to develop a mine,

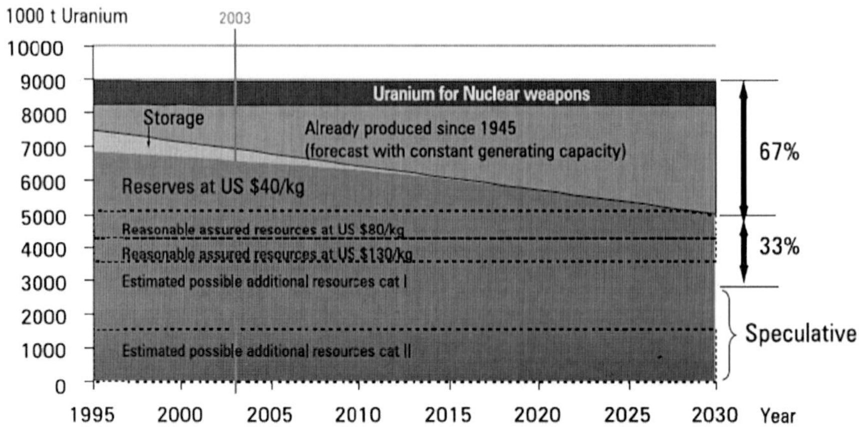

Figure 11.13: World uranium resources.

to process, to convert and to enrich the uranium as well as to build a nuclear power plant is extremely high and the impact on the environment considerable, thus making nuclear power's claim of being environmentally friendly a fake! In addition, the energy efficiency of current reactor technologies is no more than 33%, thus making electricity production highly inefficient. The age structure of the 435 nuclear reactors operating worldwide today essentially determines the future role of nuclear energy. Assuming an average reactor lifespan of 40 years, by the year 2030, 75% of the reactors installed today must be disconnected from the grid. If the number of reactors is to remain constant, 14 reactors must be built and put into operation each year throughout this interval. Worldwide, however, only around 28 reactors are under construction, and these could start operating in the next 5 to 7 years. Eleven of these reactors have been "under construction" for more than 20 years. Moreover, several countries have decided not to build nuclear plants and others have decided to gradually phase out their existing ones. Under these circumstances, it is not possible to talk of a renaissance in nuclear energy.

11.6 Contribution of all fossil and nuclear fuels[4,5]

A summary of the future availability of fossil and nuclear energy resources is shown in Figure 11.14. On the basis of what we know today, a strong decline in oil production after peak production is highly probable. The reason lies in the oil production technologies used today, which aim to exhaust the fields as quickly as possible. When peak production has been reached, a quick drop in production rates will be experienced.

Achieving peak production for oil, and subsequently for natural gas, will therefore shortly thereafter leave a noticeable gap in world energy supply, which cannot be filled by other fossil primary energy sources. Coal reserves known to us today, with a range of coverage of around 160 years, will indeed permit

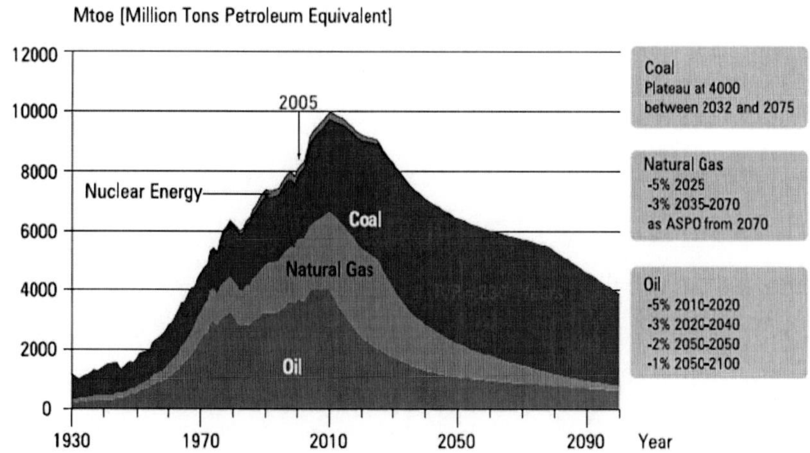

Figure 11.14: Contribution of all fossil and nuclear fission resources.

increasing production until around 2050, but in assessing coal one should take into account that the data quality is poorer than for crude oil. Since 1992, China has been reporting exactly the same reserve figures each year. In this period around 20% of the "proven" reserves have already been used up. China currently produces the largest amount of coal worldwide (almost double that of the USA). However, China's reserves are only half those of the USA. For Canada too, another major source, almost exactly the same reserve figures are published today as in 1986. On the other hand, in its report to the World Energy Council in 2004, Germany cut back its "proven" hard coal reserves by 99% (from 23 milliard to 183 million tonnes) and its lignite reserves by 85% (from 43 milliard to 6.5 milliard tonnes). Despite this sobering reality, the core statements of the IEA World Energy Outlook are:

- The energy supply over the coming 20 years will continue the trend of the past 20 years;
- There will be no restrictions on oil, gas, or coal, whether due to scarcity of resources or climate protection reasons;
- There is no reason to bring renewable energies to the market—the share of so-called New Renewable Energies (solar, wind, geothermal) will be around 2% in 2030;
- Only the share of traditional biomass usage will increase, following the trend of the past decades.

The following points are completely ignored:[4]

- Fossil energies are increasingly difficult to exploit and therefore are becoming more expensive;
- Environmental reasons will put increasing pressure on the burning of coal, oil, and gas;

- Renewable energies have shown an average growth rate of far more than 10% per year over the past 15 years, and have become increasingly cost-efficient; the price gap between conventional and non-conventional energy supplies is becoming ever smaller.

11.7 What is the solution for saving the planet?—the global energy (r)evolution

The evidence that global warming is caused by human activity and the decrease in the fossil resources are a reality despite the lack of recognition by the U.S. Administration, which refuses to ratify the Kyoto Protocol. In parallel with the Stern review,[11.4] two other major reports, "Global Energy (R)evolution, a Sustainable Energy Outlook" by EREC/Greenpeace International,[4] and "Where will the Energy for Hydrogen Production come from?—Status and Alternatives" by Ludwig-Bölkow-Systemtechnik GmbH (LBST)/European Hydrogen Association,[5] both published in early 2007, attempt to bring a solution to the consequences of the global warming by analysing future scenarios of energy use with a close focus on a range of emerging technologies that will be employed in the coming years and decades. The contribution to this debate by the two reports is fundamental. Both reach the same conclusions and identify similar scenarios.

The global energy (r)evolution scenario[4]

Two different scenarios are used here to characterize the wide range of possible paths for the future energy supply system: (1) a reference scenario, reflecting a continuation of current trends and policies ("business as usual" or BAU), and (ii) the energy (r)evolution scenario, which is designed to achieve a set of dedicated environmental policy targets.

The reference scenario is based on that published by the International Energy Agency in the World Energy Outlook (WEO) 2004.[6] This takes into account only existing policies. The assumptions include, for example, continuing progress in electricity and gas market reforms, the liberalization of cross-border energy trade and recent policies designed to combat environmental pollution. The reference scenario does not include additional policies to reduce greenhouse gas emissions. As the IEA's scenario only covers a time horizon up to 2030; it has been extended by extrapolating its key macroeconomics indicators. This provides a baseline for comparison with the energy (r)evolution scenario.

The energy (r)evolution scenario has a key target for the reduction of worldwide emissions by 50% below 1990 levels by 2050, with per capita CO2e emissions reduced to less than 1.3 tonnes per year (Figure 11.15 in order for the increase in global temperature to remain below 2 °C. A second objective is to show that this is even possible with the global phasing out of nuclear energy. To achieve these targets, the scenario is characterized by significant efforts to fully exploit the large potential for energy efficiency.[7]

---

[6] World Energy Outlook (WEO) 2004. International Energy Agency. See also WEO 2005 and WEO 2006.

[7] There is, in addition, tremendous potential for reducing energy consumption by simply cutting out the many extravagances with which we have become accustomed (this policy is

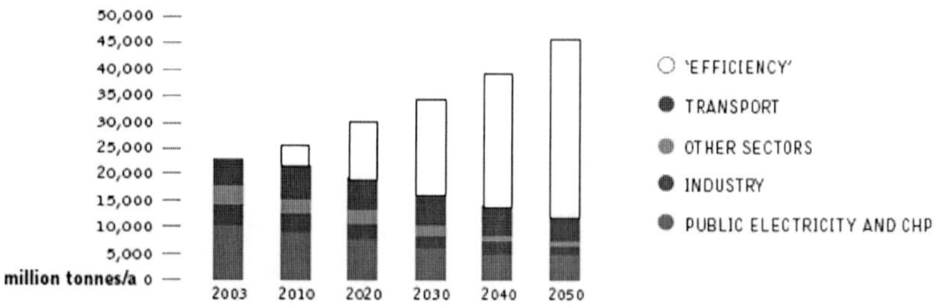

Figure 11.15: Development of global CO_2 emissions by sector under the energy (r)evolution scenario.

At the same time, cost-effective renewable energy sources are accessed for both heat and electricity generation as well as the production of biofuels.[8]

These scenarios by no means claim to predict the future; they simply describe two potential development paths out of the broad range of possible 'futures'. The energy (r)evolution scenario is designed to indicate the efforts and actions required to achieve its ambitious objectives and to illustrate the options we have at hand to change our energy supply system into one that is sustainable.

The scenarios consider global energy demand, energy and heat supply and energy and electricity generation. The development of future global energy demand is determined by three key factors:[4]

Population development: the world's population consuming energy or using energy services will increase from 6.3 milliard people now to 8.9 milliard in 2050. This continuing growth will put additional pressure on energy resources and the environment;

Economic development, for which Gross Domestic Product (GDP) is the most commonly used indicator. In general, an increase in GDP goes in parallel with an increase in energy demand;

usually referred to as CoE, "cutting out extravagances"). An excellent example of extravagance (or, in the German-speaking world, "Zuvielisation") is the vast over-endowment of urban and now even rural areas with street lighting. A few decades ago villages in rural France and Italy (for example) were totally dark at night. Now, even the tiniest commune is happy to spend public money on erecting elaborate lamps illuminating deserted pavements all through the night. In Belgium, along the numerous mororways, urban freeways etc., lamp standards are as thickly planted as trees in a forest, again burning all through the night even in the small hours of the morning, when motor vehicles (anyway obligatorily equipped with headlamps) are few and far between. The reason given for keeping the lamps on all night in cities is because of the fear of crime: i.e. they contribute to security. Matters of taste also enter: it has been a delight to observe the use of incandescent lamps in many streets in this Vake district of Tbilisi where the Workshop is taking place, aesthetically far more pleasant than the mercury or pressure sodium discharge lamps used in other European capitals (not to speak of the ordinary sodium discharge lamps, which are an abomination, aesthetically speaking); and a regret to see that some of them are already being replaced by the more energy-efficient discharge lamps. Other examples of extravagance are the outdoor gas-burning heaters placed near café tables, allowing even the scantily clad to sip a beverage in comfort in winter. The reader will doubtless be able to think of many more instances.

[8]The general framework parameters for population and GDP growth remain unchanged from those of the reference scenario.

Energy intensity: this indicates how much energy is required to produce a unit of GDP.

Both the reference and energy (r)evolution scenarios are based on the same projections of population and economic development. The future development of energy intensity, however, differs between the two, taking into account the measures to increase energy efficiency under the energy (r)evolution scenario.[9]

In terms of energy intensity, an increase in economic activity and a growing population does not necessarily have to result in an equivalent increase in energy demand. There is still considerable potential for exploiting energy efficiency measures. Under the reference scenario, it is assumed that energy intensity will be reduced by 1.3% per year, leading to a reduction in final energy demand per unit of GDP of about 45% between 2003 and 2050. Under the energy (r)evolution scenario, it is assumed that active policy and technical support for energy efficiency measures will lead to an even higher reduction in energy intensity of almost 70%.

Development of global energy demand[10,11,12,13]

Combining the projections on population development, GDP growth and energy intensity results in future development pathways for the world's energy demand. These are shown in Figure 11.16 for both the reference and the energy (r)evolution scenarios. Under the reference scenario, total energy demand almost doubles from the current 310 000 PJ/a to 550 000 PJ/a in 2050. In the energy (r)evolution scenario, a much smaller 14% increase on current consumption is expected by 2050, reaching 350 000 PJ/a.

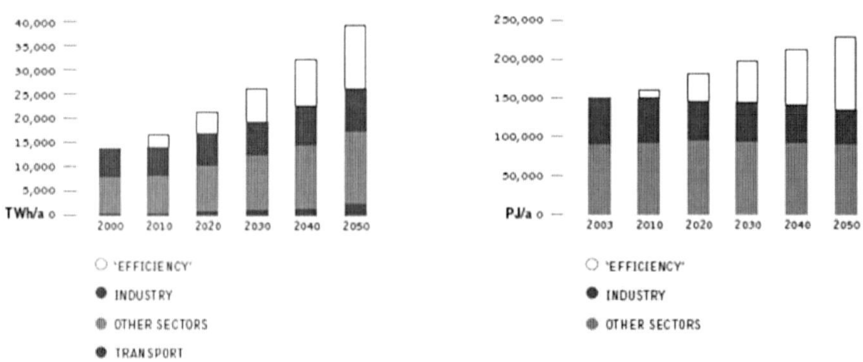

Figure 11.16: Development of global demand by sectors in the energy (r)evolution scenario. 'Efficiency' = reduction compared to the reference scenario. On the left, electricity demand; on the right, heat demand. (*Source:* EREC-Greenpeace.)

[9] But not, vide supra, cutting out extravagances (CoE).

[10] "Futu(r)e investment, a sustainable investment plan to save the climate". EREC—Greenpeace International.

[11] "Energy and Environment in the European Union 2006". European Environment Agency (EEA)

[12] "Energy Efficiency", (`http://www.iea.org/`). International Energy Agency (IEA).

[13] "Solar Generation IV", September 2007. EPIA—Greenpeace International.

Energy efficiency is essential to reduce consumption: it offers a powerful and cost-effective tool for achieving a sustainable energy future. Improvements in energy efficiency can reduce the need for investment in energy infrastructure, cut fuel costs, increase competitiveness and improve consumer welfare. Environmental benefits will also be achieved by the reduction of greenhouse gas emissions and local air pollution. The most important sectors in which energy savings can be applied are industry, buildings, appliances and transport. A few examples show where energy saving can be applied.

In the industrial sector approximately 65% of electricity consumption is used to drive electric motors. This can be reduced by employing variable speed drives, high efficiency motors and using efficient pumps, compressors and fans. The savings potential here is estimated at up to 40%.[14] The production of primary aluminium from alumina (the main constituent of bauxite) is a very energy-intensive process. It is produced by passing a direct current through a bath with alumina dissolved in a molten cryolite electrode. Another option is to produce aluminium out of recycled scrap. This is called secondary production. Secondary aluminium uses only 5–10% of the energy demand for primary production because it involves remelting the metal instead of an electrochemical reduction process. If recycling increases from 22% of aluminium production in 2005 to 60% in 2050 this would save 45% of current electricity use.[15]

In buildings, intelligent architectural design, new materials, efficient insulation and passive solar design in both residential and commercial buildings will help to curb the growing demand for active air-conditioning and heating saving up to 80% of the average energy demand.[16] For household appliances such as washing machines, dishwashers, television sets and refrigerators, energy use can typically be reduced by 30% using the best available options and by 80% with advanced technologies. For office appliances energy use can be reduced by 50–75% through a combination of power management and energy efficient computer systems. Use of standby mode for appliances is on average responsible for 5–13% of electricity use by households in OECD countries. Replacement of existing appliances by those with the lowest losses could reduce power consumption by 70%. "Low-consumption" light bulbs have now become compulsory in some countries.[17]

---

[14] In some industries, the savings have already been made. It is noteworthy that the all-electrified Swiss Federal Railways use no more electricity than several decades ago, despite a great increase in the tonnage hauled, since the modern electric motors powering their current locomotives are far more energy efficient than the old ones.

[15] A highly visible use of aluminium is in the cans in which drinks are widely sold. Everyone knows that they use far less metal (per can) than formerly, because of improvements in design and manufacturing processes. These improvements also mean that there is less to recycle. At the same time a "cutting out extravangances" (CoE) approach would deprecate the consumption of these beverages, which are perhaps not very healthy, altogether (i.e. adopting the ethos of the song "My drink is water bright &c."), automatically reducing the use of cans.

[16] Cf. Batty, W.J. Eco-design and sustainability. In: J.J. Ramsden, S. Aida and A. Kakabadse, "Spiritual Motivation: New Thinking for Business and Management". Basingstoke: Palgrave (2007).

[17] Apart from aesthetic considerations—Edison's (or Swan's) invention is sans pareil in that regard—there is also the problem that the low-consumption bulbs take some time to achieve maximum brightness after being switched on. Therefore in safety-critical situations (e.g. a staircase) they will probably be left on all the time, so the actual gain in energy saving is likely to be significantly less than as asserted by their advocates. There is also the matter of the whole manufacturing cycle to be taken into account.

In the transport sector, the use of hybrid (electric/combustion) and future fuel cell powered vehicles as well as other efficiency measures could reduce energy consumption in passenger cars by up to 80% in 2050. This will be achieved also by shifting the transport of goods from road back to rail, reverting to the pattern followed in eastern Europe prior to 1990. Changes in mobility-related behaviour patterns will however be essential.[18]

An accelerated increase in energy efficiency, which is a crucial prerequisite for renewable sources achieving a sufficiently large share of overall energy supply, will be beneficial not only for the environment but from an economic point of view. Taking into account the full life cycle, in most cases the implementation of energy efficiency measures saves money compared to increasing energy supply. A dedicated energy efficiency strategy including improvement of legislation, labelling and monitoring, therefore helps to compensate in part for the additional costs required during the market introduction phase of renewable energy sources. Several studies have demonstrated that additional costs incurred in improving efficiency are offset even in the short term by energy saving. These savings are, however, likely to be dwarfed by those achievable via CoE, but serious studies of the latter are still lacking.

Under the energy (r)evolution scenario, electricity demand is expected to increase disproportionately, with households and services the main source of growing consumption. With the exploitation of efficiency measures, however, an even higher increase can be avoided, leading to electricity demand of around 26 000 TWh/a in the year 2050 (see Figure 11.16).

Compared to the reference scenario, efficiency measures avoid the generation of about 13 000 TWh/a. Efficiency gains in the heat supply sector are even larger. Under the energy (r)evolution scenario, final demand for heat supply can even be reduced. Compared to the reference scenario, consumption equivalent to 94 000 PJ/a is avoided through efficiency gains by 2050. As a result of energy-related renovation of the existing stock of residential buildings, as well as the introduction of low energy standards and 'passive houses' for new buildings, enjoyment of the same comfort and energy services will be accompanied by a much lower future energy demand. (Figure 11.16). New advanced standards, such as the Swiss-developed "Minergie" are under consideration in most European countries. Overall development of primary energy consumption under the energy (r)evolution scenario is represented in Figure 11.17.

In the transport sector, which is not analysed in detail in the EREC/Greenpeace study, it is assumed under the energy (r)evolution scenario that energy demand will increase by a quarter to 100,600 PJ/a by 2050, saving 80% compared to the reference scenario.

Electricity generation[4,10,13]

The development of electricity supply sector is characterized by a dynamically growing renewable energy market and an increasing share of renewable electricity. This will compensate for the phasing-out of nuclear energy at the end of the life of the reactors operating now and reduce the number of fossil fuel-fired power

[18] Excessive use of passenger vehicles of all descriptions is doubtless a significant contributor to obesity, the negative impacts of which on public health are of growing concern to governments, as well as to insecurity (road accidents) and emissions of carbon and nitrogen oxides.

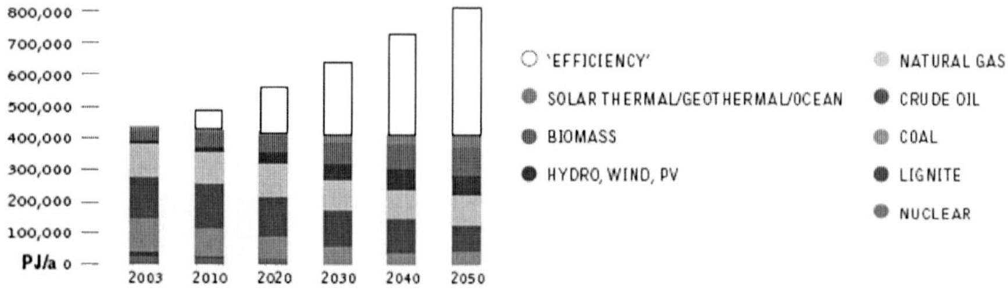

Figure 11.17: Development primary energy consumption under the energy (r)evolution scenario.

plants required for grid stabilization. The availability of the renewable energy sources is shown in Figure 11.18. These can provide 3,078 times current global energy needs. By 2050, 70% of the electricity produced worldwide is anticipated to come from renewable energy sources. 'New' renewables—mainly wind, solar thermal energy and PV—will contribute 42% of electricity generation.

The following strategy paves the way for a future renewable energy supply:[4]

- The phasing out of nuclear energy and rising electricity demand will be met initially by bringing into operation new highly efficient gas fired combined-cycle power plants, plus an increasing capacity of wind turbines and biomass. In the long term, wind will be the most important single source of electricity generation.

- Solar energy, hydro and biomass will make substantial contributions to electricity generation. In particular, as non-fluctuating renewable energy sources, hydro, solar thermal and geothermal, combined with efficient heat storage, are important elements in the overall generation mix. Co-generation systems will be used as far as possible. Figure11.19 show the mix of renewable energies that can be achieved in 2030.

- Decentralized energy systems will be created where power and heat are produced close to the point of final use, avoiding the current waste of energy during conversion, transmission and distribution.

  As shown in Figure 11.20 city centres or suburbs could become totally independent in relation to electricity and heat generation.

- The installed capacity of renewable energy technologies will grow from the current 800 GW to 7,100 GW in 2050. However, increasing renewable capacity by a factor of nine within the next 43 years requires political support and well-designed policy instruments. There will be a considerable demand for investment in new production capacity over the next 20 years. As investment cycles in the power sector are long, decisions on restructuring the world's energy supply system need to be taken now. Figure 11.21 shows global electricity generation under the two scenarios.

To achieve economically attractive growth in renewable energy sources, a balanced and timely mobilization of all technologies is of great importance. This mobilization depends on technical potentials, cost reduction and technological

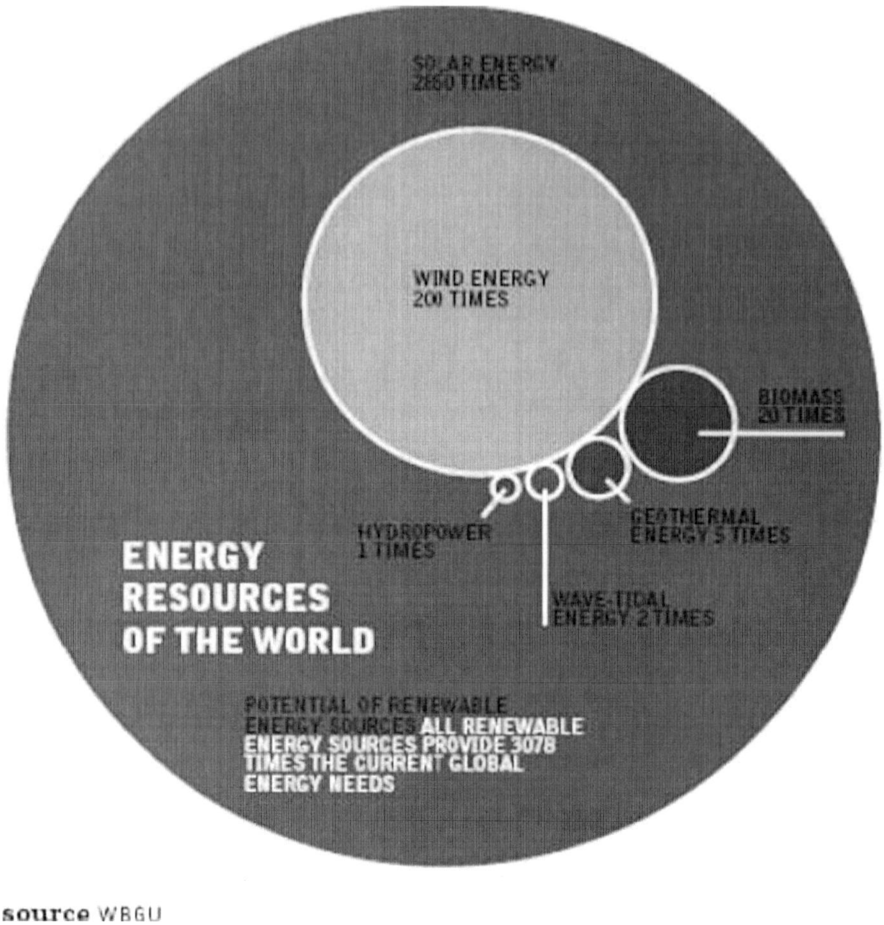

Figure 11.18: Renewable energy sources of the world.

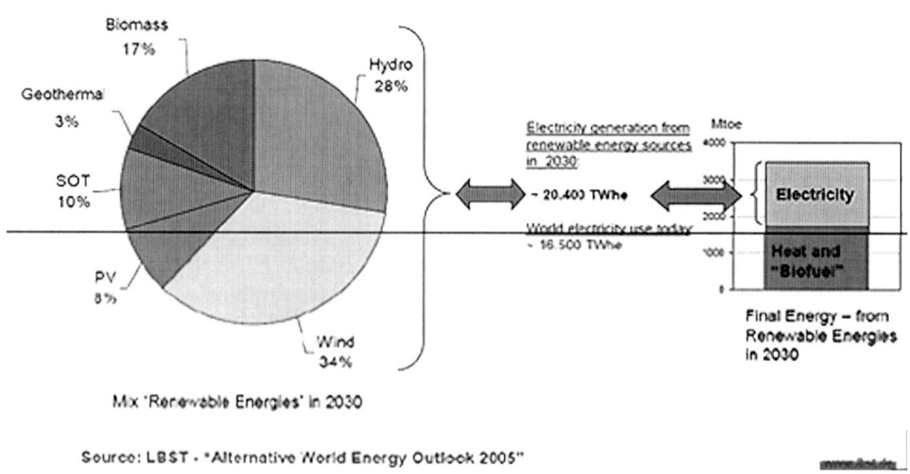

Figure 11.19: World electricity production from renewable energy sources in 2030.

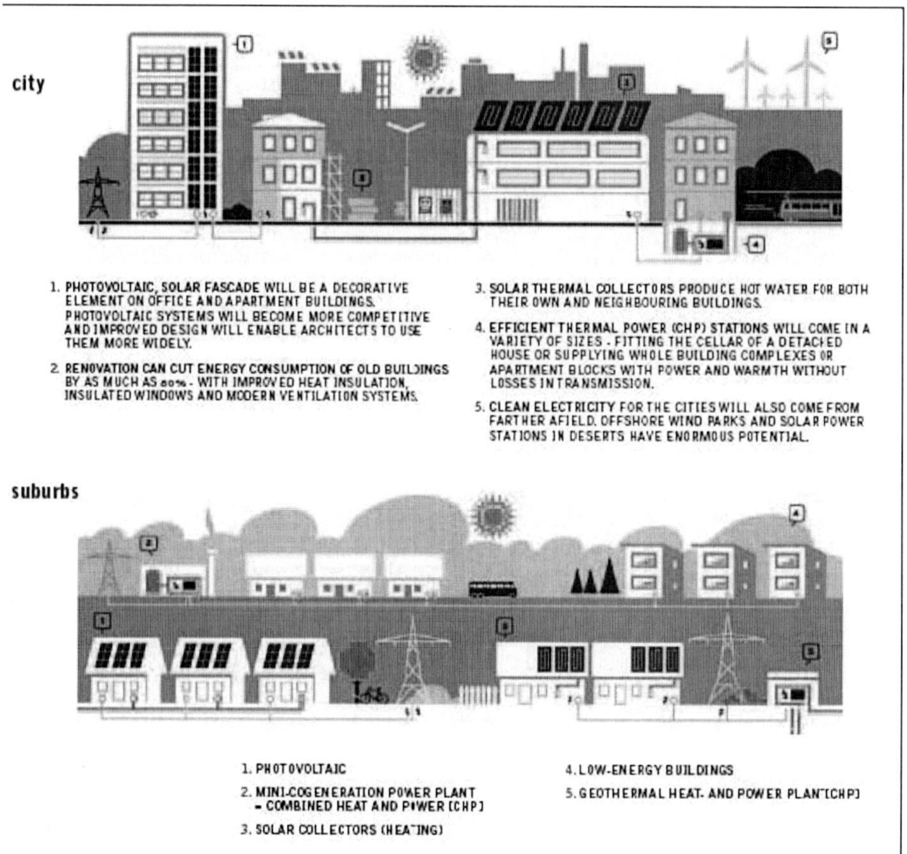

Figure 11.20: A vision of a decentralized energy future.

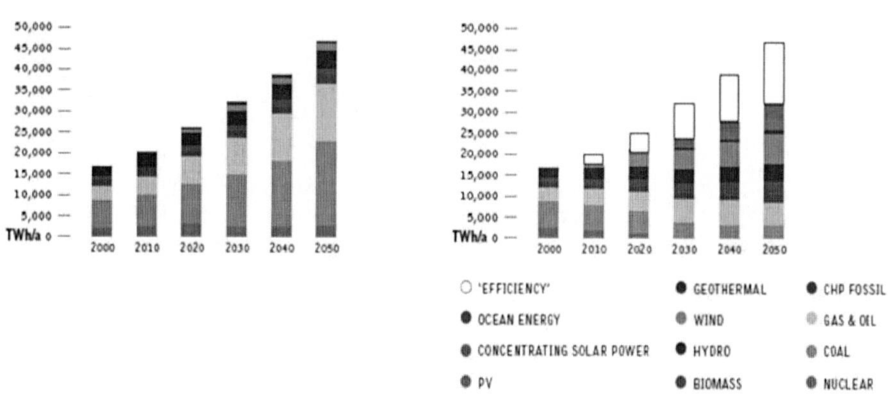

Figure 11.21: Development of global electricity generation under the two scenarios.

maturity. Note that many of the required technologies are already available. Up to 2020, hydro-power and wind will remain the main contributors to the growing market share. After 2020, the continuing growth of wind will be complemented by electricity from biomass, photovoltaics and solar thermal energy. Figure 11.22 shows a possible world energy scenario.

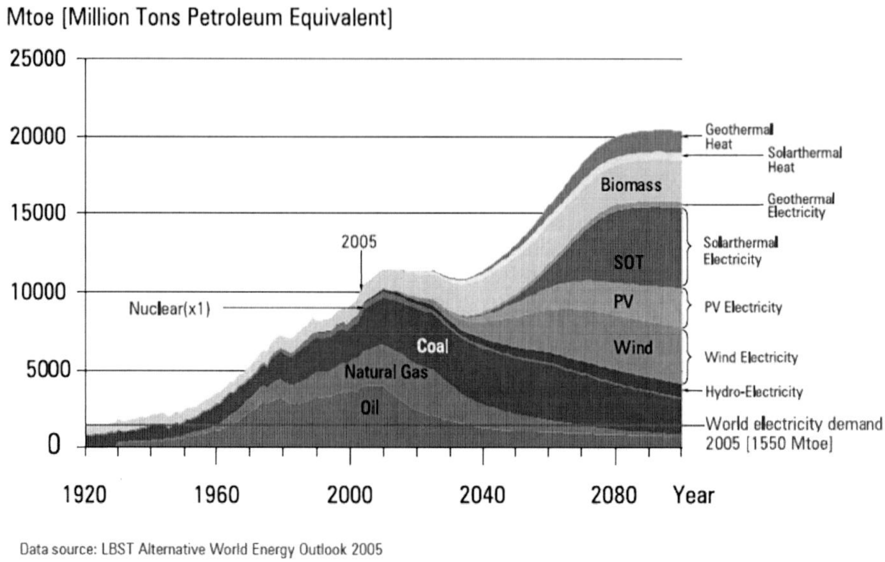

Figure 11.22: A possible world energy scenario.

Future costs of electricity generation[4,10,13,19]

Due to growing demand, we face a significant increase in society's expenditure on electricity supply. Under the reference scenario, the unchecked growth in demand, the increase in fossil fuel prices and the cost of CO2e emissions result in total electricity supply costs rising from today's 1130 milliard USD per year to more than 4300 milliard USD in 2050. The introduction of renewable technologies under the energy (r)evolution scenario slightly increases the costs of electricity generation compared to the reference scenario as shown in Figure 11.23. This difference will be less than 0.1 cents/kWh up to 2020. Note that any increase in fossil fuel prices which is foreseen in the coming years will reduce the gap between the two scenarios. Because of the lower CO2e intensity of electricity generation, by 2020 electricity generation costs will become economically favourable under the energy (r)evolution scenario, and by 2050 generation costs will be more than 1.5 cents/kWh below those in the reference scenario.

Figure 11.24 shows, moreover, that the energy (r)evolution scenario not only complies with global CO2e reduction targets but also helps to stabilize energy costs and relieve the economic pressure on society. Increasing energy efficiency and shifting energy supply to renewables leads to long term costs for electricity supply that are one third lower than in the reference scenario.

Figure 11.25 shows the trend of cost reduction for wind and photovoltaic generated electricity. The power generation costs are depicted versus the installed

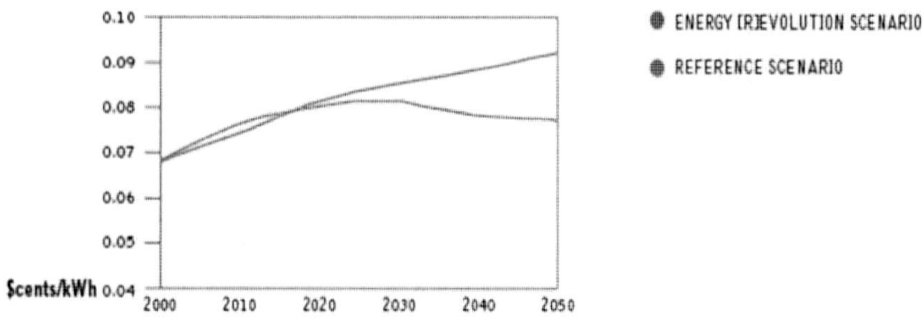

Figure 11.23: Development of global electricity generating costs under the two scenarios.

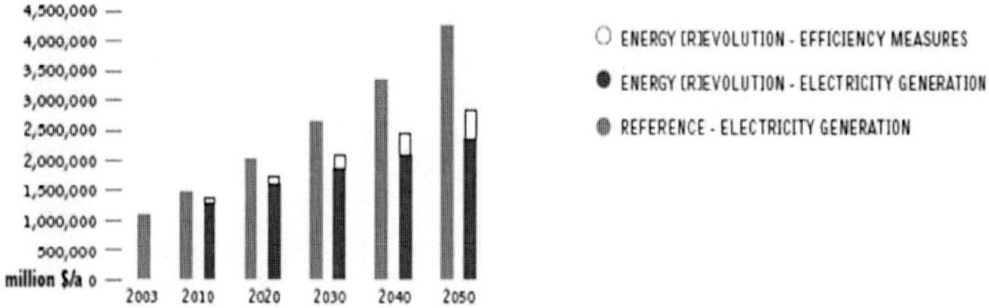

Figure 11.24: Development of total global electricity supply costs.

capacities and shown in euros/kWh_{el} based on the cumulated installed capacity. Significant cost reductions are expected, in particular for photovoltaics (PV), which will benefit greatly from the economy of scale effect where the increase in the volume of production leads to a decrease in price.

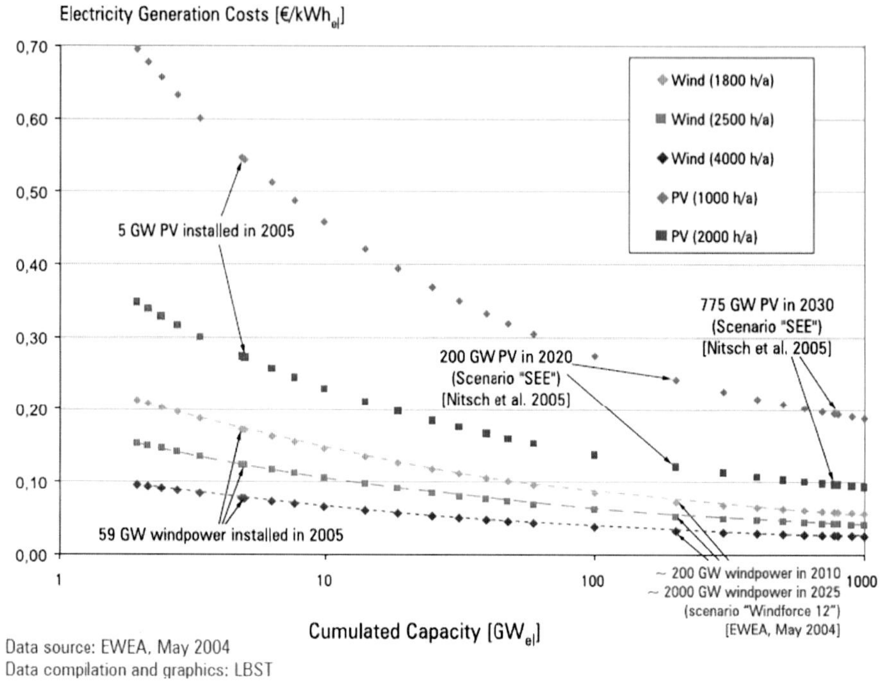

Figure 11.25: Cost reductions for renewable energies.

The report "Solar Generation IV 2007" jointly published by EPIA-European Photovoltaic Industry Association and Greenpeace International in September 2007, forecast continued growth for the solar power industry, which will generate revenues of up to 300 milliard euros by 2020 meeting 9.4% of the world's electricity demand. The report notes that total installed capacity of solar photovoltaics reached a peak of 6.5 GW in 2006, up from 1.2 GWp in 2000, with annual growth rates averaging 35% since 1998. It is projected that by 2030 the photovoltaic systems cumulative installed capacity will be 1272 GWp with an electricity production of 1802 TWh. 776 million grid-connected consumers and 2,894 million off-grid will benefit from solar electricity. Its cost is estimated between 0.07 and 0.13 euros per kWh depending on location. Cumulative CO2e savings are 6670 million tonnes. In 2040 global solar electricity output will account for 28% of global electricity demand under the energy (r)evolution scenario. Concerning wind power, the European Wind Energy Association (EWEA) and Greenpeace predict an installed capacity of around 2025 GW by 2010. It becomes clear that pursuing stringent environmental targets in the energy sector also pays off in terms of economics. The tables shown in Figures 11.26 and 11.27[19] renew-

[19]Communication from the Commission to the European Council and the European Parliament: An energy policy for Europe, Brussels: European Commission, 10.1.2007 COM(2007)

ables provide a summary of technical and economic data for electricity and heat generation for fossil fuels, nuclear and renewables.

Energy sources	Technology considered for the cost estimate	2005 Cost (€ / MWh)	Projected Cost 2030 (€ / MWh with €20-30/tCO2)	GHG emissions (Kg CO2eq/MWh)	EU-27 Import dependency		Efficiency	Fuel price sensitivity	Proven reserves / Annual production
		Source IEA			2005	2030			
Natural gas	Open cycle gas turbine	45 – 70	55 - 85	440	57%	84%	40%	Very high	64 years
	CCGT (Combined Cycle Gas Turbine)	35 - 45	40 - 55	400			50%	Very high	
Oil	Diesel engine	70 - 80	80 - 95	550	82%	93%	30%	Very high	42 years
Coal	PF (Pulverised Fuel with flue gas desulphurisation)	30 - 40	45 - 60	800	39%	59%	40-45%	medium	155 years
	CFBC (Circulating fluidized bed combustion)	35 - 45	50 - 65	800			40-45%	medium	
	IGCC (Integrated Gasification Combined Cycle)	40 - 50	55 - 70	750			48%	medium	
Nuclear	Light water reactor	40 - 45	40 - 45	15	Almost 100% for uranium ore		33%	low	Reasonable reserves: 85 years
Biomass	Biomass generation plant	25 - 85	25 - 75	30	nil		30 - 60%	medium	Renewable
Wind	On shore	35 - 175	28 - 170	30			95-98%	nil	
		35 – 110	28 – 80						
	Off shore	50 - 170	50 - 150	10			95-98%		
		60 – 150	40 – 120						
Hydro	Large	25 - 95	25 - 90	20			95-98%		
	Small (<10MW)	45 - 90	40 - 80	5			95-98%		
Solar	Photovoltaic	140 - 430	55 -260	100			/		

Figure 11.26: Table of electricity generation data for fossil fuel, nuclear and renewable sources. (*Data source:* European Commission—Communication from the Commission to the European Council and the European Parliament: An Energy Policy for Europe. Brussels, 10.1.2007 COM.) Note: Due to various uncertainties the above figures do not necessarily correspond to the figures quoted in the EREC-Greenpeace report.

11.8 The hydrogen economy[5]

Hydrogen is the cleanest fuel that can be found on the planet. It can be co-fired with fossil fuels, it can be used in fuel cells for power generation and transportation, and it can be employed in many branches of industry. Unfortunately, the production of hydrogen is high in energy requirements and this is currently mostly of fossil origin, thus making it environmentally unfriendly; in gas or liquefied form, it is difficult to store due to its explosive character. However, in the not so distant future, it will be produced exclusively with renewable energy, thus providing a totally CO2e-free clean fuel. The report "Where will the Energy for Hydrogen Production come from?—Status and Alternatives" by Ludwig-Bölkow-Systemtechnik GmbH—LBST/European Hydrogen Association[5] introduces scenarios analysing the production of hydrogen using renewable energy including wind and biomass at competitive costs.

European research and European industry are jointly associated in an important research and technology development (RTD) effort to develop new production, storage and handling technologies as well as a large number of applications such as fuel cells for electricity and heat generation to make hydrogen

1 final.

Energy sources		EU-25 market share by energy source	Market price (€/toe)	Lifecycle cost (€/toe)	GHG emissions (t CO_2eq/toe)	EU-27 import dependence 2005	EU-27 import dependence 2030
Fossil fuels	Heating gas oil	20%	525 (€0.45/l)	300-1300	3.1	82%	93%
	Natural gas	33%	230 – 340 (€20-30/MWh)		2.1	57%	84%
	Coal	1.8%	70 (€100/tce)		4	39%	59%
Biomass	Wood chips	5.7%	280	545-1300	0.4	0	?
	Pellets		540	630-1300	0.4	0	?
Electricity		31%	550 - 660 (€50-60/MWh)	550 - 660	0 to 12	<1%	?
Solar		0.2%	/	680-2320	Very low	0	0
Geothermal		0.4%	/	230-1450	Very low	0	0

Figure 11.27: Table of heat generation data for fossil fuel and renewable sources. (*Data source:* European Commission—Communication from the Commission to the European Council and the European Parliament: An Energy Policy for Europe. Brussels, 10.1.2007 COM.) Note: Due to various uncertainties the above figures do not necessarily correspond to the figures quoted in the EREC-Greenpeace report.

the cleanest fuel for the future. In the area of fuel cells a very large number of technologies are under development utilizing different concepts and materials in order to improve efficiency and reduce costs. It is expected that small portable applications will enter the market in the coming years and will help introduce the benefits of fuel cells and hydrogen to the general public. Large scale stationary and cogeneration fuel cells are experimentally employed today and should achieve a large commercial breakthrough before the end of this decade.

Transport applications will be the main driver for hydrogen demand; mass production of passenger vehicles powered by fuel cells could begin in 2010 and continue for the next 5 years to effect significant replacment of fossil fuel-powered vehicles. A new hydrogen economy will begin to compete with and replace the fossil fuel economy in the near future.

11.9 Conclusions

The Industrial Revolution has brought immense benefits to all of humanity, but if human intervention is at the root of the impact of the greenhouse emissions on climate change, it must be conceded that this same revolution is now destroying life on earth. At the same time, the progessive exhaustion of fossil energy sources will have a major impact on energy production on the planet.

The "business as usual" scenario, based on the IEA's World Energy Outlook (2004–2006) projection, is not an option for future generations. CO2e emissions would almost double by 2050 and the global climate could—assuming that current models are valid—heat up by well over 2 °C. This would have catastrophic consequences for the environment, the economy and human society. The major and very urgent issue is to reduce CO2e emissions by lowering consumption, using energy more efficiently, and making use of all types of renewable energy available and abundant on the planet (implying phasing out the "carbon society").

For the energy sector, the renewables industry and their friends[4] have—unsurprisingly—a clear agenda for changes that need to be made in energy policy to encourage a shift to renewable sources. The main demands are:

- Phase out all subsidies for fossil and nuclear energy and indexinternalizing external costsinternalize external costs. Conventional energy sources (including nuclear) receive an estimated 250-300 milliard USD in subsidies per year worldwide,[5,6,9] resulting in heavily distorted markets.
- Remove energy market distortions and internalize the social and environmental costs of polluting energy and introduce the "polluter pays" principle.
- Reform the electricity market by removing electricity sector barriers.
- Establish legally binding targets for renewable energy.
- Provide support mechanisms for energy efficiency, renewables and distributed energy.
- Guarantee priority access to the grid for renewable power generators.
- Enforce strict efficiency standards for all energy-consuming appliances, buildings and vehicles.
- Provide defined and stable returns for investors.
- Ensure worldwide compliance with the Kyoto Protocol and establish new, more restrictive targets for the next phases.

Climate change and the energy global security supply are extremely complex matters, heavily entangled in gigantic vested interests. Simple solutions are available but whether they can be put in place today and whether they would "save the planet" and the whole of humanity is another matter.[20]

[20] The author is grateful to Reinhold Wurster of LBST and Sven Teske of Greenpeace International for their input to this article.

Complexity and Security
J.J. Ramsden and P.J. Kervalishvili (Eds.)
IOS Press, 2008

Chapter 12

Complexity in environmental and meteorological research

D.N. Asimakopoulos

Department of Applied Physics, Faculty of Physics, University of Athens, Greece

Abstract. The problem of natural disaster prediction and substantiation of environmental monitoring systems to receive, store and process necessary information for the solution of relevant problems have been analysed. Some specific natural disasters have been studied focusing on the following issues. The human-nature interaction is a function of a broad complex of factors functioning both in human society and in the environment. Anthropogenic forcing of the environment should be assessed together with natural processes, in order to work out a technology to reliably predict the consequences of human activity. One of the causes of natural balance violation is environmental pollution, in particular of the atmosphere. Interactions between processes in the environment is the main mechanism for appearance of emergency situations through the fault of humans. The observed increase of the number of natural disasters is connected with the growing role of anthropogenic factors that determine the environmental conditions for the origin of critical situations. Anthropogenically-induced natural disasters include forest fires, desertification, deforestation, dust storms, floods, snow avalanches and snowslides, reduced biodiversity, etc. An optimization of risk insurance against natural disasters becomes more and more urgent with every passing year, since economic losses increase and become practically more and more poorly predicted.

12.1 "Natural disasters" as a dynamic category of environmental phenomena—climate change

Nowadays, special care is given to the complexity in environmental and meteorological research in order to understand the intrinsic dynamics of the atmospheric environment, considered as a complex system under multi-component control mechanisms. Therefore, it should be considered of first priority to correctly analyse and interpret the data of environmental phenomena and distinguish trends (caused by external effects) from the long-range fluctuations intrinsic in the data. The disasters problem serves as a good example.

The "natural disaster" is a spasmodic change in the system in the form of its sudden response to smooth changes in external conditions. At present, natural disasters are floods, droughts, hurricanes, storms, tornados, tsunami, volcanic eruptions, landslides, landslips, mud flows, snow avalanches, earthquakes, forest fires, dust storms, bitter frosts, heat, epidemics, locust invasions, and many other natural phenomena, while in the future, this list can be extended to collisions with space bodies and anthropogenic disasters—bioterrorism, nuclear catastrophes, sharp changes of the Earth's magnetic field, plague, invasion of robots, etc. (Kondratyev et al., 2002; 2006).

Recently, the frequency and scale of disasters seems to be growing. In 2001, about 650 natural catastrophes happened over the globe, with victims totalling more than 25 thousand people and economic losses exceeding 35 milliard dollars, while in 2002 only 11 thousand people perished (although the economic damage reached 55 milliard dollars). In 2003, more than 50 thousand people died because of natural disasters, with economic damage reaching about 60 milliard dollars. The USA in 2004 achieved a record in the number of tornados (562), whereas in 1995 they numbered 399. The early 2004 was characterized by an increase of emergency situations mainly of weather origin, and this year ended with a catastrophic tsunami on 26 December with enormous losses for the countries in the Indian Ocean basin. In Sri Lanka (Ceylon) alone the damage reached 3.5 milliard dollars. Over the territory of Russia, the number of natural disasters during the last decade increased from 60 to 280. Noteworthy is the fact that the type and spatial distributions of disasters are rather non-uniform (tropical storms, 32%; floods, 32%; earthquakes, 12%; droughts, 10%; and other disasters 14%; and Asia, 38%; America, 26%; Africa, 14%; Europe, 14%; and Oceania, 8%).

The principal changes that will take place in the future are forecast to be the following (Kondratyev et al., 2002; 2006):

1. Earthquakes and floods, even in several decades' time will be killing tens of thousands people in developing countries, and the developed countries will continue to suffer large-scale economic losses and braking of progress in many spheres of life;

2. Epidemics, despite the development of medicine, will, as usual, prevent the universal establishment of a healthy mode of life due to the appearance of new kinds of diseases, which may be caused by genetic engineering;

3. Aggression of people living on the territory of other peoples as well as the possible intrusion of living beings from other planets in the future can create a precedent of colonization and a principal change of the way of

life of the population on the Earth. A reduction of traditional supplies of biological food and mineral resources could engeder a change of species, which will be able to feed on solar energy or some chemical elements, of which there is plenty in the world's oceans; (for instance deuterium).

4. The impact of cosmic bodies on the Earth can cause in the future a sharp global climate change, which will lead to a global catastrophe. A comet or asteroid with a diameter of several km is able to devastate huge areas, either by direct action or due to fires, tsunami, and other extreme phenomena, as well as a change of the orbit, (bearing in mind the inherent dynamical instability of the solar system). As a result, life on Earth could stop. The probability of such an event is, however, negligible.

5. A supernova in the stellar neighbourhood of the Earth could engender the mortality of every living being on the surface due to the high radiation. But this is unlikely to occur before several million more years have passed.

6. Global glaciation can happen in the next ten thousand years as an alternative to the expected climate warming (see also Chapter 10);

7. A change of the Earth's magnetic field by inversion (or lesser change) of the poles could eliminate the ozone layer and thereby cause irreversible changes in the biosphere;

8. Anthropogenic disasters expand due to appearance of new kinds of impact on the environment and human society. They will include deviations in the social and cultural spheres, in science and in engineering. Bioterrorism will increase, problems will appear with interaction with robots, and nanotechnologies will change the structure of the energy balance of the planet, raising the efficiency of assimilation of solar energy from the present 10% to an anticipated 50% in the future.

It is convenient to consider three categories of events that affect the climate: (a) events that occur outside the Earth; (b) natural events on the surface of the Earth; and (c) human activities. It is then convenient to distinguish three separate components to the events which occur outside the Earth: namely (a) variations in the intensity of the radiation emitted by the Sun (i.e. the solar cycle—solar flares); (b) changes in the transmission properties of space between the Sun and the Earth; and (c) changes in the Sun-Earth distance. The changes in the Sun-Earth distance arise from three causes, namely variations in the eccentricity of the Earth's orbit (with a period of around 100 000 years), oscillation of the tilt of the Earth's axis (with a period of about 40 000 years) and the precession of the equinoxes (with a period of about 22,000 years). These changes cause a (slight) variation in the intensity of the solar energy arriving at the Earth and it has been known for a long time that there is a good (anti-)correlation between the intensity of sunlight reaching the Earth and the volume of ice in the polar regions, i.e. with the occurrence of the Ice Ages at intervals of around 100 000 years. The first suggestion that Ice Ages were related to the Earth's orbit around the Sun appears to have been made by Joseph Adhemar, a mathematics teacher in Paris, in 1842; he concentrated on the 22,000 year period. The theory was extended to include the changes in the eccentricity of the Earth's orbit by James Croll, the son of a Scottish crofter, who had very little formal education. He stumbled on this idea and spent his spare time in the 1860s and 1870s working on the idea; he estimated that the last Ice Age ended about 80 000 years ago. There was some interest in Croll's theory at the time; however, possibly because he was of low birth and not part of the fashionable circles of the day and because it became apparent that the last Ice Age ended only about 10 000 years ago rather than 80 000 years ago, his ideas were largely forgotten by the end of the nineteenth century. The cycles due to orbital changes are now known as the Milankovitch cycles or wobbles, after Milutin Milankovitch, a Serbian mathematician who revived and extended Croll's ideas in the early twentieth century (Pearce, 2006).

The question whether the Earth's atmosphere can become hot like that of Venus as a result of the boundless accumulation of carbon dioxide (leading to an increase of the greenhouse effect), is quite reasonable. The main reason is surface heating and positive feedback between the absorption of the outgoing heat radiation in the atmosphere by 'hot' bands of CO_2 and H_2O, for example in the middle of the atmospheric transparency window in the 8–13 μm range. For example, thermal radiation is strongly absorbed by carbon dioxide in the vibrational hot bands (100)-(001) and (020)-(001), near the 943 cm^{-1} and 1064 cm^{-1} regions respectively, and by water vapour over the entire ranges of the transparency windows. A mechanism for this positive feedback is based on the exponential temperature dependence of absorption in the hot bands as well as on the exponentially increasing CO_2 equilibrium concentration in the atmosphere due to the emission from the ocean, from the Earth's crust, or as result of biota exhalation on the one hand and the increasing of the equilibrium concentration of water vapour in the atmosphere on the other hand. Also, there is an additional positive feedback between surface temperature and methane concentration in the Earth's atmosphere. Only photosynthesis provides a natural negative feedback that can control a stability of concentration of CO_2 in the atmosphere.

The theory of catastrophes has been well developed. But its application to the description of events and processes in the real environment requires the use of the methods of system analysis to substantiate the global model of the 'nature-society' system with the use of the technical facilities of satellite monitoring. The solution of the problems lies in the sphere of ecoinformatics, which provides a combination of analytically simple, semi-empirical and complex non-linear models of ecosystems with renewable global data sets.

12.2 The 50th anniversary of the International Geophysical Year (IGY) of 1957–58; from IGY (1957–58) to IPY (2007–2008)

Almost a century after Tyndall's discovery in 1859 that changes in the concentration of the atmospheric gases could bring about climate change, and almost fifty years after Arrhenius' calculation in 1896 of possible global warming from human emissions of CO_2, a small group of scientists got together to push international cooperation to a higher level in all areas of geophysics. They aimed to coordinate their data gathering and to persuade their governments to spend extra money on research. The result was the International Geophysical Year (IGY) of 1957–58. The International Polar Year (IPY) 20072008, is a large worldwide science programme focused on the Arctic and Antarctic. In the frame of the IPY thousands of scientists from more than 60 countries will be conducting research during this two-year programme.

Under the IGY umbrella the United States and the former Soviet Union officially announced plans to launch artificial satellites as part of cooperative scientific experiments. The era of satellite meteorology began when on 4 October 1957, the Soviet Union launched Sputnik 1, the world's first artificial satellite. It was a 22 inch diameter sphere with a mass of 83 kg and four antennae projecting from it). It circled Earth once every 96 minutes and transmitted radio signals that could be received on Earth, providing the first space views of our planet's surface and atmosphere. On 3 November 1957, a second satellite, Sputnik 2, was launched which carried a dog named Laika. The United States launched its first satellite, Explorer 1, on 31 January 1958, and its second, Vanguard 1, on 17 March 1958 (Cracknell and Varotsos, 2007), see Figure 12.1.

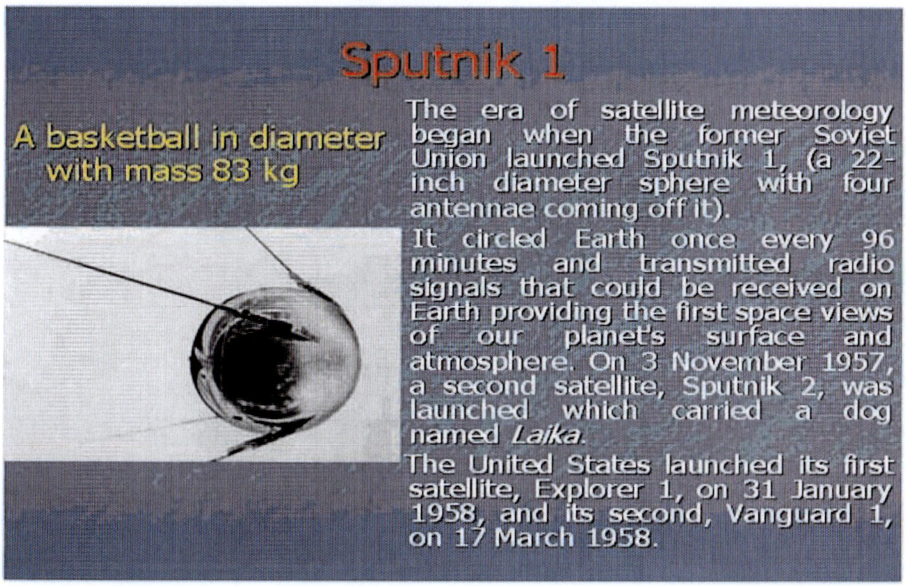

Figure 12.1: Information about the first satellites in the IGY of 1957–58.

On 1 March 2002 the European Space Agency (ESA) launched ENVISAT with various instruments on aboard. It is the largest Earth-observation spacecraft ever built (its dimensions are: 26 × 10 × 5m, and its total mass is 8211 kg), for monitoring the Earth's land, atmosphere, oceans and ice caps. It is a Sun-synchronous satellite flown at an altitude of 800 km, with inclination 98 degrees and an orbital period of 101 minutes. It has successfully reached its nominal 5-years mission lifetime, having orbited Earth more than 26,000 times. Since 1 March 2007 the ENVISAT data are available to researchers associated with the International Polar Year (IPY) in 2007–2008.

An extensive validation campaign of ENVISAT has been organized by ESA's Atmospheric Chemistry Validation Team and by the SCIAMACHY Validation and Interpretation Group through so-called Announcement of Opportunity

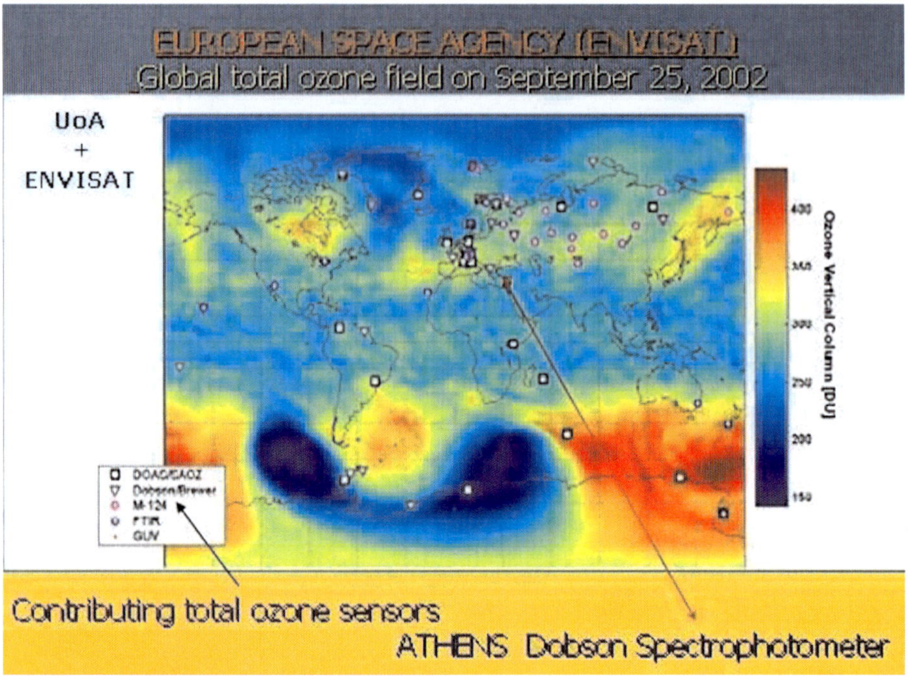

Figure 12.2: The contribution of Athens University to the ENVISAT venture.

projects. In this framework, correlative ground-based measurements have been acquired and collected into a centralized database. The University of Athens is involved in the ESA programme for the geophysical validation of the ENVISAT Atmospheric Chemistry Instruments (see Figure 12.2 and in related national activities, carrying out a variety of ground-based observations (total ozone, vertical ozone profile, solar UV irradiance, etc.).

12.3 The 2007 IPCC report

The IPCC report released on 2 February 2007 was produced by some 600 authors from 40 countries, reviewed and revised by over 620 experts and a large number of government reviewers as well as political representatives from 113 governments. Among its basic conclusions are the following (see Figures 12.3 and 12.4):

1. Warming of the Earth's climate system is unequivocal, as is now evident from observations of increases in global average air and ocean temperatures, widespread melting of snow and ice, and rising global average sea level;

2. At the scales of continents, regions, and ocean basins, numerous long-term changes in climate have been observed. These include changes in Arctic temperatures and ice, widespread changes in precipitation amounts, ocean salinity, wind patterns and aspects of extreme weather including droughts, heavy precipitation, heat waves and the intensity of tropical cyclones;

3. For the next two decades a warming of about 0.2 °C per decade is projected for a range of SRES[1] emission scenarios. Even if the concentrations of all greenhouse gases and aerosols were to be kept constant at year 2000 levels, a further warming of about 0.1 °C per decade would be expected;

4. Anthropogenic warming and sea level rise would continue for centuries due to the timescales associated with climate processes and feedbacks, even if greenhouse gas concentrations were to be stabilized.

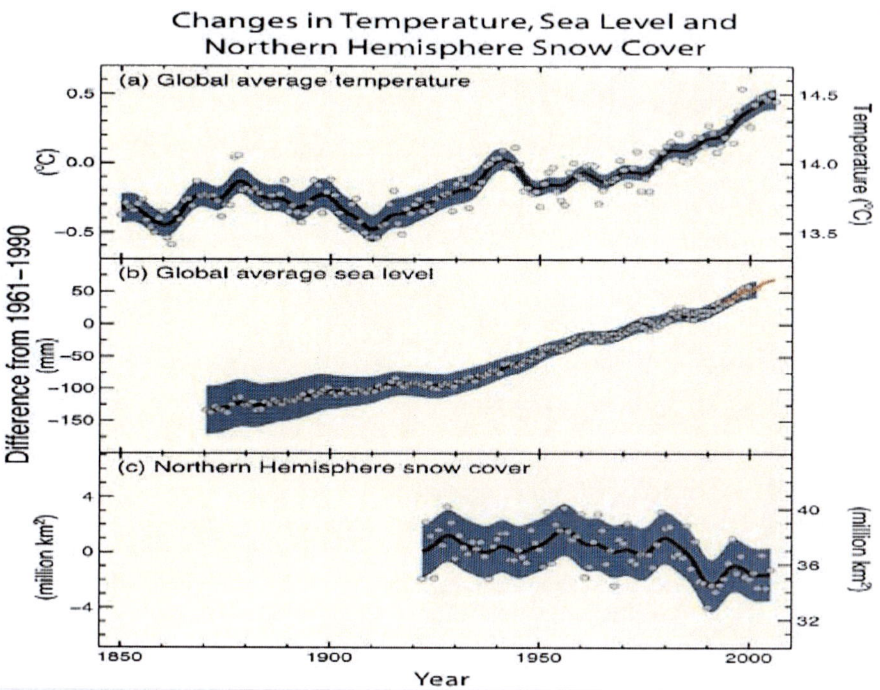

Figure 12.3: Changes in temperature, sea level and Northern Hemisphere snow cover (IPCC 2007).

Some international documents containing analyses of current ideas of climate refer to a consensus with respect to the scientific conclusions contained in these documents. This wrongly assumes that the development of science is determined not by different views and relevant discussions, but by a general agreement and even by voting! For example, an editorial in the London-based Nature[2] commented as follows on the recent release of the summary of the IPCC: "The IPCC report has served a useful purpose in removing the last ground from under the sceptics' feet, leaving them looking marooned and ridiculous". As a political or social statement it is fair enough, but it is not a scientific statement. A few centuries ago what would have been the consensus on the heliocentric nature of the solar system? Or in the 19th century what would have been the consensus on Darwin's theory of evolution, or on the age of the Earth and

[1]Special Report on Emission Scenarios.
[2]Light at the end of the tunnel. Nature (Lond.), 445, 8 February 2007

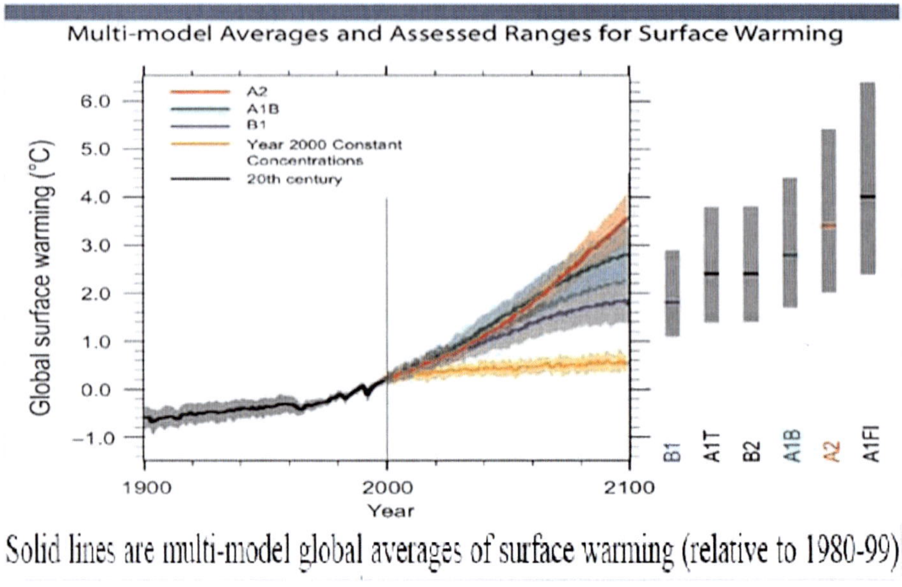

Figure 12.4: Multi-model averages and assessed ranges for surface warming (IPCC 2007).

Hutton's unconformity? Historically, it is not a matter of counting heads or counting votes (cf. Chapter 13).

Nobody doubts that the Earth's climate system has indeed changed markedly since the Industrial Revolution, with some changes being of anthropogenic origin. The consequences of climate change do present a serious challenge to the policy-makers responsible for environmental policy and this alone makes the acquisition of objective information on climate change, its impact and possible responses, most urgent. From the viewpoint of the impact on 'rapid' climate changes, of special interest are O_3, H_2O, and aerosols. However, as is well known, climate variability is both of natural (due to internal processes and external forcings) and anthropogenic origin, i.e. it possesses both internal and external variability. In order to quantify this variability, the reliability of observational data is fundamental. Without such observational data adequate empirical diagnostics of climate in all its complexity remains impossible. Yet the information concerning numerous meteorological parameters, so very important for documentation, detection and attribution of climate change, remains inadequate for the drawing of reliable conclusions. This is especially true for the global trends of those parameters (e.g. precipitation), which are characterized by a great regional variability.

A very important aspect of the climate problem consists in the recognition of anthropogenically-induced changes caused by increased emissions to the atmosphere of aerosols and aerosol-forming gases. In this connexion, highly uncertain quantitative estimates of anthropogenic impacts on global climate deserve special attention.

A principal difficulty in giving substance to the projections is the impossibility of determining agreed predictions on how the concentrations of greenhouse gases (GHGs) will evolve in future, which makes it necessary to take into account a totality of various scenarios. The huge thermal inertia of the world's oceans dictates a possibility of delayed climatic impacts of the GHG concentrations, which have already increased.

It is a very well known truth that the chaotic character of the atmospheric dynamics limits the validity of long-term weather forecasts to one or two weeks and prevents the prediction of a detailed climate change (e.g. it is impossible to predict precipitation in Great Britain for the winter of 2050). However, it is possible to consider climate projections, that is, to develop scenarios of probable climate changes due to the continuing growth of GHGs concentrations in the atmosphere. Such scenarios, if credible, may be useful for decision-makers in the field of ecological policy. The basic method to make such scenarios tangible involves the use of numerical climate models that simulate interactive processes in the climatic system "atmosphere-ocean-land surface-cryosphere-biosphere".

12.4 Global carbon cycle and climate

12.4.1 Sources of and sinks for carbon dioxide in the biosphere

Carbon dioxide is circulating in the environment among its reservoirs listed in Table 12.1. In general, the diverse compounds of carbon continuously form, change and decompose, and amidst all this diversity, natural and anthropogenic CO_2 fluxes are created by the processes of respiration and decomposition of vegetation and humus, in the burning of carbon-containing substances, in rock weathering, etc. Part of the CO_2 dissolves in the world ocean giving carbonic acid and the products of its dissociation. The content of carbon in its reservoirs and estimates of its fluxes between them is the most important problem of the analysis of the global CO_2 cycle. Numerous schemes of this cycle drawn from analysis of global interactions of living organisms and their physical and chemical media as well as estimates of carbon supplies accumulated during the historical past serve as the basis for predictions of CO_2 concentration dynamics in the Earth's atmosphere, which has recently been the subject of a hot dispute in connection with assessments of the role of CO_2 in climate warming (Kondratyev et al., 2006).

In any case, the marine environment is subject to a variety of threats within the different meanings. Together with the possible future effects of climate change discussed here, these threats include such consequences as loss or degradation of biodiversity, changes in the structure of trophic chains and the loss of habitats. According to Sherman (2003), 64 large marine ecosystems (LME) exist in the world and need to be studied in detail.

LMEs encompass coastal areas from river basins and estuaries to the seaward boundaries of continental shelves and the outer margins of the major coastal currents. These areas occupy more than 2×10^5 km^2 of coastal waters. These waters are characterized as being sinks for most of the global ocean pollution and hapless victims of overexploitation of their ecosystems. That is why the study of LMEs is the principal stage of the marine ecosystem problems on the global scale. At the present time a useful set of indicators characterizing the state of the LMEs exists (Sherman, 2003):

Table 12.1: Global carbon reservoirs (Kondratyev et al., 2003).

Reservoir	Amount of carbon/Gt
Atmosphere	720
World's oceans	38 400
Total inorganic carbon	37 400
Surface layer	670
Deep layers	36 730
Total organic carbon	1000
Lithosphere	
Carbonate (sedimentary rocks)	> 60 000 000
Kerogens	15 000 000
Land biosphere	2000
Living biomass	600–1000
Dead biomass	1200
Inland waters biosphere	1–2
Burnt fuel	4130
Coal	3510
Oil	230
Natural gas	140
Others (peat, etc.)	250

1. **Productivity indicator:** photosynthetic activity, zoöplankton biodiversity, oceanographic variability;
2. **Fish and fisheries indicator:** biodiversity, finfish, shellfish, demersal species, pelagic species;
3. **Pollution and ecosystem health indicator:** eutrophication, biotoxins, emerging disease, other health indices;
4. **Socio-economic indicator:** integrated assessments, human forcing, sustainability of long-term socio-economic benefits;
5. **Governance indicator:** stakeholder participation, adaptive management.

A modular strategy proposed by Sherman (2003) provides information for the various aspects of LME state. The productivity indicator supports data for the ecosystem parameters that are used within different items of global model. Other indicators help to make more precise the model items that parametrize the carbon fluxes within the marine ecosystem structure.

An important stage in understanding the processes of CO_2 exchange between the biosphere reservoirs is a study of the laws of the development of various ecosystems in the pre-industrial epochs, without any anthropogenic factor present (Kondratyev et al., 2006). Natural carbon fluxes between the atmosphere, oceans, surface ecosystems and inland water bodies are strongly variable both in space and in time (from year to year and seasonally). Analyses of ice cores from Greenland and the Antarctic have reliably shown variations of atmospheric CO_2 concentration in the past. Eight thousand years ago the CO_2

concentration in the atmosphere constituted 200 ppmv.[3] By the beginning of the pre-industrial epoch this estimate varied between 275 ppmv and 285 ppmv (±10 ppmv). By the year 1985 the concentration of CO_2 in the atmosphere reached ~ 345 ppmv. But in 1998 it was already 366–367 ppmv (Bolin and Sukumar, 2000). The total amount of carbon in the atmospheric CO_2 is estimated at about 700 Gt of carbon.[4] The natural CO_2 budget is estimated at $\sim 150 \times 10^9$ t C emitted annually in the processes of respiration and decomposition and assimilated in photosynthesis both on land and in the ocean, and including CO_2 dissolving in the world ocean.

Special emphasis has been recently placed on the circulation of organic and inorganic carbon in the water domain of the world ocean whose mechanisms are closely connected with the CO_2 partial pressure dynamics in the atmosphere. With some assumptions, it appears that an increase of the atmospheric share of CO_2 is followed by an increase of the partial pressure of carbon dioxide in the surface layer of the ocean at about the same rate. In this case a mechanism of the interaction of the components of carbonic acid (HCO_3^-, $CaCO_3$, CO_3^{2-}) is indicated, which, depending on the relationships between its characteristics, limits or, on the contrary, stimulates the process of either assimilation or emission of CO_2 by the ocean. The establishment of the carbon flux on the atmosphere/ocean border depends also on phytoplankton production, the relationship between the organic and inorganic shares of carbon, temperature, and hydrodynamic and other parameters of the water domain. At a certain level this flux depends also on the processes of the vertical transport of carbon with descending dead organisms, when they become bottom sediments whose contribution to the so-called biological pumping depends on the depth. In this chapter an attempt has been indicated to describe these mechanisms on a formalized level. Of course, a search for critical factors of the impact on the difference of partial pressures of atmospheric and oceanic reservoirs of CO_2 requires thorough and detailed observational studies of the ecological and geophysical processes that function in the oceanic system. In particular, the role of marine organisms in the process of transformation of calcium carbonate and its accumulation, that is, in the change of the domain acidity, has been poorly studied. All these processes have different directions and powers in the open ocean, estuaries, shallow waters, coastal zones, and river deltas.

To analyse the CO_2 dynamics in the biosphere, it is important to take into account a maximum possible number of its reservoirs and fluxes as well as their spatial distribution. It is here that numerous global models of the carbon cycle differ. The present level of sophistication of these studies does not allow one to answer the principal question about the extent of reliability of details of the database concerning the supplies and fluxes of carbon. Therefore many authors analysing the dynamic characteristics of the global CO_2 cycle rather arbitrarily utilize fragments of the databases dealing with the distribution of the carbon sinks and sources.

12.4.2 Anthropogenic sources of carbon

A key component of the global CO_2 cycle is anthropogenic emissions of the compound into the environment. The principal problem studied by most of the

[3] Parts per million by volume.
[4] 1 Gt is 10^9 (metric) tones, or 10^{12} kg.

investigators consists in the assessment of the ability of the biosphere to neutralize an excess amount of CO_2. It is here that all predictions of the consequences of the greenhouse effect are wide open to criticism. All the models of the global CO_2 cycle are based on scenarios that describe the dynamics of the extraction and burning of fossil fuels. Here a natural need appears, however, for models of the energy-economy system that require a detailed parameterization of the geopolitical structure of the world. So far, among the most widely used models of this type is a model developed by IIASA, in which the globe is divided into nine regions differing in the level of per-capita energy consumption and other parameters. The regional structure is shown in Table 12.2. With this scenario of the socio-economic structure one can attribute strategies of the development of each region and assume possible consequences for the environment of future behaviour of individual regions. Most of the similar scenarios use such indicator as the ratio of acceleration of energy consumption. This parameter varies from 0.2% to 1.5% per year. Various combinations are considered when choosing a source of energy among oil, gas, nuclear and solar energy, hydroelectric power stations, solid wastes burning. Naturally, one has to take into account demographic, technical, political, and macroeconomic factors. The size of population in most of the scenarios is assumed to grow at a rate that would reach in 2025 and 2075 respectively the levels of 7.9 and 10.5 milliard people. If all these suppositions in the scenario are assumed to be true, one can calculate the anthropogenic emissions of CO_2 and other greenhouse gases. Then it is necessary to determine the total temperature impact ΔT_Σ of these gases. Kondratyev et al. (2006) recommend the simplest additive algorithm to estimate this effect.

Table 12.2: Characteristics of the rates of growth of the economic efficiency and population in the world regions following the IIASA scenario (Mintzer, 1987).

		Population/millions in the year:		
Region	P^a (%)	**2025**	**2050**	**2075**
Australia	2.3	160	150	150
Africa	1.6	1600	2200	2700
Canada + Europe	1.6	520	540	540
China	1.9	1600	1700	1700
Latin America	1.9	720	850	900
Russia + FSU[b]	1.3	470	500	510
Central Asia	1.9	280	360	410
USA	1.2	290	290	290
South-east Asia	1.8	2600	3100	3400

[a] Annual increment of labour productivity.

[b] Former Soviet Union (Kondratyev et al., 2003).

The anthropogenic constituent in the global CO_2 cycle causes changes in the reservoirs of the CO_2 sink. The greatest changes are connected with urbanization, distorted structures of soil-plant communities, and hydrosphere pollution. The rates of the change of forest masses for pastures and cultivated lands are estimated at 5×10^4 km^2/year. Dense tropical forests are substituted with plantations at a rate of 10^5 km^2/year. This process increases the rate of desertification ($\sim 5 \times 10^4$ km^2/year), which in turn increases the amount of emitted carbon (by about 0.1 Gt C/year).

The general pattern of the present level of anthropogenic CO_2 fluxes has been rather well studied. So, due to burning of solid and liquid fuels, about 20×10^6 t CO_2 are emitted every year (with the ratio 1:1??). Burning of gas fuel contributes about 4.5×10^6 t CO_2 into the atmosphere. The contribution of the cement industry is estimated at 7.5×10^5 t CO_2. Individual regions and countries contribute to these fluxes rather non-uniformly. Table 12.3 gives some estimates of the contributions due to biomass burning.

Biomass burning in the tropics is one of the main sources of the input of minor gaseous components and aerosol particles to the troposphere. The tropics have about 40% of the global land area and contribute about 60% of the global primary productivity. The types of vegetation in the tropics are much more diverse compared to other regions. However at present the tropical forests and savannahs are being transformed into agricultural lands and pastures at a rate of about 1% per annum. This transformation is mainly caused by biomass burning, which strongly affects the chemical composition of the atmosphere, and hence the climate.

Table 12.3: Distribution of the scales of biomass burning (millions of tons of dry matter per year) (Kondratyev et al., 2003).[a]

	Sources[a]						
Region	T	S	L	B	Y	A	C
Tropical America	590	770	0	0	170	200	7.5
Tropical Africa	390	2430	0	0	240	160	9.3
Tropical Asia	280	70	0	0	850	990	3.3
Tropical Oceania	0	420	0	0	8	17	0
USA and Canada	0	0	0	0	80	250	0.5
W. Europe	0	0	0	0	40	170	0.2
Temp.-zone forests	0	0	224	0	0	0	0
Boreal forests	0	0	0 5	6	0	0	0
Totals	**1260**	**3690**	**224**	**56**	**1438**	**2017**	**21**

[a] Notation: T, tropical forests; S, savannah; L, temperate-zone forests; B, boreal forests; Y, household fuel; A, agricultural wastes; C, brown coal.

In the process of biomass burning, huge amounts of non-methane hydrocarbons (NMHC), NO_x, and many other gaseous components go to the atmosphere. As shown by analysis of the data of satellite observations, the share of the tropics constitutes about 70% of the burnt biomass, about half of which is concentrated in Africa, with an annual maximum of biomass burning observed north of the equator (in the period of the dry season). Savannahs and forests in the tropics also emit into the atmosphere a great amount of biogenous compounds. In connexion with widely spread fires in savannahs and their strong impact on the environment, Nielsen (1999) performed an analysis of special features of the spatio-temporal distribution of fires in the region where the field experiment EXPRESSO (Central Africa) was carried out from the data of AVHRR provided by NOAA satellites during the dry seasons from November 1994 to December

1997. The variability of the fires can be described by three characteristics: (1) fire probability at a given point at a given time moment; (2) probability of repeated fires at a given point during a certain time period; and (3) the spatial extent and the burning savannah temperature affecting the conditions at a given point.

Processing of satellite imagery has shown that a fire is not an accidental process. For instance, the probability of fire increases in the neighbourhood of an actual point considered. A combined analysis of the characteristics of the spatial and temporal variability of fires has made it possible to substantiate 12 typical régimes of fires as well as the dependence of the special features of the fires on those of the vegetation cover. Though there is no doubt that, as a rule, the savannah fires are caused by humans and not by other factors, specific causes of fires as a function of forms of human activity remain unclear. From the viewpoint of the temporal variability, it is expedient to classify the fires taking into account the beginning of the fire season, the rate of their development, and the duration of the fire season. In this context the following types of fires can be identified by specific dynamics of their development: fast, late or long.

The contribution of fires in savannahs constitutes more than 40% with respect to the global level of biomass burning due to which the atmosphere receives minor gaseous components, such as NMHC, carbon monoxide, methane, etc., as well as aerosols. According to the available estimates for the period 1975–1980, 40–70% of savannahs were burnt every year, and about 60% of such fires took place in Africa. In 1990 about 2×10^9 t of vegetable biomass were burnt, and as a result, 145 Tg CO_2 got into the atmosphere, which constitute about 30% compared to the anthropogenic CO_2 emissions.

Forest fires have a serious impact on the global carbon cycle. Though forest fires can occur naturally, for example, by being caused by a lightning strike, nevertheless, humans' contribution to their occurrence is constantly growing. An occurrence of a forest fire due to a lightning strike is only possible if it strikes standing wood or, in the case of a thin forest, it strikes soil covered with moss or litter. The electrical resistance of standing wood is known to be almost 100 times greater than that of the growing trees, and therefore when the lightning strikes a living tree, it does not even become charred. Therefore the monitoring of the fire danger in forests gives reliable estimates of the probability of the lightning-caused forest fires. A more complicated problem is to predict the anthropogenic causes of the forest fires. More than 90% of forest fires are known to occur within a 10 km radius of populated areas, implying that they are being caused by anthropogenic factors. Hence, the fire load on forests is strongly correlated with the spatial distribution of population density. Of course, the intensity and frequency of occurrence of the fires depend on the climate dryness in a given territory, on forests' density and on their health. A forest fire is dangerous not only because it is a source of pollutants for the atmosphere, but also because its consequences are dangerous. The fires change the forest microclimate, in particular, illumination and heating of the soil intensifies, and the hydrological régime of the territory changes. Moreover, on a territory afflicted by a forest fire the bioproductive ability of biocoenoses deteriorates and, hence, the role of this territory in biogeochemical cycles changes. It is well known that in a region with a dry climate the fire-destroyed forests are not restored naturally, and the area must be reforested. Therefore it is important to know the laws of interaction of the forest fire and biocoenosis of its territory. For instance, fires in the boreal forests contribute not more than 2% into carbon emissions into the atmosphere, but seriously affect chemical processes in the high-latitude troposphere and the radiative properties of the atmospheric,which can lead to global climatic consequences.

In general, for different reasons, biomass burning is a complex anthropogenic source of atmospheric pollution and of the global impact on the biosphere on the whole (Table 12.3). Estimates obtained by many authors show that the radiative forcing of climate determined by aerosols from biomass burning constitute about -1.0 W/m^2 (in the case of purely scattering aerosols the uncertainty of the estimates ranges between -0.3 and -2.2 W/m^2).

The reliability of the assessment of the role of CO_2 in building up a greenhouse effect depends on a detailed consideration of the global biogeochemical carbon cycle dynamics in the models and on the accuracy of the assessment of its characteristics. There are dozens of diagrams of the global carbon cycle in the form of CO_2 changes. Let us consider some of them to demonstrate their principal features and to understand the limit for necessary details of the simulation of carbon cycle compounds, beyond which it is impossible to obtain additional knowledge about this cycle and, hence, about the greenhouse effect due to CO_2. Note that all known diagrams of the global CO_2 cycle are divided into two classes: pointwise (globally averaged) and spatial (locally averaged). All the diagrams are similar in that the biosphere is divided into the atmosphere, oceans, and land ecosystems. Many diagrams divide carbon into organic and inorganic forms. As a rule, the time step of averaging all processes and reservoirs of carbon is assumed to be one year, and therefore the atmospheric reservoir is considered as homogeneously mixed-up (pointwise). The world ocean and surface ecosystems are assigned considerable detail. This detailing is based on global data-bases for these reservoirs of carbon. As a rule, the final results of such diagram-based studies are either of methodical character or they predict the atmospheric CO_2 concentrations within the limits of a certain choice of scenario for anthropogenic activity.

The scheme in Figure 12.5 (Bolin and Sukumar, 2000) gives an idea about the amounts of carbon supplies for its basic reservoirs. The estimates shown in this scheme differ drastically from the estimates given by other authors. Nevertheless, their relationships and orders of magnitude coincide in most cases. As seen, the largest carbon supply is concentrated in the world ocean. A minimum of it is in the atmosphere.

12.5 Conclusions

Nowadays, there are many atmospheric models. A serious difficulty arises regarding which is the best model to choose. As this problem of choice is crucial, there remains the possibility of comparing the climate scenarios obtained using various models. The difficulty becomes much more serious when taking into account the fact that big uncertainties in fundamental climate-forming factors (e.g. melting of the ice sheets, carbon-cycle feedbacks, the role of clouds, biogeochemical cycles) still remain. In order to reduce the level of existing uncertainties, the modelling of nature-society interactions is urgently required together with long-term, nonlinear changes in the climate system taken into account.

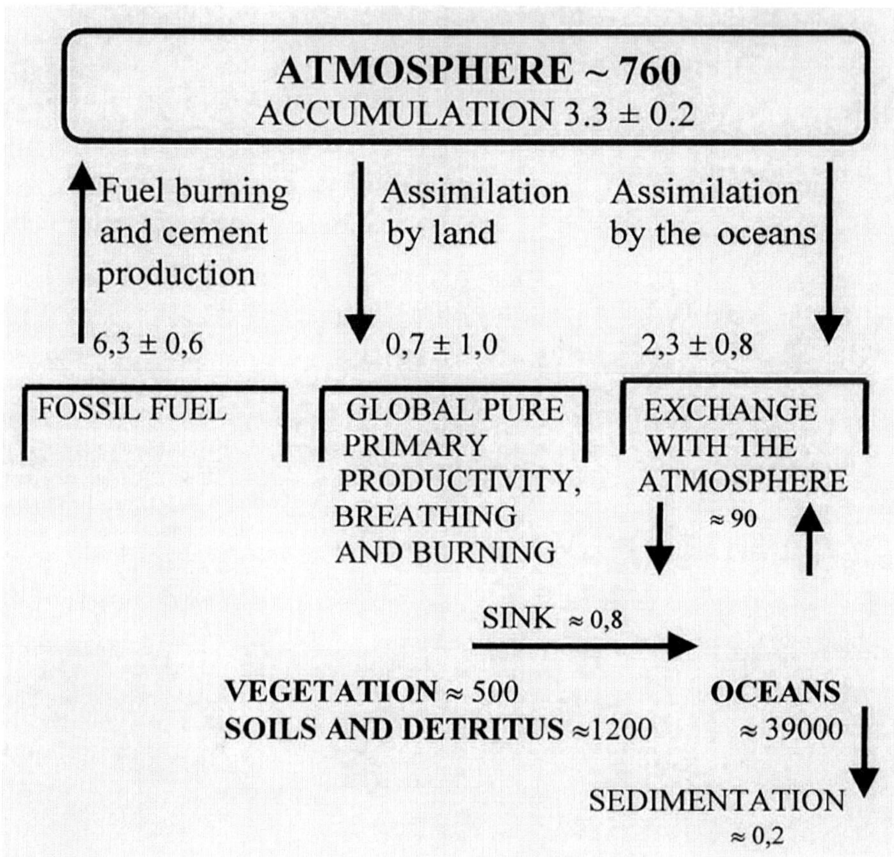

Figure 12.5: Global carbon cycle, after Bolin and Sukumar (2000). Carbon supplies are given in Gt C, and its fluxes in in Gt C/year.

12.6 References

Bolin, R. and Sukumar, R. Global perspective. In: Watson R.T., Noble I.R., Bolin R. et al. (eds). Land Use, Land-Use Change, and Forestry, pp. 23–51. Cambridge: University Press (2000).

Cracknell, A.P. and Varotsos, C.A. Fifty years after the first artificial satellite: from Sputnik 1 to ENVISAT. International Journal of Remote Sensing 28 (2007) 2071–2072.

IPCC Report: Climate Change 2007: The Physical Science Basis, Summary for Policymakers, Contribution of Working Group I to the Fourth Assessment Report of the Intergovernmental Panel on Climate Change (Website: http://www.ipcc.ch).

Kondratyev, K.Ya., Grigoryev, A.A. and Varotsos, C.A. Environmental Disasters: Anthropogenic and Natural, Chichester: Springer-Praxis (2002).

Kondratyev, K.Ya., Krapivin, V.F. and Varotsos, C.A. Global Carbon Cycle and Climate Change, Chichester: Springer-Praxis (2003).

Kondratyev, K.Ya., Krapivin, V.F. and Varotsos, C.A. Natural Disasters as Interactive Components of Global Ecodynamics. Chichester: Springer-Praxis (2006).

Pearce, F. The Last Generation: How Nature Will Take Her Revenge for Climate Change? (Eden Project Book). London: Transworld Publishers (2006).

Sherman, K. Assessment and recovery of large marine ecosystems. In: Proc. PAME, (Protection of the Arctic Marine Environment), Workshop 25–27 February 2003, Stockholm, Sweden, pp. 1–36 (2003).

Complexity and Security
J.J. Ramsden and P.J. Kervalishvili (Eds.)
IOS Press, 2008

Chapter 13

Global warming: a social phenomenon

Serge Galam

Centre de Recherche en Épistémologie Appliquée (CREA) and CNRS
École Polytechnique, 1 Rue Descartes, 75005 Paris, France

13.1 Introduction

The current situation: man is guilty

Global warming has become a world issue and everyone everywhere is dealing with it, either directly, with floods, droughts and loss of life, or by strong disturbances in the normal local weather conditions. The media are making headlines with the latest catastrophes, and politicians are addressing the issue in their speeches. But yet not much has been done. Pressure is growing from public opinion and from non-governmental organizations to take drastic measures against the well-identified cause of global warming. The scientific community is extremely active on the issue by describing detailed scenarios of the dramatic consequences of the current trend and urging governments to act immediately. The Intergovernmental Panel on Climate Change (IPCC) is monitoring world activity with thousands of climatologists. With a single voice they agree about the scientific diagnosis. During their last meeting in Paris in February 2007 they concluded unanimously that it is the increased quantity of carbon dioxide in the atmosphere which produces global warming, and they designate man as the cause. Human greed, by its exponential appetite for natural resources, is destroying the planet. At the present rate of carbon dioxide production, global warming will lead to a total catastrophe. Artists are becoming involved in this survival cause and ex-U.S. Vice President Al Gore is leading a new crusade to save the planet. Huge free concerts are taking place worldwide to raise the profile of the matter. While American President George W. Bush has been reticent to adopt countermeasures, European leaders are taking the initiative in carbon dioxide reduction, in particular led by the German Chancellor Angela Merkel and the new French President Nicolas Sarkozy. The European Union has decided on a unilateral twenty percent cut of carbon dioxide production by 2020. Many industries are taking this coming restriction very seriously and are trying to adapt without too much incurred cost. Countries like China and India are not yet engaged in reduction policies, on the contrary, they are increasing their carbon dioxide output enormously, but this fact is ignored in the quest for European righteousness. To sum up the current situation, a world danger has been clearly identified by scientists, its cause is determined precisely and the solution is clearly set. Man is guilty and must pay the price. Globalization and the consumerist way of life of modern society must be sacrificed on the altar of redemption to save our beloved and potentially martyred planet. It is in the name of Science, it is in the name of human humility.

Something sounds wrong with this so clear picture

The current unanimity of citizens, scientists, journalists, intellectuals and politicians is intrinsically worrying, since it is predicated by the generalized fuzzy feeling that man went too far in intervening in natural processes in order to reap immediate profit. The clear identification of a unique culprit, man's abuses, is too reminiscent of past ancestral reactions to collective fears provoked by natural disasters. The designation of the USA as the villain is also problematic within the current world political framework of an anti-Bush attitude. It appears that several different concerns all coalesce to identify a kind of world devil responsible for all the evils. More problematic is the insistence of presenting the unanimity of the scientific world in identifying a catastrophe caused by man's production of carbon dioxide. Claiming that thousands of scientists have voted to assert the certainty, estimated at 90%, that humans are responsible for all the observed climate changes, hides something, especially when the sole basis of these affirmations come essentially from models simulated with computers. The major real fact is the observed correlation between, on the one hand, the increase of carbon dioxide in the atmosphere, and on the other hand, the measured, albeit small, increase in global temperatures. But an observed correlation does not of course mean a single directed relation of a cause to an effect.

Remembering the past collective fears of human societies

In the history of human society, the identification of an individual responsible for all the difficulties and hardships of a society has always resulted in human destruction. Nothing good has ever ever emerged from such a unanimity of all parts of a society, quite the contrary. When the collective fears are driven against a designated culprit and monitored by those seeking a big political design for a better society, the result has been always fascism and desolation. Human society is exposing itself to great risks in this voluntary and headlong rush to "salvation", with the real possibility of destroying itself even before the global mean temperature has had time to rise significantly. But why provoke an anxiety about the proposed measures to reduce global warming by the apparent reassurance of official concern of a possible impending catastrophe?

When a scientific matter becomes emotional

The unanimity exhibited everywhere is obtained by the exclusion of any person who dares to cast doubt about the collective hypothesis. Verbal and written abuse as well funding cuts are immediately applied to those few sceptics. In particular, the much-trumpeted unanimity of the climatologist community has been obtained by the expedient of excluding those sceptical colleagues. The debate about global warming has taken on such emotional tones that what was originally a scientific debate has now become transformed into an argument typified by passion and irrationality. The degree of hostility used to quell any dissenting voice demonstrates that the current debate has acquired a quasi-religious nature and thus has become extremely dangerous. To give an illustration of what should be a scientific debate, imagine that a scientist were to question the reality of gravity. The community of scientists as a whole would be indifferent to such an opinion and just ignore it. Scientific colleagues would at the very most express some feeling of compassion toward someone who had evidently lost his head. However, this scientist would not become the victim of any kind of abuse. The violence actually exercised against scientists sceptical about the cause of global warming is an additional indicator, were any needed, that the "official" thesis of human guilt has an extremely shaky foundation. I myself became the target of virulent attacks from many different sources after I published an article in the French daily newspaper *Le Monde*, in which I stated there exists scientific unanimity about human responsibility in global warming but no proven scientific certainty.

The ancestral temptation of sacrifice

The consensual solution embodied by this assertion of human guilt is very reassuring in subduing archaic fears against human vulnerability to natural elements. First it identifies the cause of the threat without uncertainty. Second it offers a clear solution to resolve the problem and suppress the threat. Third it requires sacrificing the current standard of life, which for many, is synonymous with conspicuous consumption and abuse. Moreover, it is in accord with our records of the past. Throughout history, it is found that when facing uncontrolled natural elements, our ancestors had the tendency to persuade themselves that they were the cause. Always they associated big and small natural catastrophes with God's anger against mankind's sins. God was upset and exerted the deserved punishment through the violence of nature. And for many millennia, human beings believed that they could stop this violence with redemption in the form of animal and human sacrifices. Fortunately, the growth of scientific understanding has taught us that there was no foundation to this custom. And yet all of a sudden, against all expectations, this ancient and archaic system of beliefs is resurgent once more with fresh vitality. The incredible paradox is that scientists in the name of science are actually encouraging such behaviour. And just as in ancient times, the new prophets are announcing the end of the world. Again, it is our greedy and profligate ways of life that are responsible for this imminent end. And again, the prophets are demanding sacrifices in order to pacify nature. Fortunately this time, they are not demanding that we sacrifice our lives, but instead that we sacrifice our way of life, including technological progress and scientific research.

13.2 No present scientific certainty about human guilt

To analyse the various aspects of the global warming debate it is essential to come back to the supposed certainty of the scientific proof stating that man is guilty. All media and journals assert the scientific proof by quoting especially the February, 2007 meeting of the IPCC held in Paris where 2500 scientists voted in favour of a human cause. Here stands a major confusion between what is a political decision and what is a scientific proof. In the case of a political decision, unanimity and the number of votes are essential ingredients in weighing the validity of the decision taken. On the contrary, science has nothing to do with either unanimity or number of voters. One might recall that the consensus of scientists regarding erroneous "truths" has often been used to oppose the acceptance of genuine new discoveries. A scientific proof can be discarded by the scientific community for some time, cf. the famous examples of Galileo and Einstein. Hence, if one insists so much on the very broad consensus backing the "scientific proof" of human guilt for global warming, that in itself may demonstrate that the asserted "proof" is absent. One must be very clear about this matter. At present, contrary to what has taken place during recent years, there exists no scientific certainty about human culpability concerning global warming. There is only the strong conviction of thousands of scientists that it is so. This is not an insignificant matter in making priorities for the research objectives but it should not, in any case, be an argument to forbid parallel research in other directions. The debate must stay wide open within the community of climatologists. The matter is simply not yet resolved scientifically, even if politically it appears to be. To make the issue at stake more precise, imagine that for some incredible reason 10 000 physicists from all over the world unanimously voted that gravity does not exist. Evidently such a vote would not alter the reality of gravity's existence by one iota. On the other hand, such a vote could convince millions of people that they could safely jump off a high place, and as a consequence be seriously injured or die. This is the root of the present danger where a supposed truth is purported to be scientific. It may be necessary to reassess how science is presented. If one has some novel proof of a phenomenon (which, unlike mathematical proof, essentially rests on an overwhelming accumulation of evidence from repeated experiments), one simply says or writes, "X et al. have demonstrated (or proved) that ...". One never says, "the scientific community of the world, united in conclave, have unanimously decided that ...". It is in the domain of politics and sociology that consensus is accepted in order to justify a choice, precisely because there does not exist the possibility of proof (or incontrovertible accumulation of evidence) in favour of that choice. Thus, the position of the scientists, to be more precise those belonging to the community of climatologists who have created a unanimous view, engaged in political lobbying, and utilized the media to secure for themselves larger and larger financial subsidies, is particularly disturbing from the viewpoint of the free development of research. The effective elimination of "dissidents" from this community is the first cause for alarm.

When the charge of proof is up to the defendants

Serious climatologists will recognize that consensus does not as such establish a proof and that doubt is always possible. But in return for this scientific open-mindedness and in order to discard nonsense claims, they will state that to establish a credible doubt, it must be sustained by a proof or overwhelming solid evidence in favour of the non-guilt of mankind. It must be clearly recognized that up until now, such proof cannot be given. There exists no proof of the innocence of mankind. But here stands a fallacious reversal of what should be proved. It is not the duty of the sceptics to have to bring a proof of whatever it is they are sceptical as long as they are not stating anything but their doubt about some claimed new truth. Rather, it is up to the scientists making the new assertion who must bring the corresponding proof, in this case of human guilt. The rules of the debate have been inverted. Guilt has been erected as the truth, and it is up to the defendants of the opposite view to bring proof of the absence of guilt. It is an absurd trap in which to fall, and which distorts the entire debate. This adroit deception has a pernicious effect. The respective roles of the opponents have been surreptitiously inverted, and all further real inquiry into the matter is now subject to a barrier in the shape of an automatic accusation of superfluity. Man has been declared guilty simply because, at the present time, no other bearer of guilt has been found, and as mentioned, there are some superficially attractive reasons for ascribing guilt to man.

The misleading use of probability

Those climatologists convinced about global warning stipulate, in order to remain rigorous, that they are 90% certain of human guilt, but quickly add that this is essentially the same as 100%, and emphasize that it would be almost criminal to wait for a certainty of 100% before acting. This reasoning is offered as an elementary application of the so-called precautionary principle, and vaccination against infectious disease is frequently given as an example of the application of this principle. Unfortunately, there is no link between the problem of vaccination and global warming. The use of a probabilistic concept in this latter context leads to a serious and dangerous confusion. The use of the notion of probability in order to evaluate a risk is based on the existence of a collection of identical alternative events, the realization of any one of which is largely random. The probability of meeting an infected person, and of being infected, is an example of a situation to which probabilistic concepts may be legitimately applied. The evaluation of the risk is only reliable when the statistics describing the event (in this case, infection) are sufficiently large. Probability theory then allows the calculation of what could be the result of the event of meeting somebody by chance. Thus one can legitimately talk about the probability of being infected in a certain region, or indeed of the probability of winning the lottery. In the case of the risk of infection, if the vaccine exists and has no undesirable side-effects, there is likely to be a good case for vaccinating oneself, even if the probability of being infected is much less than 90%. On the contrary, to use the notion of probability in order to define the degree of confidence in the diagnosis of a unique problem may lead to dramatic errors. In order to discover the truth about a specific unique problem, one has to somehow aggregate a large number of indications, many of which are very different from each other, each one revealing only one part of the overall truth. Unlike the repetition of the same event, these different indications have very different statistical weights. Some seem major, other minor. One can gather a very large number of them, all pointing in the same direction (or perhaps not). Progressively, a truth is deduced in accord with all the available indications, but without necessarily being the truth. There is no question here of a mathematical proof, nor of a unique and incontrovertible relation of cause and effect. Until such proof or incontrovertible demonstration has been accomplished, some new indication found from some previously unsuspected or not investigated source has the potential to annihilate the entire conviction constructed up to that point, and to itself form the basis of the definitive establishment of the real truth. The example of a person accused of a crime well illustrates the subtlety of the process of proof (in the non-mathematical sense) of guilt. One may possess 99% of the indications, yet a single additional fact whose veracity is not in doubt can, at the last minute, exonerate the accused person. Each case is unique. It is meaningless to apply statistics in such cases, and to attempt to do so leads to dangerous arbitrariness. Numerous judicial errors have resulted from this fallacy. In the case of a political diagnosis of a unique situation, choice is made according to a conviction established on the basis of a certain number of indications, and not by a 'proof'. The Bush administration was persuaded, to a very high degree, of the presence of weapons of mass destruction in Iraq, yet the conviction was wrong. What happened thereafter is well known. This does not of course imply that every strong conviction is necessarily wrong. In the case of the climate, we are facing a unique situation concerning the the Earth. No statistics are possible, and hence to obtain an incontrovertible demonstration is essential in order to avoid committing an error with irreparable consequences. If decisions are to be taken, with the present status of scientific knowledge, they must be taken as political choices, and not as a consequence of scientific fact.

What is the scientific basis for human guilt?

Three empirical facts, the increase of the global temperature of the planet, the increase in the carbon dioxide concentration in the atmosphere, and an increase in the production of carbon dioxide by humanity form the main basis upon which thousands of climatologists assert the "undeniable" conclusion of human culpability. By comparing the graphs of global temperature with time, and of the increase of quantity of carbon dioxide in the atmosphere with time, they postulate a cause-effect correlation between the two phenomena. But to correlate them in a unique relation of cause and effect is an erroneous simplification that leads to premature judgments. The two effects may influence each other reciprocally, and they could also be produced separately by other independent factors that cannot be individually identified in the extremely complex global context of climatology, which is still far from being understood. On the basis of this postulate, climatologists have constructed models capable of reproducing the climate in the past, and then, on the basis of numerical simulations carried out on the computer, they can run the models into the future and make predictions. Now, these models are intrinsically mere approximations to reality, but they are not themselves the reality. How is it possible to be certain not to have neglected some factor, considered to be insignificant today, but which tomorrow may turn out to be essential in the evolution of the climate? Models promoted 15 years ago are not the same as the models in use today and the current models will in turn also become obsolete. While the use of models is an amazingly useful tool for scientific investigation, it should be always emphasized that any model contains an enormous dose of uncertainty that depends on the knowledge available at the moment of its construction. Hence, inevitably, a new discovery can at any instant invalidate the model. A model—I myself construct models in my work—should serve to orient research, but should not become a substitute for the reality that it attempts to describe. In addition, one should distinguish between precise results established reasonably reliably for a local set of circumstances, and their generalization to a global context, especially when this context is constructed on the basis of models that by their very nature embody an a priori vision of the phenomenon and whose only justification is their ability to reproduce when simulated on a computer a certain number of empirically established results. The current models used in climatology may turn out not to be false, they even may be valid, but they are not the reality. Therefore their predictions should be considered with caution and with a measure of doubt, particularly before indicating the way in which political and economical decisions should be implemented on a global scale.

Why not reduce carbon dioxide emissions anyhow?

People could argue that there is nothing bad in drastically reducing the anthropogenic carbon dioxide emissions in the framework of combating waste and pollution. This is acceptable, but the strategic question is how such a reduction will be implemented. It would be an enormous error to accede to a unique system of thought such as that currently emerging from the combination of political demands for the reduction of carbon dioxide emissions with anti-scientific ideology calling for a 'return to nature', advocating a halt to development, and a moratorium on investment in technology and expenditure on scientific research. In case the current climate changes have natural causes, focusing our entire efforts on a drastic reduction of anthropogenic carbon dioxide emissions, implying a suppression of our advanced technologies, could leave us defenceless in the face of a newly hostile climate, and could simply accelerate the disappearance of the human race, quite the opposite of the intended outcome. On the contrary, it would appear to be essential to intensify real scientific research, free of fallacies, aimed at understanding our universe better in all aspects in order to assist mankind to adapt to global warming. If the rate of global warming turns out to be as rapid as is sometimes forecast, the development of new technology and new scientific understanding will be our only chance of survival. Therefore, the reduction of pollution should be undertaken while optimizing human development and rationalizing our energy consumption without needlessly destroying our mode of life, and could become a trigger to substantial technological innovations. To make a mistake by prematurely selecting human development as the cause of global warming and drastically suppressing technology could be a fatal for us as a species. On the other hand, to make the mistake of prematurely asserting that global warming is natural, will at least allow us to formulate appropriate remedial actions, regardless of the root cause, i.e. even if it is anthropogenic. Major climate changes have taken place on Earth in the past and will doubtless happen again in the future, accompanied at each epoch by the disappearance of tens of thousands of species, and without any human intervention at all. This will happen with certainty in the future, so it is better to be ready for it.

13.3 Social warming worse than global warming!

To render effective the study of climate and the associated range of possibile actions, one must also understand the dynamics of human nature in the face of collective fears, otherwise a collective fear may produce a social instability, whose immediate consequences could be far more dramatic than the ones eventually caused by global warming. Several advance warnings of some "social heating" can already be identified. Although they are minor, they might well be the first indicators of a more significant collective disaster. The history of human societies in crisis should be revisited in this context. One first step could be the creation of a "world observatory" that would dispassionately list and study the collective fears appearing all over the planet, whether they are baseless or not. It should include the loss of climatic regularity, technological development in general, terrorism and globalization. Through the work of this observatory one could perhaps avoid needless social catastrophe, driven by fear, at the highest levels of political institutions. Some kinds of rules would have to be invented to prevent the coalescence of archaic, yet legitimate, fears in the majority of the population being manipulated by the the social intentions of an active minority promulgating for instance sacrificial expiation. Whether the intentions of that minority are well-meaning or unscrupulous is not in fact the primary coinsideration here. Available data on the history of human fears and their consequences could at least prevent a re-enactment of some of the appalling scenarios of the past.

The priority against global warming

The threats and current disturbances created by changing terrestrial climate make achieving an understanding of what is really going on highly urgent. And the only chance to achieve such a goal is to cool down the debate on global warming by replacing it with scientific debate. It is at this stage more than legitimate to raise doubts among the public, politicians and scientists, concerning a matter that should properly be the subject of dispassionate scientific research. To cast such a doubt should not in any way be taken as a tentative attempt to thwart the reduction of pollution and the waste of natural resources. Pollution, global warming and globalization are different problems, although certainly connected, and one should take action on each front separately. Our survival will ultimately depend on it. In particular, if global warming is natural, the situation will become much more painful than expected. In that case there exists no guarantee that we can ultimately do anything about it. Moreover, even the elementary steps to be taken to protect ourselves are not in that case clearly definable, which in itself engenders an intolerable existential anguish. But to use a scapegoat to calm down such an unbearable stress would be not only dangerous for the immediate future, but will jeopardize the long-term future.

13.4 Conclusion

To sum up, the social and human aspects of global warming imply that caution should be taken to prevent opportunistic politicians, now more and more numerous, to subscribe to the proposed temptation of a sacrifice scenario in order to reinforce their power by employing re-emerging archaic fears. Let us keep in mind that in a paroxysm of fear opinions can be activated very quickly among millions of mobilized citizens, ready to act in a unique direction against one enemy: it suffices then only to point out the target of their fears. Such phenomena are being studied within the new emerging field of sociophysics, in particular, the dynamics of minority opinion spreading and the nature of rumour propagation.

13.5 Bibliography

S. Galam, *"Pas de certitude scientifique sur le climat"*. Le Monde, 07 février 2007.

V. Maurus, *"Hérésie"*, (Chronique de la médiatrice) Le Monde, 18 février 2007.

P. Dumartheray, *"Climat, si on nous mentait"*, 24 Heures (Lausanne), 26 mars 2007.

S. Galam, *"La preuve scientifique n'est pas faite"* (interview par C. Baïotti), Auto-Moto, No 144, pp. 108–110, mai 2007.

S. Galam, *"Climat: culpabilité et tentation sacrificielle"*, Revue 2050, No 5, pp. 83–90, juillet 2007.

S. Galam, *"Minority opinion spreading in random geometry"* European Physical Journal B 25 (2002) 403–406.

S. Galam, *"Modelling rumours: the no plane pentagon French hoax case"*, Physica A 320 (2003) 571–580

S. Galam and F. Jacobs, *"The role of inflexible minorities in the breaking of democratic opinion dynamics"*, Physica A 381 (2007) 366–376.

Part IV

The Technology of Security

Complexity and Security
J.J. Ramsden and P.J. Kervalishvili (Eds.)
IOS Press, 2008

Chapter 14

Complex technology: a promoter of security and insecurity

Jeremy J. Ramsden

Cranfield University, Bedfordshire, MK43 0AL, UK

Earlier in this book we have developed the idea that insecurity can be traced to a spectrum of causes, ranging from natural (such as earthquakes), hence essentially beyond human control, to the motivation of an individual malefactor. In most of these causes, the feeling of insecurity emanating from their perception is well-founded in the shape of concrete threats to actual safety: the earthquake or volcanic eruption may destroy, and the malefactor may kill or injure. Technology, which is such a distinctive product of human civilization, has played an enormous rôle in concretely combating these threats to safety, and hence removing the feeling of insecurity. In many cases we are even confronted by the paradox that the resulting feeling of security is greater than that actually warranted by the provision of the technology.

Examples will be easy enough for the reader to find. Modern geology allows volcanic eruptions to be predicted with reasonable accuracy, and hence appropriate evacuations can be carried out or, better still, people can be discouraged from living in zones at risk from the consequences of an eruption. The technology of anti-earthquake building design has reached amazing levels of sophistication in Japan and Taiwan. At the level of threats perpetrated by individuals, there is both the technology of individual self-defence—details are surely superfluous—and the institutions of human society, including the provision of police forces and elaborate legal systems, both of which nowadays make extensive use of heavily technology-based forensic services: in view of the failure of many crimes to be solved, and the failure of many cases brought to court to reach a satisfactory conclusion, the paradox of security exceeding the actual contribution to safety appears to be operating here too. One might indeed speculate that the level of success of these institutions is, by some process of intrinsic regulation, kept at just that level that allows an acceptable degree of security to be felt by the population, without incurring enormous expense.

One aspect of these developments in which technology has played an indispensable part has been the application of high-speed data processing, especially using the digital computer. Certainly some aspects of forensic deduction, such as the comparison of DNA sequences, would be unthinkable without the assistance of the digital computer. Other aspects of security, such as the desire to preserve a sustainable, functional ecosystem (this is often called "environmental security") are essentially merely complex multi-objective optimization problems, to which so-called 'soft' computing approaches can often be applied.[1] Typically some proposed human action, such as the construction of an airport or an oil refinery on a hitherto more or less pristine piece of estuary wetland, needs to go ahead (otherwise people could not travel, or they would have no fuel to power their vehicles), while minimizing the adverse impact on the rest (non-human part) of the biosphere. The main goal of the decision-making exercise is usually to minimize loss of biodiversity. Once the desired goals have been defined, it may be considered to be a purely technical problem to deduce where the Pareto front lies,[2] requiring the technology of the digital computer to enable solutions to be found in a reasonable time. Both the choice of goals, and the choice of which solution among those lying on the Pareto front, are however ethical decisions. A particular difficulty is that some of the goals, such as "maximizing the amount of concrete that can be used", or "maximizing the amount of seasonal employment for construction", or even simply "maximizing the amount of personal profit", may be very present in the minds of players, but are not included in the list of criteria used to construct the Pareto front, hence the whole process of multi-objective optimization becomes distorted. This is rather an aspect for consideration in Part V of this book, however. What is also clear is that technology is both part of the problem and the solution: the technology of aircraft and internal combustion engines creates a demand for the airport or the oil refinery in the first place; other technologies can then at least alleviate the adverse consequences of their construction. In other words, the problem of minimizing the environmental impact of some construction project falls into a greater multi-objective optimization problem, in which the goals are human development and the preservation (or, indeed, augmentation) of biodiversity. It is however generally impracticable to apply any kind of formal multiobjective optimization procedure to this higher-level problem: whereas all the issues surrounding a major construction project could fall within the zone of action of a single authority (which may be a national government), who will apply some kind of optimizing process even if they do not formally construct the Pareto front), optimizing human development and biodiversity involves the whole planet, and any actions have to rely on the weak intrinsic tendency of the world to regulate its existence.[3]

[1] See for example A. Brintrup et al., Evaluation of sequential, multi-objective, and parallel interactive genetic algorithms for multi-objective optimization problems. J. Biol. Phys. Chem. 6 (2006) 137–146.

[2] The plot of the objective functions whose nondominated vectors are in the Pareto optimal set is called the Pareto front. 'Pareto optimal' means that no criterion can be increased without causing a simultaneous decrease in at least one other criterion. 'Pareto front' encapsulates the idea that there is a set of possible solutions, rather than a single solution.

[3] The difficulties of 'solving' these kinds of problems are compounded by the multiple

Max Frisch is said to have remarked that "Technology is a way of organizing the universe so that man does not have to experience it." In our technological age it is easily forgotten how not so very long ago many of our surroundings evoked feelings of fear and despondency. Edward Whymper was exhilarated by his scrambles among the Alps, but noted (with some surprise) that among the natives the mountains were regarded as terrible and threatening objects, and most had little ambition to climb them. Nowadays, we counter the fear of getting lost by global positioning systems, the fear of the dark by electric lighting, the fear of death following an accident by helicopter rescue services and well-staffed and -equipped intensive care units in hospitals, and so on. Each of these advances brings of course some concomitant disadvantages, but most of the latter seem to be of a secondary nature, such as excessive casual tourism in previously solitary and unspoilt tracts of nature, "light pollution" impeding the work of the astronomer, and a tendency for people to take excessive risks in sporting activities. Indeed, the 17th century quotation (cf. Chapter 2) "The way to be safe, is never to be secure" reflects a profound wisdom.

Other disadvantages of advancing technology are definitely less secondary. The machinery of warfare seems to keep up very well with advances in 'peaceful' technology (indeed, the desire to make war has often been a significant driver of technological advance). The amazing variety of household artefacts that make our lives ever more convenient have required a heavy price to be paid in the form of industrial pollution. A similar argument applies to the convenience of personal motorized transport. It is as if a kind of socio-technical equivalent of Lenz's law is operating to limit the output of all the efforts of mankind. Thus, the use of credit cards means that their bearer cannot be robbed of cash, but on the other hand credit card fraud has reached mammoth proportions.[4] Hence, although technology often increases both security and safety, it can equally well have the opposite effect.

Moving towards a more specific relationship between technology and complexity, technology often implies vastification, which is a kind of complexification. Thus, technology lies behind the enormous increase (especially in the UK, but France is now pursuing a similar policy, and doubtless other countries will follow suit) of video surveillance cameras in buildings and on streets, but the security forces (i.e. police) are typically overwhelmed by the amount of information flooding in from all these cameras. Similar arguments apply to other automatic alerting systems, such as batteries of sensors in nuclear power plants, water works and chemical factories.

More profoundly, "complexity begets complexity". This is simply a manifestation of a well known principle in biology, according to which speciation to fill empty ecological niches creates fresh niches, and hence the opportunity for further speciation. An example of this in technology might be the ever-increasing capability of microprocessing hardware, whose existence has engendered the creation of new software, enabling more sophisticated operating systems to be

timescales involved. A national government may for example be confronted by the dilemma that by not constructing a new airport it will preserve regional biodiversity, but will in the medium term be economically disadvantaged relative to neighbouring countries that have constructed airports, and will therefore be weakened in endeavours to exercise responsible stewardship of the planet on a global platform.

[4]There is however a qualitative difference, notably in that the cash robbery may endanger life or limb (cf. Chapter 21), whereas credit card fraud does not, although it may be inconvenient.

designed, but these operating systems and their associated applications mostly work to the limit of the available hardware, so there is usually a disappointingly small real increase in user functionality. 'Simple' operations (from the viewpoint of the user) require many more (hidden to the user) processor operations, but often only because of more elaborate graphical user interfaces and, hence, the potential acceleration of the functional operation due to the greater hardware capability is vitiated, and from the viewpoint of the user there may even have been a slowdown. Even the argument of the software developer that the result is more aesthetically pleasing is nugatory, because, as we all should know, "*de gustibus et coloribus non disputandum.*"

Sometimes this kind of complexification can have serious practical consequences. In the automobile industry, motor-car designers are considering the introduction of sophisticated braking systems with the capability, for example, of autonomously completing a sharp braking action of the part of the driver in order to bring the vehicle to a complete stop, inspired by actual observations showing that accidents sometimes occur because of drivers prematurely relaxing pressure on the brake pedal in a situation requiring an emergency stop. But such a braking system could bring a vehicle to a complete stop in the fast lane of a motorway following sharp braking to avoid a collision with a slower car that had just pulled out into the fast lane in front of the driver! This inability to foresee all possible consequences of a novel design feature is one reason why aircraft designers are more conservative than their automotive counterparts, the consequences of some unforeseen action in the air being more likely to be immediately fatal. The general problem is that vastification of a system quickly brings it up to the "complexity ceiling", beyond which it is no longer possible to explicitly test behaviour for all combinations of input parameters, simply because the parameter space has become so vast;[5] clearly modelling can play an important rôle in exploring space that it is no longer feasible to explore physically, but it must be the right kind of modelling (cf. Chapter 6).

Most of the other chapters in this Part deal with quite practical matters. We have already mentioned the psychological importance of having effective anti-earthquake technology for buildings. A similar argument applies to protection from bomb attacks. One recalls the policy of the Swiss government, dating from the time of the Second World War, making it a legal requirement for every building to have adequate underground shelter for all its occupants. The knowledge that these shelters existed, even though they were very rarely used, doubtless contributed to maintaining morale, and hence high productivity at work, and so forth. On the other hand, recalling the remarks about the balance between cost and effectiveness made at the beginning of this chapter, excessive emphasis on security can actually have a demoralizing effect, because it implicitly promulgates the view that catastrophe is inevitable. It seems that the correct balance has in the past been found essentially by the collective instinct of society, and it is gratifying that the recent introduction of proper quantitative methods for evaluating the appropriate amount of expenditure to mitigate risk have shown that in many cases the actual level is appropriate.[6]

---

[5] Some general strategies for overcoming this difficulty have been put forward by W. Banzhaf, G. Beslon, S. Christensen, J.A. Foster, F. Képès, V. Lefort, J.F. Miller, M. Radman and J.J. Ramsden, From artificial evolution to computational evolution: a research agenda. Nature Reviews Genetics 7 (2006) 7290–735.

[6] P.J. Thomas et al., The extent of regulatory consensus on health and safety expenditure.

One of the most striking manifestations of technological advance in human development is personal mobility. The desire to travel seems to be strongly innate in a significant proportion of mankind. Some of the most daring and heroic exploits in the history of civilization have been accomplished in the course of territorial exploration, whether at sea or on land. Examples are too numerous and well known for it to be necessary to mention any specific cases. One may suppose that the adventures of a few exceptionally entrepreneurial individuals reflected a general desire, present with less intensity but nonetheless universally present within the population. The very fact that we, as human beings, can move around using our legs makes mobility an intrinsic attribute of humanness, and through our mental processes we are able to enormously amplify the possibilities of exploitation of this attribute. Nevertheless, the exigencies of sheer survival for the subsistence peasantry that constituted the bulk of the population for many centuries would have prevented any practical expression of the desire to move around, other then between cottage and field, or to and fro behind the plough, except for that adventurous minority who embarked upon great voyages of exploration.

The other exception to the enforced immobility of the bulk of the population (frequently reinforced by a legal framework that formally prohibited the serf from moving away from his land) was the movement of large armies. Even the excesses of modern mass tourism pale into insignificance beside the 600 000 people who marched in Napoleon's Grand Army to Moscow in 1812. The primary motivation for the great system of roads constructed throughout Europe by the Romans was military. Naturally they turned out to be of great commercial value as well, although they tended to fall into disrepair during the Dark Ages. The loss was especially acute in Britain, where land travel remained difficult for centuries. This perhaps explains the tremendous enthusiasm that greeted the invention of the railway, especially when combined with steam-powered locomotives. A dense network of lines was rapidly constructed—10 000 km by around 1850, only 50 years after the authorization of the Surrey Iron Railway (the world's first public goods railway, authorized in 1801 and completed in 1803) by Act of Parliament,[7] and the first steam locomotive built by Richard Trevithick (in 1803). At that time (1850) an amazing speed of almost 120 km/h had been reached by the steam locomotive "Great Britain" running on broad gauge (7 feet, 2.14 m) track, and the system as a whole was carrying about 100 000 passengers annually. Following the invention of practical internal combustion engines by Otto, Benz and Diesel, automobiles gradually replaced railway trains as an object of fascination in the public mind. Indeed, the popularity of the private motor-car has been one of the most striking socio-technical phenomena of the 20th century.[8] In densely populated Britain for example, nowadays a

Part 2: Applying the J-value technique to case studies across industries. Trans. IChemE B 84 (2006) 337–343.

[7]Private railways were already in use for conveying ore and coal in the 17th century.

[8]It is impossible here to do more than sketch in the most superficial way the grounds for making this assertion of *striking* popularity. One very important piece of evidence has been the seeming equanimity with which the enormous mortality due to motoring accidents has been acquiesced in.[9] It is very frequently reported, e.g. on radio news bulletins, that a handful of people have died in an earthquake or in flooding in some remote part of the world, while the doubtless far greater number of people killed in motoring accidents on the same day in the capital city of the country concerned goes unmentioned. This would be inexplicable were it not for the extraordinary hold on the human mind that the motor-car seems to have acquired.

constant stream of motor traffic passes through most places. People think nothing of driving 50 km to work, for entertainment, or for shopping. The result of this tremendous personal mobility has been an almost total loss of *Ortsinn*, the sense of place.

Keeping in mind the theme of this book, complexity is present here in a twofold fashion. Firstly, there is the complexity of the technology itself—as is immediately apparent upon raising the bonnet of any modern motor-car. Secondly, there is the complexity that this technology has brought society—in its vastly increased interconnectivity (to which other technologies, such as the telephone and the Internet, have also contributed), both purely social and in the sense of the intricate web of interdependencies necessitated by the practical realization of complex technologies, comprising schemes of education and training for ensuring the transmission and development of high-level knowledge, and complex supply networks for bringing the products of specialized component technologies together for final assembly.

How does this affect security? Firstly, as we have already pointed out, technology allows man to combat many natural and highly effective sources of danger. This is opposed by the feature that many technologies are intrinsically dangerous, not least because of the tremendous 'amplification' of human power that is associated with many of them. Here I wish to draw attention to another result of technology, namely the loss of *Ortsinn* and the concomitant loss of security.

Ortsinn is the sense of place. Clausewitz emphasized it as a quality very important in military contexts; this has dramatically changed in the era of the intercontinental ballistic missile (ICBM), which allows war to be waged essentially remotely, and if nuclear submarines carrying nuclear warheads are taken into consideration even the missile launchers do not need to be localized in a particular place.

So it is with society in general. As long as mobility was restricted, for example by bad roads, people were obliged to spend most of their time in the same place. In a community such as a village or town, one met the same people every day; the community was effectively self-sufficient for many of its needs. These inherent strong associations effectively ensured a high level of security. Crime was mostly associated with itinerant people. One could trust people with whom one had dealings almost every day; the position of someone who provided substandard goods or services would be intolerable in such a community. Legislation to ensure adherence to trading norms was scarcely necessary. But the appetite for manufactured and exotic goods, which perforce came from elsewhere (unless the factory happened to be in the town where one lived), necessarily eroded Ortsinn.[10] An efficient transportation network was firstly required to distribute centrally manufactured goods throughout the country; nowadays high personal mobility means that most retail purchasing activity is itself concentrated in great impersonal centres (supermarkets and hypermarkets). Quality (insofar as

[10] A striking modern example of such a development is Moscow ice cream, whose quality was universally appreciated in Soviet days. Even some months before the end of the Soviet Union, and especially around Moscow's railway terminals, many private vendors sprang up, in competition with the official state product, selling ice cream so inferior that no one would buy it twice. This private business was however profitable because so great is the daily influx of visitors to Moscow that they were able to flourish solely through the business afforded by first-time buyers.

it exists) is ensured by legislation. While it is obviously impossible for a village artisan to make a personal computer, hence we really have to accept the loss of *Ortsinn* for sophisticated manufactured goods (and all that implies in terms of maintaining a corps not only of people sufficiently well trained—in centralized establishments—to maintain the technology, but also people capable of making the inductive leaps needed to develop it further), a similar impossibility does not apply to food supply. In most countries it would still be possible for the majority of food to be produced and consumed locally, which, for the reasons outlined above, would be a guarantee of quality—one would generally know the supplier personally. Only in great urban centres is such an arrangement impracticable—and it is of course in those centres where the legislation promoting the agro-industrial complex has been enacted in the first place.[11] Thus even staple items of diet, such as bread, milk and meat, are nowadays typically transported hundreds of miles to the household where they are consumed.

The contrasts and tensions brought about by this transformation have been especially prominent in Europe and North America. In the latter, there is a remarkable contrast between the excellent quality of seasonal locally grown produce sold in the so-called "farmers' markets", and the low quality of the produce (e.g., fruit and vegetable mainly grown in California and Florida) sold all the year round in supermarkets, and emanating from vast, highly mechanized farms making extensive use of fertilizers, pesticides and plants genetically modified to resist pests and pesticides. In Europe, legislation imposed upon the member states of the European Union (EU) has almost eliminated the equivalent of the farmers' market, and it is practically impossible to find high-quality produce anywhere.[12]

One of the great dangers of the EU's Common Agricultural Policy is the astonishing loss of diversity among crop plants. Whereas a hundred years ago over two thousand varieties of apples were sold commercially in France, nowadays only about four varieties are legally permitted to be sold, all of them very inferior and none of them indigenous.[13] Similar (but mostly less striking) losses of diversity have taken place in the varieties of carrots, peas, potatoes etc. grown commercially (Table 14.1). In some cases, in theory at least, the farmer is allowed to plant any variety he chooses, but only varieties chosen from a very small list, sometimes containing only a single variety, are eligible for the battery of financial subsidies routinely paid to farmers in the member states of the European Union; hence by electing to plant some superior variety not on the list, the farmer, who of course doubtless has a family to support, forfeits his subsidies. Often the only available seeds, even to the amateur horticulturalist, are for those varieties on the "authorized list".[14] Bringing the discussion to the

[11]Here we may note that many present-day European countries have an incongruously large capital, a legacy of former times when they formed the centres of much larger systems, but where nowadays a truly disproportionate fraction of the population lives (around 20% in the case that of Austria, England, Hungary etc.—figures which are however dwarfed by Argentina, where about a third of the population lives in Buenos Aires, and without any imperial past).

[12]This was already lamented in the 1980s by Freddy Girardet, proprietor and chef of "Chez Girardet" in Vaud, which at the time had the reputation of being the finest restaurant in Europe.

[13]This policy is scheduled to be consummated by 2010 within the European Union.

[14]A number of organizations, some private, some state-administered, have built up seed banks of ancient indigenous varieties of food crops, so there is some hope that should policy swing to a more enlightened view of the importance of diversity, it could be recreated.

country where this Workshop is being held, it should be clear that the consequences of European Union membership for Georgian agriculture, noted for the excellence and variety of its produce, will be catastrophic.[15] Since 1945 about 95% of indigenous wheat varieties are no longer cultivated in Greece; in South Africa a single variety of sorghum (introduced from Texas) has replaced all others; approximately 99% of commercial Turkish crops are sown with imported seed.[16] Worldwide, the number of comestible plants (and one should bear in mind that in excess of 90% of man's alimentation is vegetable; more than half of our calories come from cereals; about 70% of crop production is from maïze, rice and wheat) is estimated at between 10 000 and 80 000, of which perhaps only about 3000 are really used in a significant way. It is estimated that 10 000 years ago, when the world population was probably only about 5 million (i.e., about 25 square kilometres per person), about 5000 varieties of plants were grown for food. Nowadays, with a population of about 5000 millions, only about 150 varieties are used on a large scale.

Table 14.1: The number of varieties furnishing the bulk of commercial crops (examples). Data are from 1970. The situation has been exacerbated since then.

Crop	Number of varieties	Percentage of harvest	Region
All	29	90	world
All	24	99	world
Winter wheat	2	40	USA
Potatoes	4	72	USA
Peas	2	100	USA
Apples	10	100	France
Apples	1	70	France

Any loss of diversity is dangerous because it limits the capacity of a system to recover from a perturbation, i.e. its resilience (cf. Chapter 8). It appears that the stability of being food supply is at present only secured by extensive use of fertilizers and pesticides, and the system looks extremely vulnerable to any unusually severe perturbation, or an unusual combination of lesser perturbations.

Brief reference to the development of biofuels in order to reduce dependence on fossil fuels has been made in Part III. The increasing diversion of crops from food to fuel is likely to put further pressures on diversity, since it will be necessary to further increase the output per unit cultivated area in order to maintain food supplies, implying increased dependence on a small number of specially developed (genetically engineered) high-yielding, thus vulnerable, crops. The present situation regarding the possible benefits from biofuels is murky indeed, advocates and critics both clamouring for attention (see Figure 14.1), and clear-thinking objectivity falling prey to the usual political distortion. A commonly heard argument is that the current vast population of the Earth means that all practically available land needs to be used for cultivating food, and there is thus

[15] Cf. J.J. Ramsden, Georgia and the EU. Georgian Times, 17 April 2007.

[16] T.M.P. Duggan, A permanent solution for Turkish agriculture. Turkish Daily News, 29 September 2000.

no reserve for growing biofuels. On the other hand, the European Union has for several years actively encouraged farmers to leave land uncultivated (and even pays them a subsidy for doing so), in order to combat food overproduction.

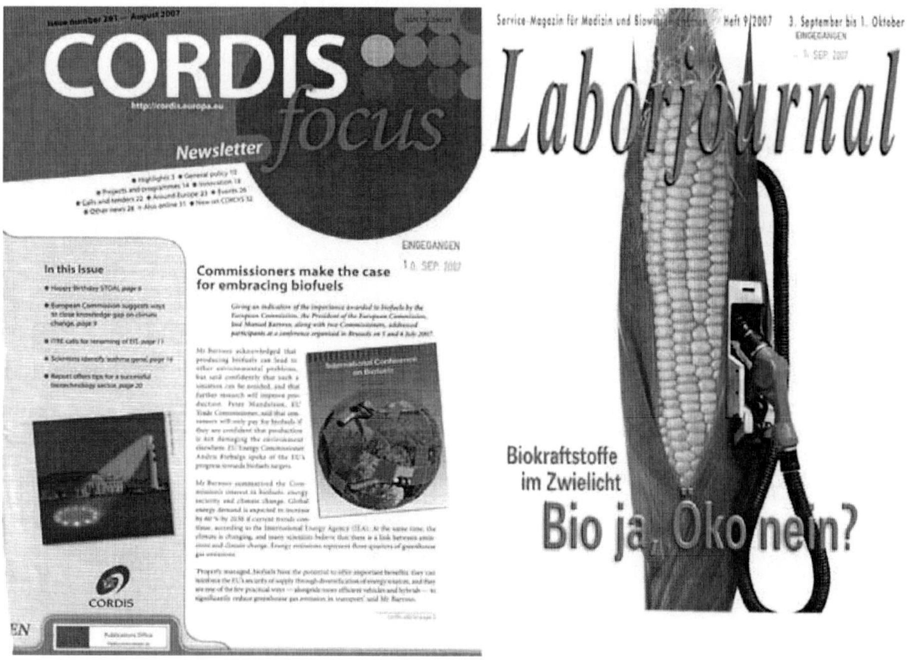

Figure 14.1: An illustration of two contrasting views of biofuels. On the left, the cover of Cordis focus, an EU newsletter (August 2007); on the right, the cover of Laborjournal, a German bioscience magazine (September 2007).

The European Union's Common Agricultural Policy (CAP) has shaped farming and food production practically ever since the inception of the Union (or its predecessors).[17] It is of course necessarily backed up by a logistic network that brings its low-quality mass production to supermarkets throughout Europe. This has also brought a new kind of problem of security: because the whole system has become impersonal, and the buyer no longer knows the producer, food security (in the sense of trusting its quality) has now to be assured by an elaborate system of legislative guarantees (for example, the life history of each cow from which prepared steaks originate must now be available at the point of sale). Unfortunately, widespread common venality works to ensure that very real health risks remain.[18] Indeed, the more complex the legislation, the easier it is for loopholes to be found by the unscrupulous (a 'soft' analogy of complexity begetting complexity in technology). In view of the evident limitations of technological solutions, one hesitates to promote yet more technology, but one solution to these problems would be to quantify the characteristics of

[17]Its principal creator, Sicco Mansholt, seems to have been inspired by the Soviet *kholkoz* or collective farm, and by the dictatorially-pursued agro-industrialization that took place in Romania under Nicolae Ceausescu.

[18]See for example M. Buchner, M. Loeckx and B. Salomon, Gesundheitsrisiko Schweinefleisch—die kriminellen Praktiken in der Tierhaltung. Vienna: Czernin Verlag (2001).

produce to a far greater extent than is possible or even feasible at present (e.g. to ensure that conditions of growth, culture, harvesting and subsequent processing are stated at the point-of-sale, vastly supplementing the very meagre and incomplete information currently available). A further development would be biosensors that can monitor at any point, including at the point-of-sale, key parameters such as the content of residual pesticides.[19]

There is certainly no shortage of challenges for personal food quality sensors. While contamination of foodstuffs with pesticides or hormones is illegal, other forms of adulteration are officially sanctioned. Thus, olive oil labelled "extra virgin" might well contain up to about one third of oil that is not exactly that;[20] and "pure" chocolate is allowed to contain up to 5% of vegetable fats other than cocoa butter. This last distortion (introduced on 1 July 1995 by the Swiss federal government) is particularly invidious, since not only has consumer trust being broken, but the formerly renowned quality of Swiss chocolate has since then sunk towards mediocrity. For a few years thereafter, Spanish and Italian chocolates ranked as the best, but their future is uncertain because of a decision by the European court of justice in 2001, as a result of an action brought by other European Union member states, that no longer requires chocolate adulterated with vegetable fats other than cocoa butter to be labelled "substitute chocolate" in Spain and Italy.[21] The motivation for replacing some of the cocoa butter with fats of different origin is unfortunately none other than human greed (in this case of the chocolate manufacturers), even though in this case the gain seems miniscule compared with the loss in quality. It is however difficult to assess the actual effects on chocolate sales caused by these changes, since other factors, including global trends in taste, doubtless also have an influence.

Clearly some of these issues surrounding the modern food chain impinge on health. As an attribute of a biological system, health is certainly a complex matter, and at the same time is probably the major source of concern, and hence of insecurity, among the population. It is fitting, since this Workshop was held in Georgia, to pay tribute to phage therapy—the use of harmless (to humans) bacteriophage to kill infectious pathogenic bacteria—that was pioneered in the Eliava Institute of of Bacteriophage, Microbiology and Virology (IBMV) located not very far from where the meetings were held. This is an example of therapy that brings safety but not security—there has been an extraordinary reluctance to use it, especially in Western Europe, seemingly because of some kind of intuitive dislike of inoculating a person with viruses. Nevertheless, the approach appears to be medically sound and may turn out to be of significant value for combating the growing problem of bacterial resistance to antibiotics.

Although governments worldwide as well as supranational bodies such as

[19]The numbers of food-borne illnesses vary quite widely among countries: in recent years there have been about 25 cases annually per 100 inhabitants in the USA, but only 3–4 in the UK, and only about 1 in France. The numbers going to hospital and possibly hospitalized are much smaller, only about one per 1000 inhabitants regardless of country, and only about one per 100 000 actually died, regardless of country.

[20]Within the European Union, Commission Regulation (EC) No 1019/2002 (containing approximately 4000 words) attempts to more strictly define different kinds of oil and blends thereof; in principle "extra virgin" olive oil may not contain any refined oil. To what extent this regulation is complied with is another matter. Even the regulation contains some distortion e.g. 'cold' (as in pressing) is defined as less than 27 °C, which the man in the street would rather refer to as tepid.

[21]C-12/00 and C-14/00. Later in that same year the international Codex Alimentarius was modified to accommodate the new definition.

the World Health Organization (WHO) have expressed grave concerns about the possibility, even probability, of a major epidemic of some infectious disease in the near future, threats to humanity from the microbial world are far from new, as has been admirably documented in the case of fungi, for example.[22] It appears to be somewhat unthinking to simply class the activities of microbes as threats. Microbes live in symbiosis with us as part of our common ecosystem, and undoubtedly have a useful rôle to play in human development. By calling them "nature's censors", Sir Albert Howard intended to point out precisely this;[23] we can be sure that their elimination would be undesirable. The real threats come from intensive farming and excessive urbanization, and the environmental degradation that tends to accompany both: alas, all characteristics of early 21st century civilization.[24]

Another characteristic of early 21st century civilization, particularly in the Western world, is the abrupt abandonment of smoking tobacco as a form of pleasurable relaxation. The sheer inconsistencies regarding the current state of legislation (and this seems to apply to most European and North American countries) concerning tobacco, the so-called "drugs of abuse" (cannabis, cocaine, heroin etc.) and alcohol beggar belief. Tobacco is subject to a punitive purchase tax (which must still be a significant source of revenue for the government, despite the diminishing number of smokers), and draconian penalties are imposed upon people caught smoking where it is forbidden (I believe the fine for smoking on the upper deck of a London 'bus is £1000, i.e. 2–3 weeks' wages of an average employee in Britain). On the other hand, the drugs of abuse are not taxed at all (but it is still illegal to consume them), yet in this case it is the purveyors who are subject to the heaviest penalties (i.e. more likely imprisonment than mere fines). The least consistency is found regarding alcohol. In some countries (e.g. Japan and many Eastern European states) consumption of alcohol by the drivers of motor-cars is strictly forbidden, whereas in others moderate consumption is tolerated.[25] Excessive alcohol consumption appears to be a major cause of violent crime (as well as a major source of illness), yet beyond a fairly punitive level of taxation, little is done by governments to restrict or discourage its consumption. Indeed, some European Union research projects seem to be aimed at positively encouraging it.

No paper has been contributed to this Workshop on the topic of health (as a security issue) and complexity, although clearly it is highly pertinent;

[22]Large, E.C., The Advance of the Fungi. London: Jonathan Cape (1940).

[23]"Insects and fungi are not the real cause of plant diseases but only attack unsuitable varieties or crops imperfectly grown. Their true rôle is that of censors for pointing out the crops that are improperly nourished and so keeping our agriculture up to the mark. In other words, the pests must be looked upon as Nature's professors of agriculture: as an integral portion of any rational system of farming." A. Howard, An Agricultural Testament, Ch. 6. London: Oxford University Press (1943).

[24]But which was already recognized some time ago: "In their one-sided chase after quantity, experimental station workers are not only misleading practice, the are doing the greatest possible disservice to the true cause of agricultural research. They have failed to insist on the effective return to the soil of the waste products contributed by the plant, the animal, and the community ... There is no need to trouble so much about yield. It is surely unnecessary to lumber up still further the world's congested markets with produce that none can buy." Sir Albert Howard, quoted by E.C. Large, loc. cit., p. 434.

[25]It is also notable that there are striking differences between countries in the numbers of accidents leading to injury or fatality committed by motorists. Whether these differences can be linked to certain differences in legislation, cultural differences, or even genetic differences, appears to be as yet unknown.

hopefully this deficiency will be made good in the future. The emphasis on the maintenance of variety as an important guarantee of the safety underpinning security applies in a very profound way to health. Disease appears to be one of nature's instruments for maintaining that variety within a population. Even allergy, which seems to be so superfluous an affliction to those who suffer from it, has its value: the effective functioning of our immune system requires, inter alia, an amazing variety of antibodies to be maintained, and it seems to be almost inevitable that the elaborate and almost unbelievably complex—in terms of both the variety of its components and their intricate interconnexions—machinery that is responsible for both the innate and adaptive immune response must sometimes overreact to stimuli,[26] and hence its elimination might be neither possible nor desirable. Clearly there is a vast unexplored field here.

Mention of the close link between technology and warfare has already been made, but at least in the Western world, for the last 50 years the bulk of the population has had little contact with the state of warfare, leaving aside isolated terrorist actions. Nevertheless, technology impinges on personal security in several very direct ways. The handgun is undoubtedly a product of technology, and has greatly evolved from the cumbersome pistols with which the highwayman would threaten his hapless stagecoach victims, who seem to have been as a rule so terrified that the highwayman probably rarely had occasion to actually attempt to fire them, perhaps just as well, from his point of view, because of the high probability of a misfire. Nowadays however a small revolver can be easily concealed on the person, and while there is a long tradition of death and serious injury caused by the use of firearms in personal disputes in the USA, the recent and rapid increase of such incidents in the British Isles is a new phenomenon. While public violence was not absent from earlier epochs, disputes were more likely to be settled using fisticuffs; the technology of weaponry that had been so successfully developed for warfare (such as the English longbow) hardly involved items to be carried around on the street as a matter of course. Members of the public clamour for more severe restrictions on the acquisition of firearms every time such an incident is reported, yet it is the general use of automobiles that is partly responsible for the enormous increase (throughout the 20th century) in crimes of a violent nature. It will probably be hard to find an example of such a crime that did not involve the automobile in one form or another. Furthermore, it was already pointed out in 1927 that the automobile "seems to have created a disregard for the safety of others, and even for the value of human life. In the year 1924—in the comparatively brief space of 12 months—5030 people were killed in motor-car accidents ... obviously a large proportion of these accidents must be due to carelessness, which can only be classed as criminal."[27]

Finally, the ever-increasing sophistication of complex technology leads to moral and ethical questions relating to security (this topic evidently bridges the themes discussed in this Part and those in the final one of this book). Apart from the rights and responsibilities of robots that is much discussed by the protagonists of artificial intelligence (but need not be further discussed here since clearly its application is still some way off), there is the already very real issue of responsibility for actions carried out by machines under human control. At a petty level, "blaming the computer", rather than its human programmer, or

[26] This might be considered as a manifestation of living on the "edge of chaos".

[27] Anon., Where Freedom Falters, p. 282. London: Charles Scribner's Sons (1927). The data applies to the USA, the population of which was about 110 million at that time.

whoever mis-entered data, is an everyday occurrence, yet one which can cause real distress and inconvenience to the person whose word of honour is mistrusted because of the conflict with what he or she has said and the mechanical output of the computer. Given that we are all, as members of a technologically sophisticated civilization, party to the widespread presence and use of technology, this naïve and even ignorant attitude seems as incomprehensible as it is to be deplored. Government ministers of education are fond of talking about the need for literacy and numeracy among schoolchildren, but what about "technical literacy"?

The other issue of morality and technology concerns the extensive use of surveillance. Since computer-based pattern recognition techniques are still very far from being able to analyse visual records automatically, camera-based surveillance can only have an effective impact on security if a cripplingly (to the rest of the economy) large proportion of the working population is able to devote their attention to examining the records. For a while, of course, the mere presence of the cameras may be reassuring to the law-abiding citizen and hence contribute to security, and by discouraging crime may also contribute to safety, but as it becomes generally realized that the surveillance is ineffective, its contribution to both security and safety will vanish. If, on the other hand, it does become effective (for example, through the implementation of powerful algorithms for image analysis), the question *Quis custodiet ipsos custodies?* arises. It has not yet been answered in the context of surveillance but clearly will have to be tackled, especially in view of the increasing doubt cast on the continuation of the Hobbesian social contract.

This Introduction has especially focused on some of the issues that could not be covered by the individual papers devoted to technology. Most of the chapters that follow in this Part are devoted to the topic of *networks*, which have such a prominent place in contemporary technological civilization. One contribution and deals with the problem of protecting people in underground structures from terrorist explosions, and another with the way in which plants can contribute to environmental security by gently neutralizing noxious chemicals.

In conclusion, we see that technology is intimately related to security. While one of the early motivations of technology may well have been the desire to render an implacably hostile nature more benign, it has turned out that many of the applications of technology to the natural world have ended up wreaking uncontrollable havoc, such as in the dust bowl of North America, or the Aral basin in central Asia, and hence palpably degrading security.

Furthermore, there is a vastly ramified web of of indirect ways in which technology influences security both positively and negatively. For example, the technologies of printing and papermaking (and of the motors which provide the power to drive them) can indeed be used to disseminate elevating and beautiful pictures and words, but can also be—and indeed are—used to disseminate terrorist manifestos and tasteless or even immoral literature. And the inextricable connexions between technology and commerce have not been favourable to the predomin/.ance of beneficial applications of technology. This was already perceived, and eloquently expressed, over a century ago when industrial development was in full spate:[28]

[28] A. Weir, The Historical Basis of Modern Europe, pp. 393–394. London: Swan, Sonnenschein, Lowrey (1886).

> Boulton, who made of Watt's invention a commercial success, possessed great business abilities, but they ministered solely to a delight in industrial generalship. Roebuck, who at an earlier period was the guardian of the embryo engine, also essayed to organize industry on a large scale, but his dominant motive was the practical application of scientific knowledge ... [the engineers] spent their whole energy on devising and superintending the removal of physical obstacles to society's welfare and development ... the thought of making man's dwelling place more commodious cast into insignificance anticipations of personal enrichment.
>
> But the system assumed a new character as soon as it was made available for the general public. The elevation of society was lost sight of in a feverish desire to acquire money. Beneficial undertakings had been proved profitable; and it was now assumed that a business, so long as it was profitable, did not require to be proved beneficial. The sophism suited vulgar inclinations, and its authority became a principal force in the social dynamics of modern Europe.

Therefore, we must approach the subject of technology's rôle in human development with very mixed feelings. They can be no ground for supposing that the application of technology is inevitably and irrefutably beneficial; the best we can realistically hope for is that the negative effects that seem to inevitably follow in the train of any otherwise admirable advance do not exceed in magnitude the positive ones.

Modern technology is above all exemplified by the information revolution (information technology, IT), also referred to as the "digital era". Digitization means representing (encoding) everything by a finite number of states—and thus ultimately by a list of binary numbers (that is, zeros and ones). While this digital representation is very necessary for ensuring the faithful transmission of the numbers in an electronic (or other) circuit—in other words, "zero" in turn does not have to be represented by exactly zero volts, but any voltage up to a certain threshold, which might be 0.5 V, will do; and similarly "one" can be represented by any voltage within a range from one to two volts—and hence what might in practice be the incessant transmission of information (such as trillions of additions carried out by a computer) can be accomplished without the accumulation of errors, the 'finitization' of potentially infinite nuance is an immense impoverishment. Fundamentally, the cardinality of the continuum is greater than that of the natural numbers (integers). The continuum, forming the canvas for all human invention, is necessarily infinite, and is therefore qualitatively different from the finite set of integers that constitutes any digital representation. Therefore, it was no surprise that early digital recordings of music on compact discs (CDs) could be clearly aurally distinguished from analogue (continuum-based) recordings on gramophone records—curiously enough the latter, despite the presence of some noise, always seemed more authentic to the listener. Similarly, television "digitized" (discretized) images into 405 lines, introducing a distortion that would have been perceived, on purely aethetic grounds, as being beyond all reasonable bounds of acceptability had it not been for the amazement attending the technical feat of transmitting moving pictures.[29] Figure 14.2 gives an example of the distortion of the original

[29] 405 lines (the Marconi-EMI system) was the standard introduced in 1936 in the UK, which

introduced by course-grained digitization.

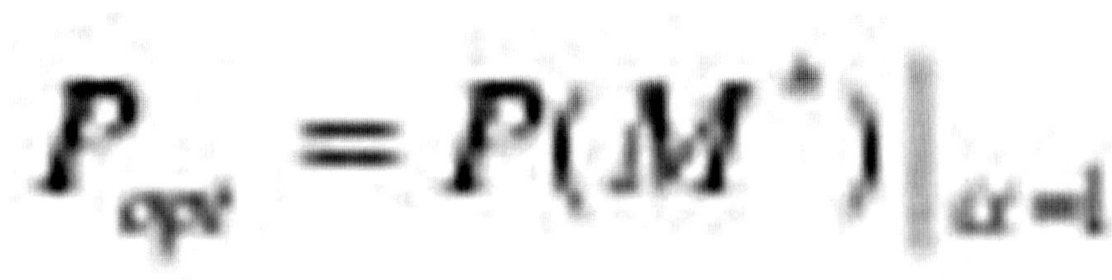

Figure 14.2: Equation (18.3) in Chapter 18 as originally supplied by the author as a bitmap.

Digitization—defined as the representation of something using a finite set of states—is now found everywhere, from the requirement when sending something by courier to choose the description of the contents from a predefined list (no member of which ever seems to correspond exactly to what one is sending!), to the very widely used questions with multiple-choice answers for examinations, and a finite set of "tick boxes" used in making medical diagnoses. (This last example is particularly inadequate for representing the complexity of any human state.) The vast increase in the information that is collected and must be processed necessitates the use of a high speed digital computer,[30] hence digital representation (for the reason mentioned above), but the representation itself, and hence even more so the results of the processing, often sufficiently distorts what is represented so as to vitiate the whole purpose of gathering the data and processing it (e.g., for carrying out a medical diagnosis). It is ironical indeed that just as nanotechnologies now permit us to look at the world with unprecedentedly fine resolution, we are representing it with increasingly coarser resolution. It is perhaps also symbolic of our digital age that a musical instrument like the theremin, capable of representing a continuum of pitch, is now scarcely heard any more.

Furthermore, there are many ways in which digitization impinges adversely on security. On the one hand it is now possible to store, access, and process (using high-speed digital computers) almost unimaginably vast quantities of (digitally encoded) data about individual persons. On the other hand, even with this power, it is necessary to adopt a severely coarse-grained representation of each person (for example, is/is not a terrorist—but as is clear from Chapter 3, this will completely fail to detect the passive supporter) to keep the amount of information within the limits of what can be handled in practice. In most cases, this coarse-grained representation will be such a severe distortion that the whole exercise merely clogs up society, irritates all those whose plans are subject to friction and delay as a result, and achieves very little if any increase in safety.

pioneered television broadcasting. When France later (1948) introduced her own system, significantly better, indeed for many purposes acceptable, image quality was achieved with Barthélémy's 819 lines. It persisted until 1967. Alas, the more lines there are, the more expensive the technology, and when the time came (1964) to unify the different national systems under the umbrella of an international standard, a lower, cheaper resolution of 625 lines was selected. Perhaps it was already realized then that the aesthetic value (content) of broadcasts would in future scarcely merit anything better!

[30] In German, elektronische Datenverarbeitung (EDV).

Moreover, as Wirth has pointed out,[31] "it is as a rule impossible to ascertain whether a digital storage symbol has been manipulated, unlike with paper-based storage, in which some clues generally remain." Given that, due to the general loss of *Ortsinn*, and general globalization, documentary evidence of the persona is more than ever relied upon, rather than personal knowledge by the official, as is possible in a city state, the impact of undetectable manipulation of this evidence will inevitably become increasingly problematical.

In one sense at least, however, technology does appear to exert an unambiguously beneficial effect, namely in quickening the working of man's mind when he is involved with technology. Despite the fascination and complexity of nature, the inner mechanisms of which far transcend the intricacy and complexity of man-made technological artefacts, there is no doubt that the engineer has a quicker intellect than the shepherd; indeed most agriculturists have become, if not engineers, at least users of engines and they seem to be thoroughly imbued with their allure. Furthermore, the industrialization of England led to a great flowering of institutes dedicated to the improvement of the mind, the likes of which had not previously existed in rural communities. Besides, we are all inextricably associated with our pervasive technology. It is not a meaningful option to cut oneself off from it, Thoreau-like, and live in the woods; as part of human society we are all partakers and accomplices in the development of even more technology, and though individually we may make more or less use of it, we could not survive without it.

[31] W. Wirth, The end of the scientific manuscript? J. Biol. Phys. Chem. 2 (2002) 67–71.

Complexity and Security
J.J. Ramsden and P.J. Kervalishvili (Eds.)
IOS Press, 2008

Chapter 15

The problems of protecting people in underground structures from terrorist explosions

E. Mataradze,[a] T. Krauthammer,[b] N. Chikhradze,[a] E. Chagelishvili[a] and P. Jokhadze[c]

[a] *G. Tsulukidze Mining Institute, Tbilisi, Georgia*
[b] *Centre for Infrastructure Protection and Physical Security (CIPPS), University of Florida, USA*
[c] *Georgian Technical University, Tbilisi*

Abstract. This chapter examines the methodological framework for designing facilities for the protection of people and underground facilities from internal explosions. The basic requirements to be fulfilled by protective systems, considering the characteristics of shock wave propagation in the limited space of underground structures, are established. A structure for an automatic protective system is proposed with a module for identification of the explosion and a blast energy absorber at its centre. Research carried out by the CIPPS (University of Florida) and the Tsulukidze Mining Institute (Georgia) was aimed at increasing the reliability of blast identification in tunnels and developing design parameters for a shock absorber.

15.1 Introduction

The analysis of acts of terrorism shows that the illegal use of force against people is neither spontaneous nor impulsive. Such acts are planned in advance and are targeted at civilians and civil (non-military) sites. Besides, terrorists aspire to achieve the maximum destructive effect by using the explosive charges available to them. This makes underground structures one of the most attractive targets for terrorists.

The growing development of underground space, typical of the ongoing urbanization in this century, raises the topicality of the maintenance of a high level of safety in underground structures. This problem involves the task of the minimization of the consequences of accidental and terrorist explosions in subway systems, transport tunnels, underground explosive storage facilities and other underground sites.

Terrorist attacks on underground structures are carried out both through surface and internal explosions. Reliable protection from surface explosions can be achieved through the proper selection of the design depth of the underground structures and the support parameters of tunnels taking account of the expected dynamic load upon explosion. The design of protection systems for underground structures to cope with internal explosions raises harder challenges.

Underground structures and long corridors with limited cross-section are most vulnerable to internal terrorist explosions because of blast confinement in the limited space. Anti-terrorist preventive actions, such as the inspection and control of the places where people gather with the aim of detection and destruction of explosives, will not always produce desirable results. The notorious terrorist explosions in the London underground railway system (in July 2005, when 56 people died and over 700 received injuries), and in Moscow (in February 2004, when 39 died and 134 people were injured; in August 2004, 49 died and over 300 people were injured) point to the urgency of developing effective systems for protecting people.

A large number of scientific centres are currently involved in comprehensive research aimed at developing a means of localization of accidental and terrorist blast effects in underground structures. The research focuses on two main tasks. The first implies elaboration of methods for reducing the blast effect by using solid or perforated concrete barricades, and absorbent materials [1,2]. Despite their effectiveness, such passive methods fail to protect people in close proximity to the point of explosion in relatively dangerous blast-affected zones. The second, growing, research effort is aimed at developing automated systems of protection, which are activated after the explosion and discharge of various agents in the protected zone.

The first studies aimed at the development of an automated system for blast energy localization in underground openings were undertaken in the 1950s to address the growing threat of methane explosions in coal mines. The earliest protective systems, applied in mines located in several countries (e.g. UK, Germany, Russia, etc.), had a very simple design, and consisted of open, wooden or plastic containers that were filled with 40–50 litres of water, placed on shelves attached to walls in the underground structure. The walls were equipped with a rope mechanism for overturning the containers that was operated by the explosion of an electric detonator. The latter was activated by a signal from a photo-electric sensor that detected the methane explosion. A water and/or dust barrier was thus formed in the underground structure. Experience has shown that passive devices of this type failed to effectively localize the blast.

At present, most of the coal-producing countries apply more advanced protective systems, which unlike the earlier ones are provided with a mechanism for the discharge of water or the distribution of blast absorbing powder [3]. Systems of this kind have been developed and applied in various mines; e.g.

in the Ukraine (AVP-1), Germany (BVS), UK (Graviner), Russia (ACVP-LV), the USA, etc. Intensive research has continued in recent years to develop a system for protection from dust explosions in industrial sites and plants producing chemicals, plastics, textiles, pulp and paper, pharmaceuticals, and carrying out milling operations [4].

The existing automated blast protection systems used in coal mines and other industrial sites have similar designs. They contain a blast identification module (usually an optical sensor and an electric device generating a trigger signal), and a blast energy absorber that contains a blast-suppressing agent dispenser and a device to eject the material into the protected media. Analysis has revealed the following major disadvantages of the existing blast suppression systems:

- Lack of reliability of the effectiveness of a blast identification device in complex underground openings, especially during long-term operation;
- Low speed of the blast energy absorber activation (the time of formation of blast-absorbing media by these systems is about 50–150 ms; i.e. the protective systems fail to absorb blast pressure at a distance of 30–90 m from the blast site if the propagation velocity of the shock wave has a typical value of 600 m/s. Faster blast waves will increase these distances;
- Inadequate discharge of a blast-absorbing agent, required for reducing the excess pressure and temperature to an acceptable value, as well as the inability to meet the geometric constrains of an underground structure.

15.2 Structure of a protective system

The goal of the research is to design a new type of system for protection against unpredictable explosions. The functions of the protective system are as follows:

- Reduction of excessive pressure on the shock wave front;
- Protection from gas and dust resulting from the explosion;
- Prevention of fire after an explosion;
- Immediately transmit information regarding the place and time of the explosion to rescue services.

The protective system consists of a blast identification module and a blast energy absorber for shock wave mitigation. The identification module includes sensors and microprocessors equipped with the software for blast identification (blast vs non-blast signals) installed in the protected zone of a tunnel (Figure 15.1).

Effective operation requires the following conditions to be met: (i) the system shall operate and provide protective media before the arrival of a shock wave to the specified zone of the underground structure; and (ii) the protective barrier shall ensure the reduction of the over-pressure ΔP_1 to an acceptable value ΔP_2 (Figure 15.2).

Unfortunately, the nature of terrorist attacks does not allow one to devise a protective system that would fully exclude the development of unacceptable

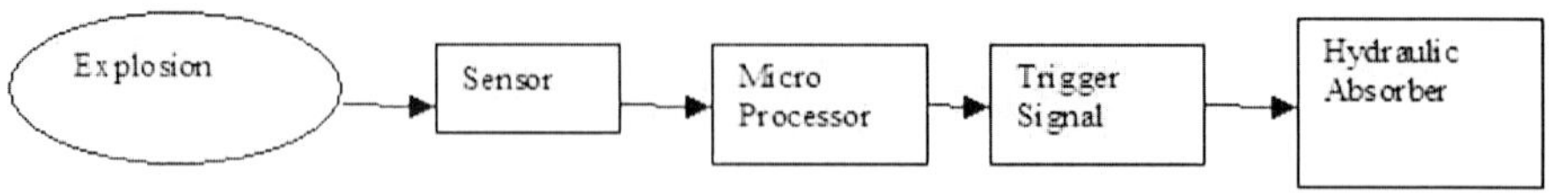

Figure 15.1: Parameters for designing a protective system.

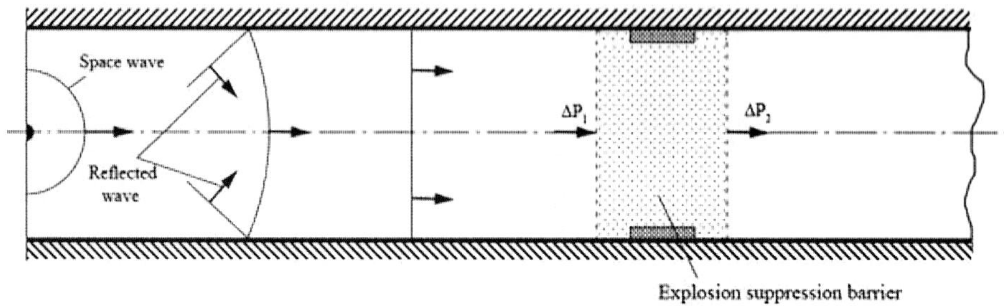

Figure 15.2: Principle of operation of a protective device.

overpressures. However, the proper selection of parameters for a protective system may effectively restrict the area of dangerous overpressure propagation and hereby enhance the safety of people and equipment in the underground structure.

15.3 Parameters of influence

The parameters determining the working efficiency of the protective system can be divided into two groups:

1. Parameters of external impact, i.e. characteristics of the effect of blasts in underground structures;

2. Internal parameters of the protective system, namely its time, energy absorption and design characteristics.

The ability of the system to respond in time and adequately to the external impact determines the protection effectiveness. The main parameters of the external impact are as follows: overpressure on the shock wave front, the positive phase duration of the shock wave, the impulse and shock wave velocity. The protective system has the following parameters: time characteristics (the time of activation from the moment of detonation and the duration of detonation); and energy absorption characteristics (amount and speed of the discharge of the blast suppression agent) (Figure 15.3). The protective system parameters are determined according to the preliminarily established parameters of external impact. It is also important to ensure that the system is impact-resistant and that it does not impede the operation of the tunnel.

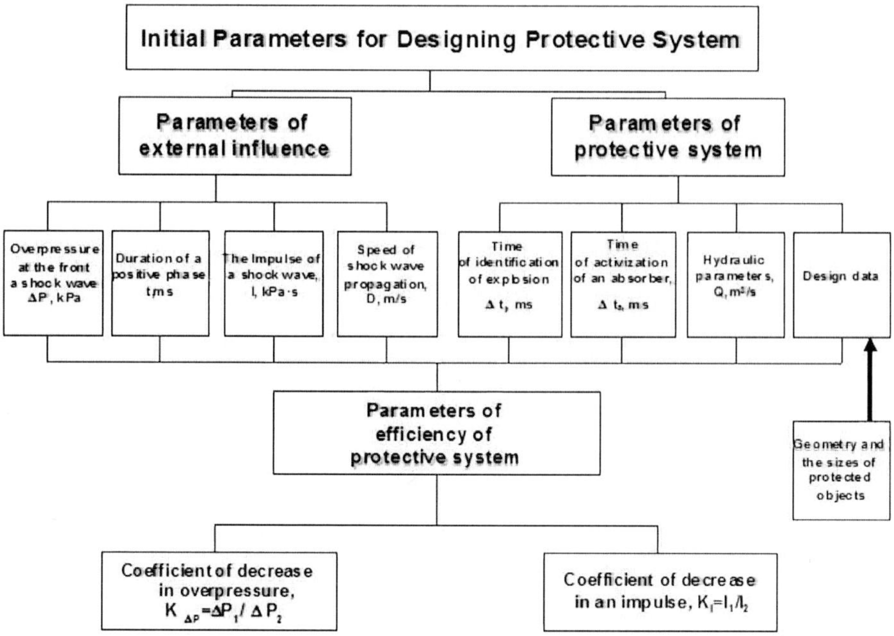

Figure 15.3: Parameters for designing a protective system.

15.4 External impact parameters

The analysis of blast effects in underground structures shows that shock overpressure, positive phase duration, and impulse and shock wave velocities depend on charge weight q, distance from the charge R, cross-sectional area of the underground opening S, and ultimately on the normalized charge weight $q/(RS)$. An analysis of the parameters of a shock wave in an underground structure due to a terrorist explosion is presented in Table 15.1 [5].

Various studies and regulations provide different assessments of the threat to humans from various blast overpressures, as shown in Table 15.2.

15.5 Main parameters of the protective system

Time parameters of the system. Synchronization of the operation of the hydraulic absorber with the time of the shock wave arrival is achieved when:

$$R/D \leq \Delta t \tag{15.1}$$

where R is the distance of the charge from the absorber, D is the shock wave propagation velocity and Δt is the elapsed time from the moment of detonation to the moment the protective system is activated, i.e.

$$\Delta t = \Delta t_{\mathrm{id}} + \Delta t_{\mathrm{v}} + \Delta t_{\mathrm{c}} \tag{15.2}$$

where Δt_{id} is the time required for blast identification, Δt_{v} is the time required for pyrotechnic valve activation, and Δt_{c} is the time required for water curtain formation, depending on the water discharge rate.

Table 15.1: Blast effect characteristics.[a]

		t					
		R/m					
$q/(RS)$	DP	10	20	D	v	T	ρ
0.01	110	0.2	3.8	470	180	365	2.10
0.05	390	1.3	4.9	700	445	500	3.35
0.10	570	18.4	55.3	805	552	585	3.90
0.15	790	19.7	59.2	930	675	690	4.40
0.20	1000	20.6	62.0	1040	772	787	4.89
kg m^{-3}	kPa	ms	ms	m s^{-1}	m s^{-1}	K	kg m^{-3}

[a] DP is shock overpressure, t is the positive phase duration, v is the speed of movement of the compressed air layer, T is the temperature at the shock wave front, and ρ is the air density at the shock wave front. Units are given in the bottom row.

Table 15.2: Assessment of the threat to humans from various blast overpressure levels.[a]

Excess pressure/kPa					
190	69 ... 76	55	24	16	8.3 ... 5.9
Lethal outcome	Lethal outcome or serious injury	Lethal outcome or serious injury of ears and lungs	10% probability of injury of ears and lungs	Lethal outcome or serious injuries less likely	No lethal outcome nor serious injury
400	120	65	35	13	10
Probability of injury (%)					
100	75	50	35	5	0

[a] Sources: upper part: DOD 5154.45 [6]; lower part: [7].

It is obvious that the effectiveness of a protective system can be achieved for low values of time duration. The latter should not exceed several milliseconds (e.g. under the shock wave propagation velocity $D = 700$ m/s, the localisation of the harmful effect of an explosion at a distance up to 15 m can be achieved when $\Delta t = 21.4$ ms). Therefore, the main task of the current research is to minimize the intervals Δt_{id}, Δt_{v} and Δt_{c} to ensure the reliability of the protective system effectiveness. Another important time parameter of the system is the system operation duration after its activation. The effectiveness of the system is achieved when the duration of the system of activation $\geq \tau$, where τ is the time of arrival of the shock wave.

15.5.1 Energy absorption parameters of the system

The major parameter is to discharge a sufficient volume of energy absorption agent for the formation of a shock wave absorption curtain in a few ms. The number N_{v} of simultaneously operating valves in a system can be determined based on the rate of absorption agent discharge per unit area of cross section in an underground structure, and expressed as:

$$N_{\text{v}} = \sum Q/Q_{\text{v}} = Q_{\text{u}}S/Q_{\text{v}} \,, \tag{15.3}$$

where $\sum Q$ is the total volume of water required for the formation of an energy absorption curtain, Q_{v} is the volume of energy absorption agent discharged from one valve, Q_{u} is the absorption agent discharge volume per unit area, determined according to the blast parameters, and S is the cross sectional area of the underground opening.

15.6 Methods of identification of blasts in underground openings

There are several methods that can be used for the identification of blasts, based on the analysis of seismic and electromagnetic radiation generated by blasts, as presented in Table 15.3 [8].

The present authors have conducted experiments in the underground test opening of the Tsulukidze Mining Institute to select blast identification methods, while addressing the parameters of the protective system. During the experiments, electromagnetic pulses (EMP) and optical signals, generated during the explosion of condensed charges in tunnels, were recorded simultaneously. The results were used for establishing blast identification criteria required for the development of the identification software [9].

15.7 Hydraulic shock energy absorber with a pyrotechnic element

A wide range of analytical and experimental studies have been conducted recently to determine the interaction of water barriers and shock waves [10–13]. The results of these research works have shown that wave mitigation can be achieved by forming a highly dispersed (droplets) liquid in a zone of shock wave

Table 15.3: Comparative analysis of blast identification methods.

Methods	Advantages	Disadvantages
Seismic	1. Allows identification at large distances Allows identification in underground openings with irregular shapes 2. No need for precision orientation of sensors	1. Data multiprocessing is time-consuming 2. Needs preliminary determination of wave conductor characteristics 3. Explosion conditions affect the identification reliability 4. Affected by temperature
Electromagnetic pulse (EMP)	1. Less processing required for identification 2. Provides for fast identification 3. Allows combination with optical measurements	1. Has spurious electromagnetic effects 2. Signal attenuates with distance 3. Needs precision orientation of sensors
Optical	1. Requires less processing for identification 2. Allows fast identification 3. Allows combination with EMP measurements	1. Pre-explosion conditions affect identification reliability 2. Has spurious optical effects 3. Needs precision orientation of sensors
Combined	1. Identification reliability	1. Complexity of system 2. High cost

propagation. The following are the basic design parameters of the hydraulic shock energy absorber: droplet diameter, spray velocity and volume of water sprayed upon the activation of the absorber, as well as the amount of time between the operation of the absorber and the arrival of the starting signal.

Analyses of the experiments reported by [10] and [11] have shown that the most effective explosion-mitigating water spray systems are those generating either droplets less than 10 μm in diameter or in the range 200 μm–1 mm. Various approaches have been adopted to determine the necessary volume of sprayed water. For instance, this value is determined according to the weight of the explosive charge [13] or the area of the cross section of the underground opening under protection [1]. Until recently, no truly justified calculating method has been proposed to establish the amount of water that would simultaneously take into account the anticipated charge of the explosive and the lateral dimension of the underground structure. Further research is needed to find a proper solution to this complex problem. The spray velocity for the absorber with a pyrotechnic element is established by taking into account the pressure developed upon activation of a gas generator and the design of the spray elements of the absorber. The time of operation depends on the type of electrical initiator and the characteristics of the gas-generating material.

The hydraulic energy absorber consists of a main pipeline and separate segments and/or branches of pipelines, adapted to the specific geometry of the protected structures, and installed at the appropriate intervals. The segments are equipped with pyrotechnic fast acting valves with spray-forming elements that, at the command of the trigger signal, can produce tailored water barriers along the whole perimeter of specific segments of the structure (Figure 15.4).

The design of the absorber allows one to select appropriate hydraulic and time-operational parameters that will enable the simultaneous activation of the outlet valves so that the protective system can generate stable water-dispersing media and thus ensure effective blast energy suppression. Benchmark testing of the operating element of an absorber with a pyrotechnic device and membrane valves has been conducted in the experimental station of the Tsulukidze Mining Institute. The operating element of the absorber with a pyrotechnic device and membrane valves had the following properties: internal pipe diameter, 200 mm; length, 1.0 m; diaphragm sizes, 0.03×0.8 m. Propellant RST-4K, pyroxylin blasting powders NBT-3 12/TP, 07/32 and 4/1, and blasting powder "Sokol" were used as gas-generating materials in the tests.

The hydraulic energy absorber consists of a main pipeline and separate segments and/or branches of pipelines, adapted to the specific geometry of the protected structures, and installed at appropriate intervals. The segments are equipped with pyrotechnic fast-acting valves with spray-forming elements that, at the command of the trigger signal, can produce tailored water barriers along the whole perimeter of specific segments of the structure (Figure 15.4).

The design of an absorber allows the selection of appropriate hydraulic and time-operational parameters that will enable the simultaneous activation of the outlet valves so that the protective system will generate stable water-dispersing media to ensure effective blast energy suppression. The outcome of the tests have revealed that the membrane valve operated 1.5–2.0 ms after the trigger. The proposed design has additional advantages, such as the ability to more effectively control the water ejection velocity and volume per unit underground opening cross sectional area.

Figure 15.4: Layout of the location of a blast absorber in an underground opening.

15.8 Conclusions

- Due to their sluggish time characteristics, existing blast protection systems made of a blast identifier and an absorber fail to respond adequately under current operational requirements and are, therefore, unreliable for the protection of people and underground facilities from internal terrorist explosions;
- The reliability of blast identification in underground structures can be achieved by means of the application of a combined technique implying simultaneous recording of electromagnetic impulses and of blast-generated optical signals;
- Effective operation of a blast energy absorber is achieved by the use of a pyrotechnic device and membrane valves, enabling the operation of the system within 1.5–2.0 ms from the trigger signal, and providing effective control of the energy absorption agent ejection velocity and volume per unit underground opening cross section area.

Acknowledgment. This research was carried out with the financial support of the NATO Science for Peace (SfP) Programme.

15.9 References

[1] Gurin, A., Malyi, P. and Savenko, S. Shock Waves in Underground Openings. Moscow: Nedra (1983) (in Russian).

[2] Berger, S., Sadot, O., Melamud, G., Anteby, I., Kivity, Y., Britan, A. and Ben-Dor, G. Attenuation of shockwaves by barriers in tunnels and corridor-type structures. Proc. 2nd Intl Conf. on Design and Analysis of Protective Structures, pp. 71–77. 2006, Singapore.

[3] Jigrin, A. and Gorlov, A. Protection of underground openings from the explosion of methane and/or coal dust, Vzryvnoe Delo, No 95/52 (2005) pp. 115–120 (in Russian).

[4] Hattwing, M. and Steen, H. (eds). Handbook of Explosion Prevention and Protection. Wiley VCH (2004).

[5] Mataradze, E., Krauthammer, T., Chagelishvili, E. and Chikhradze, N. System for protecting people and underground facilities from internal terrorist explosions. Proc. 2nd Intl Conf. on Design and Analysis of Protective Structures, pp. 185–194. 2006, Singapore.

[6] Adushkin, V., Kogarko, S. and Lyamin, A.G. Calculation of safe distances under blasts in the atmosphere. Vzryvnoe Delo, No 75/32 (1975) pp. 82–94 (in Russian).

[7] Baker W. E., Cox P.A., Westine P.S., Kulesz J.J. and Strehlow R.A. Explosion Hazards and Evaluation. Amsterdam: Elsevier (1983).

[8] Mataradze, E., Chagelishvili, E., Mikhelson, R., Krauthammer, T. and Butskhrikidze G. Identification of explosions in underground structures. Proc. Intl Conf. on Measuring Microsystems for Environmental Monitoring, pp. 132–137. 2004, Tbilisi.

[9] Mataradze, E., Krauthammer T., Chagelishvil E., Chikhradze N., Khoperia Z., Jokhadze P. and Bochikashvili P. Frequency analysis of electromagnetic radiation from small explosive charges in tunnels. Proc. 2nd Intl Conf. on Design and Analysis of Protective Structures, pp. 195–199. 2006, Singapore.

[10] Thomas, G. On the conditions required for explosion mitigation by water sprays. Trans. Inst. Chem. Engng B 78 (2000) 339–354.

[11] Bjerketvedt, D. and Bjorkhaug, M. Experimental Investigation of the Effect of Waterspray on Gas Explosions. Christian Michelsen Institute Report for the UK Department of Energy, OTH 90 316. London: HMSO (1991).

[12] Van Wingerden, K. Mitigation of gas explosions using a water deluge. Process Safety Progress 19 (2007) 173–178.

[13] Chor Boon Ng and Choung, K. Large scale water mitigation tests in Sweden. Proc. 2nd Intl Conf. on Design and Analysis of Protective Structures, pp. 147–156. 2006, Singapore

Complexity and Security
J.J. Ramsden and P.J. Kervalishvili (Eds.)
IOS Press, 2008

Chapter 16

Degradation of anthropogenic contaminants by higher plants

G. Kvesitadze[a] and E. Kvesitadze[b]

[a] *Durmishidze Institute of Biochemistry and Biotechnology*
[b] *Georgian Technical University, Tbilisi*

16.1 Introduction

The elimination of contaminants from the environment by micro-organisms of different taxonomic groups is a property developed through evolution, which constitutes an increasingly important element of the ecosystem of the Earth. Until recently, plants, which still occupy about 40% of the world's land area, were considered as organisms having only a limited potential for contaminant sequestration within cell organelles. Analysis of experimental data over the last two decades has revealed the great ecological potential of plants and their nature as a complex cleaning system. Functional processes in higher plants can lead to the complete detoxification of anthropogenic contaminants; the enzymes of the plant carry out oxidation and conjugation; and some plant varieties have the ability to accumulate huge amounts of heavy metals and deposit them in cellular structures. There remain uncertainties related to the multistage degradation process of the contaminants, but we highlight the interchangeability of enzymes participating in oxidative degradation of organic contaminants in higher plants, and stress the importance of the phenoloxidase enzyme, hitherto unknown, in the remediation processes.

The human contribution is now comparable to natural contaminants, such as the emission of poisonous gases during a volcanic eruption and earthquakes, swamps, and the synthesis of toxic compounds by lower (micro) organisms and higher plants. As a result of urbanization, the growth of industry and transport, production of chemicals for agriculture, military activities, etc. the concentration of anthropogenic toxicants entering the environment, especially in some regions, exceeds recognized permissible standards. In spite of difficulties in quantitative, as well as in qualitative estimation, the anthropogenic contribution is measured in milliards of tons annually (see also Chapter 10). Most dangerous among these contaminants are those with greatest persistence, bio-accumulation, and toxicity. The huge amounts of these hazardous substances, or toxic intermediates of their incomplete transformations, are accumulated in different niches of the biosphere, significantly affecting the ecological balance. Recently ecological technologies aimed at minimizing the flow of toxic compounds have been developed (Tsao, 2003; Kvesitadze, 2006), but the flow of toxic compounds into the biosphere is still increasing. The international character of this problem is determined by such factors as global migration of contaminants (migration between soil, air and water; geographical; biotic; etc.).

Nevertheless, the plant kingdom assimilates toxic compounds, removing them from the environment, naturally providing long-term protection and preventing their environmental dispersal, and represents the main natural agent for the maintenance of the ecological balance.

Nature's antipollutant arsenal includes about 400 species of plant worthy of especial attention, divided into two categories: hyperaccumulators, which absorb such toxic metals as arsenic, lead and mercury; and root-level killers, which break down organic pollutants such as petroleum below the ground (see also Chapter 8). Evolution likely favoured these traits: one discourages predators by dosing them with lethal metals; the other allows flora to adapt to harsh soil.

The first phytoremediation study was conducted in 1948, when Italian scientists noticed dense accumulations of nickel in the alyssum vine. Uses date even further back. Miners in the 19th century found ore deposits by looking for fields thick with the copper-loving mustard plant. Now, in a neat demonstration of ecoefficiency, some green plants are being smelted to recapture ores trapped in their tissues.

The international character of this problem is determined by such factors as global migration of contaminants (migration between soil, air and water, geographical, biotic, etc.). Nevertheless, the plant kingdom assimilates toxic compounds, removing them from the environment, naturally providing long-term protection and preventing their environmental dispersal, and represents the main natural agent for the maintenance of the ecological balance.

The seriousness of the problem of contamination remediation is now recognised through the emergence of the environmental lobby associated with the concerns on climate change. The contribution of phytoremediation is perceived to be growing within the complex system comprising the ecology of the planet. Forbes magazine (Helman, 2001) has reported that "The concept of using trees and plants to purify polluted land, a process called phytoremediation, is beginning to take root in America's USD 30 billion-a-year remediation industry. Cleaning the U.S.'s 217,000 polluted sites with traditional techniques would cost an estimated USD 187 billion and require more landfills. Plants can do it with less labor, noise and waste. The cost savings can be tremendous. Mechanical pump-and-treat remedies to keep underground petroleum plumes or toxic leachate from seeping into groundwater cost an annual USD 1.7 million per hectare. Trees can do it for USD 500,000."

16.2 Plants and remediation pathways

Plants are now recognized as an important ecological tool and in order to properly evaluate their detoxification potential. The following ecobiological characteristics apply:

- Higher plants simultaneously contact three main ecological niches: soil, water and air.
- Well-developed root systems of higher plants determine the soil-plant-microbial interaction, representing a unique integrated process, significantly affecting the overall plant metabolism.
- The large assimilating surface area of plant leaves (adaxial and abaxial) significantly exceed the aboveground surface beneath the plant, and permit the absorption of contaminants in large quantities from the air via the cuticle and stomata.
- The unique internal transportation system operates in both directions, distributing all the penetrated compounds throughout the entire plant.
- The autonomous synthesis of vitally important organics generates extra energy during the prolonged remediation process.
- Enzymes exist for catalysing oxidation, reduction, hydrolysis, conjugation and other reactions of the multistage detoxification process.
- There is large intracellular space to deposit heavy metals and conjugates.

In order to penetrate into a leaf, the xenobiotic (contaminant) should pass through the stomata, or traverse the epidermis, which is covered by a film-like wax cuticle. Generally, stomata are located on the lower (abaxial) side of a leaf, and the cuticular layer is thicker on the upper (adaxial) side. Gases and liquids penetrate through the stomata into the leaves. The permeability for gases depends on the degree of opening of the stomatal apertures (4–10 nm), and for liquids, on moistening of the leaf surface (which depends in turn on the surface tension of the liquid and the morphology of the stomata). The majority of toxic compounds penetrate into a leaf as solutions (pesticides, liquid aerosols, etc.). It was established for the leaves of zebrine (*Zebrina purpusii*), that the crossover surface tension of their lower surface is 25–30 dyne/cm (for comparison: the surface pressure of water is 72.5 dyne/cm, and for ethanol 22 dyne/cm) (Schonherr, 1972). Liquids with a surface tension less than 30 dyne/cm have a constant angle of contact with the surface of a leaf and instantly penetrate into the stomata. Liquids with surface tension greater than 30 dyne/cm penetrate into the stomata without moistening the leaf surface.

The pathways of lipophilic organic contaminant penetration were shown by the absorption of gaseous hydrocarbons in hypostomatic leaves (Ugrekhelidze, 1976). The leaves of the field maple (*Acer campestre*), wild Caucasian pear (*Pyrus caucasica*), vine (*Vitis vinifera*) and narrow-leaved oleaster (*Elaeagnus angustifolia*) were placed in an atmosphere containing ^{14}C-methane or [1,6-^{14}C]-benzene, and contact with labelled hydrocarbon occurred only from one side of the leaf. The total radioactivity of the non-volatile metabolites formed showed that the absorption of gaseous alkanes and vapours of aromatic hydrocarbons is processed not only through the stomata (most preferred), but also through the cuticle. Similar results have been reported for a number of herbicides (α-naphthylacetic acid, 2,4-dichlorophenoxyacetic acid(2,4-D), picloram and derivatives of urea), applied in soluble form to leaves (Sargent, 1972; Sharma, 1970). The abaxial side of a leaf, rich in stomata, absorbs the organic substances more intensively than the adaxial side. These results demonstrate the active participation of stomata during the absorption of toxic compounds, see Figure 16.1 (Ugrekhelidze, 1976).

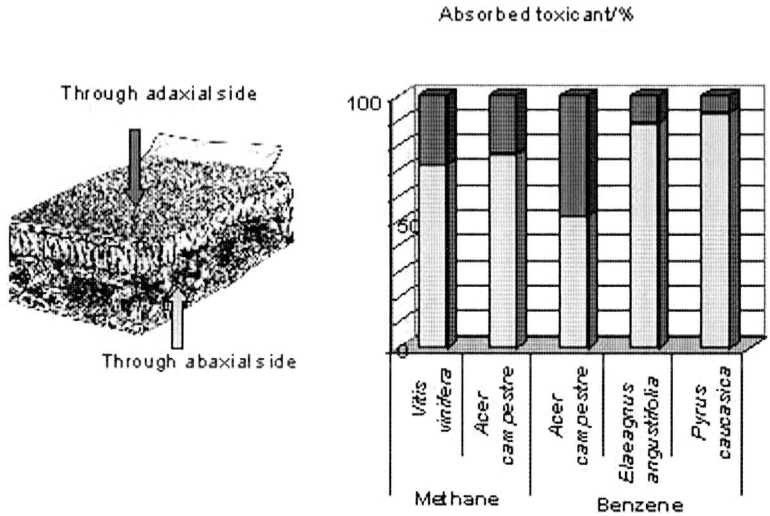

Figure 16.1: Absorption of ^{14}C-methane (specific radioactivity 1 mCi/mL) and [1-6-^{14}C] benzene (specific radioactivity 4.9 mCi/mL) by the hypostomatic leaves of plants. Concentration of methane in the air is equal to 1.5% by volume; 8 h exposure under illumination. Concentration of benzene in the air is equal to 2 mg/L, 4 h exposure in darkness.

The contaminant penetration into the roots essentially differs from the leaves. Substances pass into roots only through cuticle-free unsuberized cell walls. Therefore, roots absorb substances much less selectively than leaves. Roots absorb environmental contaminants in two phases: in the first fast phase, substances diffuse from the surrounding medium into the root; in the second they gradually distribute themselves and accumulate in the tissues. The intensity of the contaminant absorption process depends on the contaminant solubility, molecular mass, concentration, polarity, pH, temperature, soil humidity, etc. (Kvesitadze, 2006; Korte, 2000).

There is now experimental data demonstrating that plants are able to activate a definite set of biochemical and physiological processes to resist the toxic action of contaminants by the following mechanisms:

- Excretion
- Conjugation of contaminants with intracellular compounds and further compartmentalization of the conjugates into cellular structures

- Decomposition of environmental contaminants to standard cell metabolites or their mineralization. Commonly, plants gradually degrade organic contaminants to avoid their toxic action on their cells. Different plants differ by up to four orders of magnitude in the contaminants assimilating potential, suggesting a classification as strong, average and weak assimilators of different structure contaminants. For instance the most active assimilators uptake nearly 10 mg of benzene per 1 kg of fresh biomass per day, while the assimilation potential of the weak absorbers is measured in hundredths of mg (Ugrekhelidze, 1997).

The fate of a contaminant depends on its chemical nature, external temperature, variety of the plant(s) and phase of vegetation, etc. The simplest pathway through the plant is excretion. The essence of excretion is that the toxic molecule undergoes no chemical transformation, and being translocated through the apoplast, is excreted from the plant. This pathway of xenobiotic (contaminant) elimination is rather rare and takes place at high concentrations of highly mobile (phloem-mobile or ambi-mobile) xenobiotics.

In the great majority of cases, contaminants absorbed by and penetrating into the plant cell undergo enzymatic transformations leading to an increase in their hydrophilicity—aprocess simultaneously accompanied by decreasing toxicity. Successive phases of contaminant transformations are shown (in accordance with Sandermann's (1994) green liver concept) in Figure 16.2.

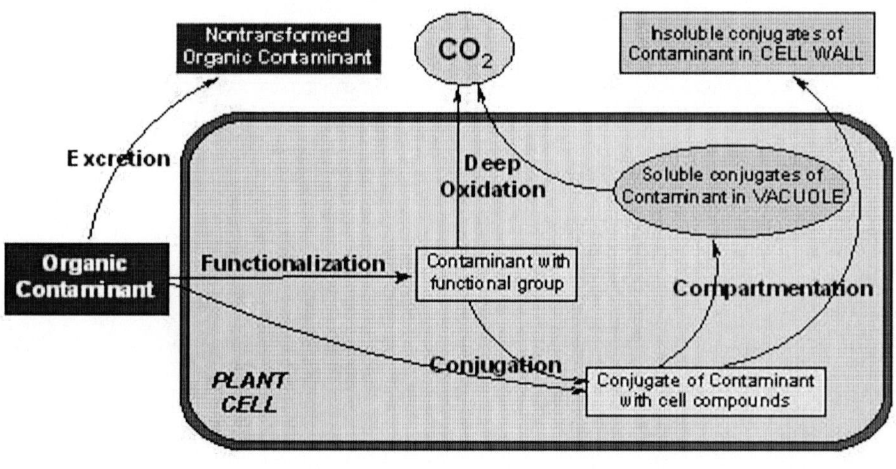

Figure 16.2: The main pathways of organic contaminant transformation in plant cells.

Functionalization is a process whereby a molecule of a hydrophobic organic xenobiotic acquires a hydrophilic functional group (hydroxyl, carboxyl, amino, etc.) as a result of enzymatic oxidation, reduction, hydrolysis, etc. Due to the introduction of the functional group, the polarity and corresponding reactivity of the toxicant molecule is enhanced. This promotes an increase of the intermediate's affinity to enzymes, catalysing further transformation.

Conjugation takes place as a basic process in phytoremediation and consists of the formation of chemical coupling of the contaminant to the endogenous cell compounds (proteins, peptides, amino acids, organic acids, mono-, oligo-, polysaccharides, lignin, etc.), so forming peptide, ether, ester, thioether or other type covalent bonds. Intermediates of the contaminant, initial transformations or the contaminants themselves possessing functional groups capable of reacting with intracellular endogenous compounds, are all susceptible to conjugation.

Commonly, immediately after penetration by the contaminant, the main part of the toxicant molecules undergo conjugation and only a small amount is deeply degraded (0.1–5% depending on structure). Conjugation is a widespread defence mechanism in higher plants, especially in cases when the contaminant penetrates into the plant cell and the concentration of the contaminant exceeds the plant's transformation (decomposition) potential. Increased amounts of deep degradation to regular plant cell metabolites, or CO_2 and water, is achieved in the case of linear, low molecular weight structures of contaminants (Ugrekhelidze, 1976; Kvesitadze, 2006). The toxicity of the conjugates compared to parent compounds is decreased due to binding with non-toxic cellular compounds. Conjugates are kept in the cell for a certain duration without causing visible pathological deviations from cell homeostasis. The conjugate formation also gives the plant cell extra time for the internal mobilization, and the induction of enzymes responsible for further transformation. Relatively quickly after the termination of the incubation of the plant with the contaminant, conjugates are no longer found in the cells. Some attempts have been made by the authors (unpublished data) to estimate different plant (soybean, ryegrass) cells' potential to accumulate conjugated benzene in the case of toxicant saturation. In spite of incomplete information, it was suggested that for genetically unmodified plants, it could be, as a minimum, several molecules of contaminant conjugates per plant cell. Although conjugation is one of the most widely distributed pathways for plant self-defence, it is neither energetically nor physiologically particularly advantageous for the plant process. Firstly, the formation of conjugates leads to the depletion of vitally important cellular compounds, and secondly, unlike deep degradation, the formation of conjugates maintains the contaminant basic molecular structure, and hence results only in a partial and provisional decrease of its toxicity.

Compartmentation in most cases is the final step of conjugate processing, and results in temporary (short or long) storage of the conjugates in defined compartments of the plant cell. Soluble conjugates of toxic compounds (coupled with peptides, sugars, amino acids etc.) are accumulated in the cell structures (primarily in vacuoles), while insoluble conjugates (coupled with lignin, starch, pectin, cellulose or xylan) are moved out of the cell via exocytosis into the apoplast and accumulated in the cell wall (Sandermann, 1994). The compartmentalization process is analogous to mammalian excretion, essentially removing the toxic part from metabolic tissues. The major difference between detoxification in mammals and plants is that plants do not have a special excretion system for the removal of the contaminant conjugates. Hence they use a mechanism of active transport for the removal of the toxic residues away from the vitally important sites of the cell (nuclei, mitochondria, plastids, etc.). This active transport is facilitated and controlled by the ATP-dependent glutathione pump (Martinova, 1993) and is known as "storage excretion" (Coleman, 1997). The pathway of toxic compound processing i.e.,

- functionalization
- conjugation

- compartmentalization, is well illustrated by the processing of anthropogenic contaminants. One such example demonstrating the transformation of organochlorine pesticides is the hydroxylation of 2,4-D followed by conjugation with glucose and malonyl residues and deposition in vacuoles (Sandermann, 1987), see Figure 16.3.

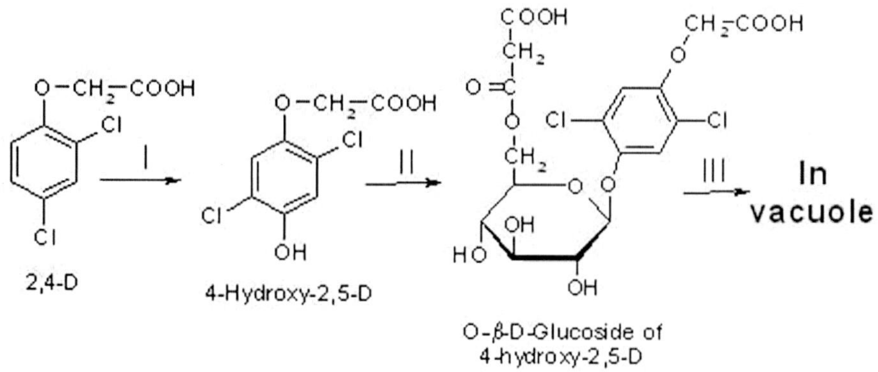

Figure 16.3: How 2,4-D is transformed for deposition in vacuoles.

16.3 The rôle of enzymes

Anthropogenic organic toxicant decomposition processes are closely related to many aspects of higher plant cellular metabolism. In prolonged and multifunctional detoxification processes, quite a few enzymes are actively involved. According to their catalysed reactions they directly or indirectly participate in the detoxification process. Ttransformations of contaminants during functionalization, conjugation and compartmentation are of an enzymatic nature. It is remarkable that due to their unusual flexibility in the absence of xenobiotics, these plant cell enzymes catalyse reactions typical of regular plant cell metabolism. The following enzymes directly participate in the transformation process of anthropogenic contaminants:

- Oxidases, catalysing hydroxylation, dehydrogenation, demethylation and other oxidative reactions (cytochrome P450-containing monoöxygenases, peroxidases, phenoloxidases, ascorbatoxidase, catalase, etc.).

- Reductases, catalysing the reduction of nitro groups (nitroreductase).

- Dehalogenases, splitting halogen atoms from halogenated and polyhalogenated xenobiotics.

- Esterases, hydrolysing ester bonds in pesticides and other organic contaminants. Conjugation reactions of contaminants in the plant cell are catalysed by transferases: glutathione S-transferase (GST), glucuronosyl-O-transferase, malonyl-O-transferase, glucosyl-O-transferase, etc. Compartmentation of contaminant intermediates and transformation-conjugates takes place under the action of the ATP-binding cassette (ABC) transporters (Eckardt, 2001). Depending on the structure of the contaminant, some other enzymes may also be involved in their degradation.

Prolonged cellular decomposition of contaminants involves the participation of enzymes that provide the plant cell with the extra energy needed for the induction of the enzymes, and for the vitally important secondary metabolites. Enzymes involved in these and similar processes indirectly participate in the detoxification of the contaminants. The correlation between the penetration of the organic contaminants (alkenes, aromatic hydrocarbons, polycyclic aromatic hydrocarbons) in the plant cell and the corresponding changes in the activities of the enzymes participating in the energy supply (malate dehydrogenase) and nitrogen metabolism (glutamate dehydrogenase, glutamine synthetase), is highly affected by xenobiotic concentration, exposure time and mode of illumination (Kvesitadze, 2006).

Ecologically, the most advantageous pathway for organic contaminant transformation in plants, is deep oxidative degradation. In higher plants, the following enzymes are mainly responsible for this process: cytochrome P450-containing monooxygenese, peroxidase and phenoloxidase. The characteristics of these enzymes, responsible for the degradation of different organic contaminants, are tabulated in Figure 16.4.

Enzyme	Physiological function	Existence in cell	Localisation	Specificity to toxicants	Limiting factors	Stability
Cytochrome P450 containing monooxygenase	Participation in a number of intracellular synthesizing reactions	Small amount, inductive nature	Endoplasmatic reticulum, cytosole	Very high affinity to nonpolar toxicants	NADPH, NADH	Labile, inactivatin g during substrate oxidation
Peroxidase	Hormonal regulation, lignification, response to stress, removing of peroxides	Large amount, inductive nature	***Cell wall, vacoules, cytosole, tonoplasts, plastids, plasmalemma***	Affinity to aliphatic compounds	Hydrogen peroxide or organic hydroperoxide s	Stable
Phenoloxidase	Oxidative transformation of phenols, lignification, cell defence reactions	Large amount, presents in latent form too, inductive nature	Chloroplasts, cell wall, cytosole, tonoplasts	Affinity to aromatic compounds	Endogenous phenols	Stable

Figure 16.4: Table of plant oxidative metalloenzymes.

Cytochrome P450-containing monooxygenases (EC 1.14.14.1) are mixed-function enzymes located in the membranes of the endoplasmic reticulum (microsomes) (Robineau, 1998). The monooxygenase system contains a redox-chain for electron transport; the initial stage of electron transfer is a NADPH-cytochrome P450 reductase (EC 1.6.2.4); the intermediate carrier is cytochrome b5, and the terminal acceptor of electrons is cytochrome P450. When NADPH is used as the only source of reductive equivalents, the existence of an additional carrier, a NADH-dependent flavoprotein, is required. NADH may also be oxidized by the NADPH-dependent redox system. In the latter case cytochrome b5 is not required (Hansikova, 1994). The cytochrome P450-containing monooxygenases use NADPH or NADH reductive equivalents for the activation of molecular oxygen and incorporation of one of its atoms into lipophilic organic compounds (XH), which results in the formation of hydroxylated products (XOH) (Schuler, 1996). The second atom of oxygen is used for the formation of a water molecule, see Figure 16.5.

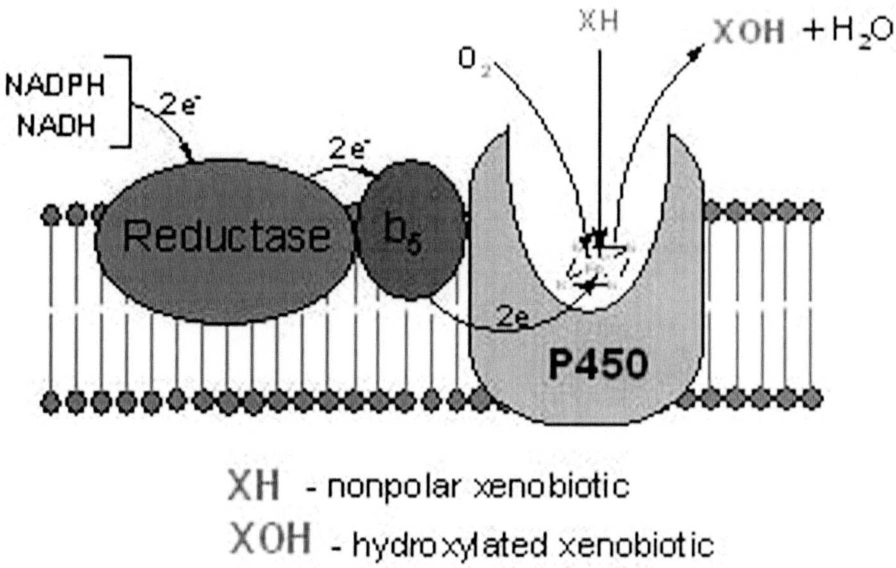

Figure 16.5: Microsomal monoöxygenase system.

Plant cytochrome P450-containing monoöxygenases play an important role in the hydroxylation of organic contaminants (Sandermann, 1994). The enzymes participate in the C- and N-hydroxylation reactions of aliphatic and aromatic compounds, N-, O-, and S-dealkylation, sulfo-oxidation, deamination, N-oxidation, oxidative and reductive dehalogenation, etc. (Schuler, 1996; Morant, 2003). The resistance of plants against herbicides is mediated by their rapid intracellular transformation into hydroxylated products and subsequently conjugated to carbohydrate moieties in the plant cell wall. For example, the N-demethylation and ring-methyl hydroxylation of the phenylurea herbicide chlorotoluron in wheat and maize are cytochrome P450-dependant processes (Fonne, 1990; Mougin, 1990). For some phenylurea herbicides in the Jerusalem artichoke, cytochrome P450-mediated N-demethylation is sufficient to cause significant or complete loss of phytotoxicity (Didier, 2002).

Peroxidase. In higher plants, peroxidase activity increases in response to stress. Among the multiple functions of this enzyme, one of the major ones is the protection of cells from oxidative reactions typical of all photosynthetic plants. The great catalytic versatility of peroxidase is its predominant characteristic, and, therefore, no single rôle exists for this multifunctional enzyme. Peroxidase is defined by the following reaction:

$$RH_2 + H_2O_2 \rightarrow 2H_2O + R \qquad (16.1)$$

where R symbolizes the compound being oxidized. The peroxidases catalyse a number of free radical reactions. Alternatively, the compound that is directly oxidized by the enzyme further oxidizes other organic compounds, including xenobiotics. According to the current hypothesis, the great majority of organic contaminants in plants are oxidized by peroxidases (Stiborova, 1991). This notion is based on the wide and ubiquitous distribution of this enzyme in plants (the isozymes of peroxidase in green plants occur in the cell walls, plasmalemma, tonoplasts, intracellular membranes of endoplasmic reticulum, plastids and cytoplasm), and the high affinity and wide substrate specificity of plants peroxidases to organic xenobiotics of different chemical structures. In the literature the participation of plant peroxidases in hydroxylation reactions of xenobiotics has been widely discussed. For example, peroxidases from different plants are capable of oxidizing N,N-dimethylaniline (Shinohara, 1984), 3,4-benzpyrene, 4-nitro-o-phenylene diamine (Wilson, 1994), 4-chloroaniline (Laurent, 1994), phenol, aminoflourene, acetaminophen, diethylstilbestrol, butylated hydroxytoluene, hydroxyanisoles, benzidine, etc. (Sandermann, 1994); horseradish (*Armoracia rusticana*) peroxidase oxidizes tritium-labelled [C_3 $^3H_3$]TNT (Adamia, 2006).

Phenoloxidases. This group of copper-containing enzymes (other names are tyrosinase, monophenol monoöxygenase, phenolase, monophenol oxidase, etc.) are widespread within the plant cell organelles catalysing both monoöxygenase and oxygenase reactions: the o-hydroxylation of monophenols (monophenolase reaction) and the oxidation of o-diphenols to o-quinones (diphenolase reaction) (Sanches, 1994). Currently, accepted enzyme nomenclature classifies hydroxylating phenol oxidase as monophenol monoxygenase (EC 1.14.18.1) and o-diphenols oxidizing phenol oxidase as catechol oxidase (EC 1.10.3.1). Plant phenol oxidases appear to be a group of specific enzymes, oxidizing a wide range of o-diphenols, such as DOPA (dihydroxyphenylalanine), catechol, etc, but which are unable to convert m- or p- diphenols to the corresponding quinones (Rompel, 1999). The active centre of phenol oxidases contains two copper atoms and exists in three states: 'met', 'deoxy' and 'oxy'. A catalytic cycle for the phenoloxidase also may involve a non-enzymatic reaction, with participation of the o-quinone intermediate (Rodriguez, 1992), see Figure 16.6.

Phenoloxidases actively participate in the oxidation of aromatic xenobiotics. Phenoloxidase from spinach, analogously to many other plants, oxidizes aromatic xenobiotics (benzene, toluene), by their hydroxylation and further oxidation to quinone (Ugrekhelidze, 1997). In a number of the cases, if the xenobiotic is not a substrate for the phenoloxidase, it may undergo co-oxidation in the following manner: the enzyme oxidizes the corresponding endogenous phenol by forming quinones or semi-quinones or both, i.e. compounds with a high redox potential. These compounds activate molecular oxygen by forming oxygen radicals, such as the superoxide anion radical ($O^{2-\bullet}$) and the hydroxyl radical ($^{\bullet}OH$) (Guillen, 1997), which gives the compounds the capacity for further oxidation of the xenobiotic. The formation of these radicals enables phenoloxidase to participate in contaminant degradation processes, see Figure 16.7:

Analogously, nitrobenzene is oxidized to m-nitrophenol, and the methyl group of [C_3 $^3H_3$TNT (Adamia, 2006) by phenoloxidase from the tea plant. The information confirming participation of this enzyme in the oxidative degradation of xenobiotics in higher plants is sparse (Ugrekhelidze, 1997), despite the fact that participation of phenoloxidase should definitely be expected. Lac-

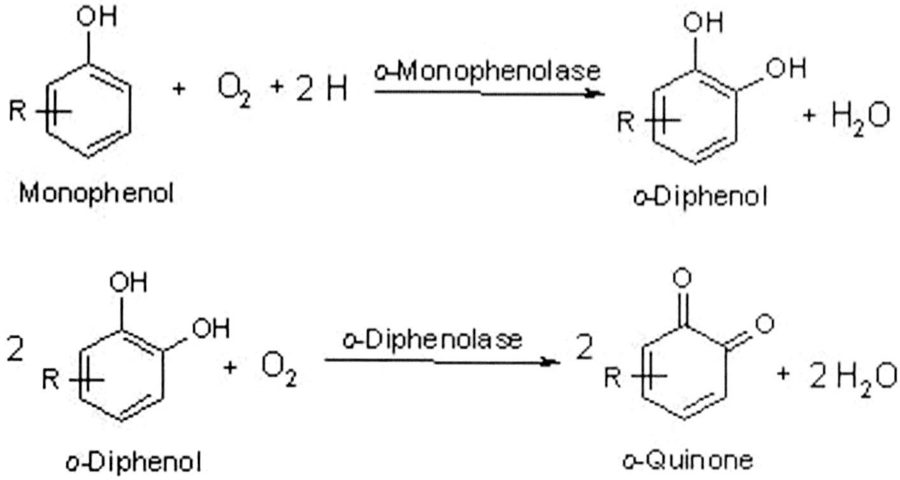

Figure 16.6: Action of phenoloxidases on monophenol and o-diphenol.

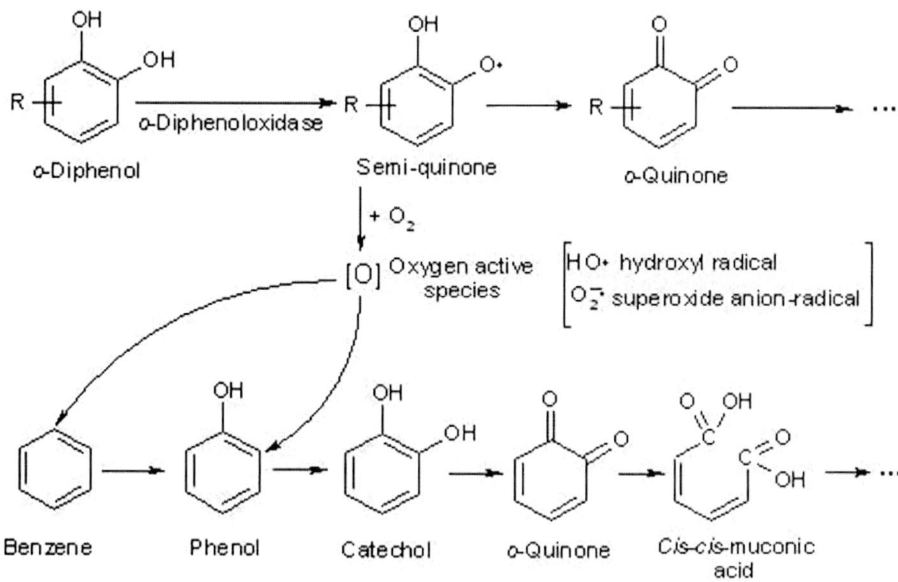

Figure 16.7: Enzymatic oxidation of o-diphenols (top line) by phenol-oxidase and non-enzymatic co-oxidation of benzene (bottom line).

cases of basidial fungi, analogous to higher plant phenoloxidase, have been better explored. Laccase degrades different aliphatic and aromatic hydrocarbons (Colombo, 1996), and actively participates in the enzymatic oxidation of alkenes (Niku, 2000). Crude preparations of laccase isolated from the white rot fungus *Trametes versicolor* oxidize 3,4-benzopyrene, anthracene, chrysene, phenanthrene, acenaphthene and some other PAHs (Collins, 1997). The intensity of oxidation of these antroprogenic contaminants is increased in the presence of such mediators as: phenol, aniline, 4-hydroxybenzoic acid, 4-hydroxybenzyl alcohol, methionine, cysteine, reduced glutathione, and other compounds that are substrates of laccase (Johannes, 2000). These data indicate that in the cases of fungal laccase and plant o-diphenoloxidase, the oxidation of hydrocarbons is carried out by a co-oxidation mechanism (Ugrekhelidze, 1980; Ugrekhelidze, 1997).

Apparently, metalloenzymes differing in their localization in the plant cell organelles, structural organization, mechanisms of action, and substrate specificity, allow plants: firstly, to oxidize a wide spectrum of organic contaminants including aromatic structures; and secondly, to regulate functional interchange of these enzymes during contaminant oxidative degradation (caused by the inability or decreased potential of any of them to carry out further oxidation of structurally unsuitable intermediates).

Deep degradation of organic xenobiotics is a multistage, mainly oxidative enzymatic process and only insignificant amounts of the toxic molecules undergo direct degradation; the majority are conjugated with endogenous secondary metabolite contaminants (more than 80%) accumulated in vacuoles and apoplasts, and their further transformation takes place with some delay. The emission of $^{14}CO_2$ (up to 5% in the case of labelled linear contaminants) from plant cells indicates that the formation of conjugates and their compartmentalization is followed by deep oxidation of the toxic parts of their molecules (Ugrekhelidze,1986; Ugrekhelidze, 1997; Chrikishvili, 2005).

Based on the experimental data the most rate-limiting stage of the whole process of xenobiotic transformation seems to be the initial hydroxylation of nonpolar contaminants.

16.4 Degradation processes

The transformation of small molecular weight aliphatic xenobiotics, such as methane, in the tea plant (*Thea sinensis*) proceeds by the formation of fumaric acid. The transformation of ethane, propane and pentane leads to the formation of low molecular mass compounds largely composed of di- and tri-carbon organic acids. Labelled fumaric, succinic, malonic, citric and lactic acids are identified in plant leaves exposed to these low molecular mass alkanes, with most of the radioactivity incorporated into succinic and fumaric acids. The absence of oxalic acid directly indicates that the ethane in plants is oxidized mono-terminally. The oxidation of ethane at one terminal carbon atom leads to the formation of acetyl-CoA, which in turn participates in the Krebs cycle, Figure 16.8 (Durmishidze, 1968).

At the terminal carbon atom, oxidation (of propane) forms propionic acid, successively undergoing further b-oxidation and resulting in the formation of malonyl-CoA, and decarboxylation by formation acetyl-CoA. The formation

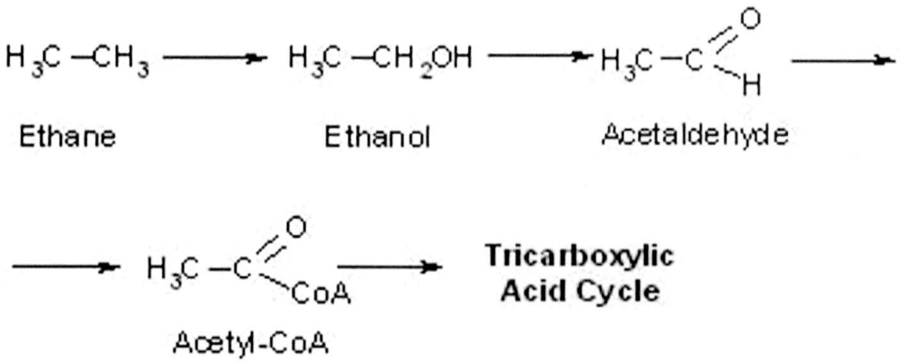

Figure 16.8: Transformation of ethane in higher plants.

of low molecular mass compounds such as monocarbonic acids suggests that propane and pentane could be oxidized mono-terminally, by intermediate incorporation into the Krebs cycle or by forming valeric acid (Ugrekhelidze, 1976).

Long chain alkanes are subjected to similar transformations. For instance, after 40 min of incubation of leek leaves with an emulsion of exogenous [^{14}C] octadecane in water, 9.6% of the total label is detected in esters, 6.4% in alcohols, and 4% in organic acids (Cassagne, 1975).

The most significant input in the understanding of the detoxification process of plants has been the discovery nearly 40 years ago of their ability to transform (oxidatively decompose) benzene and phenol via aromatic ring cleavage. As a result of this degradation, carbon atoms of the contaminant are incorporated into organic acids and amino acids. Similar data were reported for nitrobenzene, aniline, toluene, a-naphthol, and benzidine transformation in plants (Durmishidze, 1974; Mithaishvili, 2005; Jansen, 1969; Tkhelidze, 1969; Ugrekhelidze, 1976). The oxidation of benzene and phenol by crude enzyme extracts of plants forms muconic acid as a result of ring cleavage, with catechol as an intermediate (Durmishidze, 1969): see Figure 16.9.

Further oxidation of muconic acid results in the formation of fumaric acid. Labelled muconic and fumaric acids are found in plants exposed to labelled benzene or phenol. Cleavage of the aromatic ring in endogenous substrates proceeds by the transformation of 3,4-dihydroxybenzoic acid into 3-carboxymuconic acid (Tateoka, 1970). Phenoxyalkyl-carboxyl acids containing four and more carbon atoms in their side chain often undergo oxidation in plants. For instance, 2,4-dichlorophenoxybutyric acid is oxidized by the formation of 2,4-D (Hawf, 1974; McComb, 1978; Taylor, 1978).

Finally, contaminant degradation proceeds to standard cell metabolites or mineralization. The plant cell, in degrading the xenobiotic, not only avoids its toxic action but also utilizes its carbon, nitrogen, and other atoms for intracellular biosynthetic and energy requirements. The totality of such transformations is the essence of the plant's detoxification process. Direct complete xenobiotic degradation in a plant cell is, however, accomplished only slowly depending on concentration.

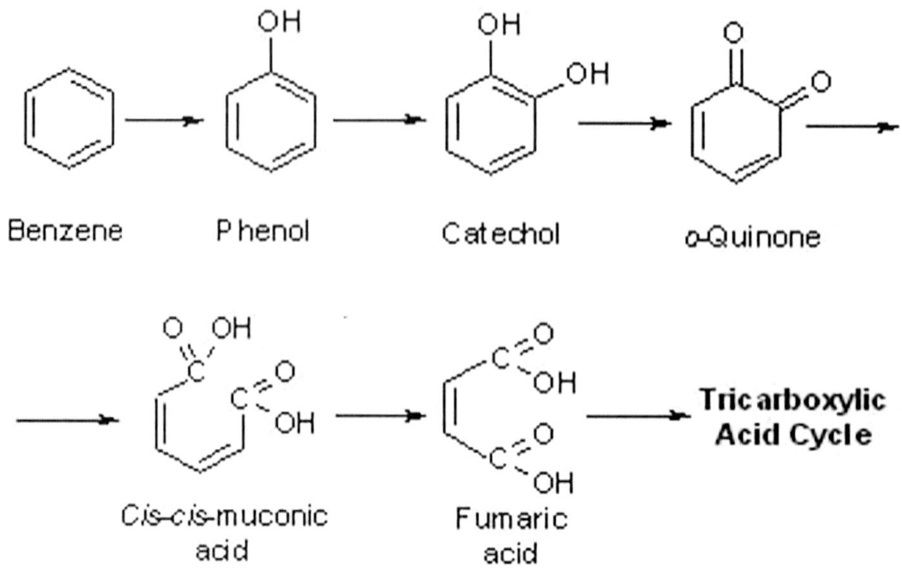

Figure 16.9: Oxidative degradation of benzene in plant cells.

16.5 Plant ultrastructure dynamics due to xenobiotics

To evaluate the ecological potential of plants, demonstration of their response to contaminants at the level of cell ultrastructure is the most precise indication of a plant's capacity to detoxify. Penetration of small amounts of contaminants leads to structurally invisible but measurable perturbations of cell metabolic processes such as the induction of enzymes, inhibition of intracellular metabolic processes, and change in the levels of regular secondary metabolites. At larger concentrations, contaminants provoke clearly noticeable perturbations of cell ultrastructural organization. It has been shown that the complex of changes and alterations in the main metabolic processes of plant cells elicited by organic pollutants (pesticides, hydrocarbons, phenols, aromatic amines, etc.) are connected with the perturlations of cell ultrastructural architecture. The sequence and depth of destruction of plant cell organelles are determined by the variety of the plant, the chemical nature, concentration and exposure duration of the action contaminant, etc. (Buadze, 1998; Zaalishvili, 2000). In experiments with a number of various higher plants exposed to different ^{14}C-labelled toxic compounds, the penetration, movement and localization of the contaminants in the plant cells has been observed to engender changes in the ultrastructural organization. The effects of toxic compounds on cell ultrastructure, depending on its concentration, may be classified into two types:

- metabolic: the contaminant is digested by the plant in spite of the mobilization of the plants internal protection mechanism

- lethal: leading to indigestion and plant death.

Figure 16.10 (Zaalishvili, 2000) shows maize root apex cells exposed to ^{14}C-nitrobenzene action, its penetration across the plasmalemma and its localization in subcellular organelles. Studies of the penetration of ^{14}C-labelled xenobiotics into plant cells indicate that in the early stages of exposure (5–10 min), labelled compounds are detected in the cell membrane, in the nucleus and nucleolus (in small amounts), and, seldom, in the cytoplasm and mitochondria. As a result of prolonged exposure the amount of label significantly increases in the nucleus, in the organelle membranes, in tonoplasts, and in vacuoles (Zaalishvili, 2000), i.e. the xenobiotic becomes distributed in most of the subcellular organelles, but ultimately there is a tendency for the contaminants to accumulate in vacuoles.

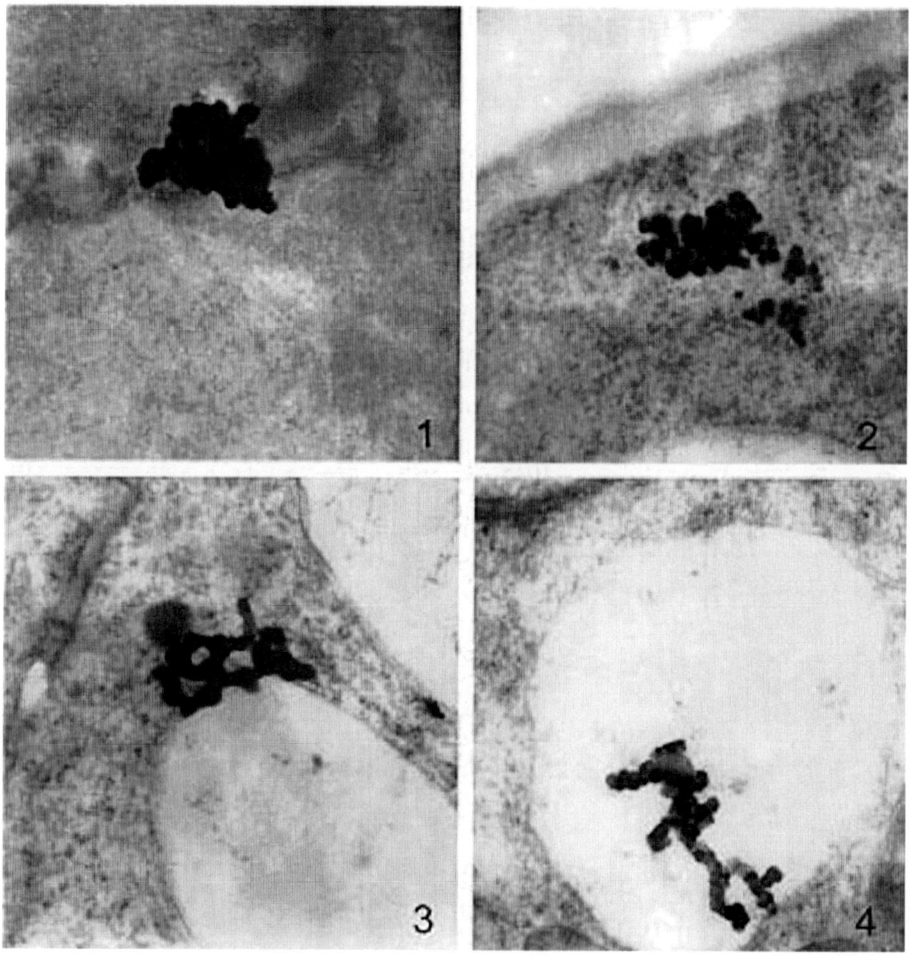

Figure 16.10: Electron micrographs showing the penetration and movement of ^{14}C-labelled nitrobenzene (0.15 mM) in a maize root apex cell. The xenobiotic penetrated through the plasmalemma (1, × 48 000), moved to the cytoplasm (2, × 36 000), and thereafter translocated into vacuoles (3, × 50 000; 4, × 30 000).

The general picture of the evolving action of organic contaminants on plant cells is the following:

- Initially, changes in the configuration of the nucleus become noticeable. Simultaneously inhibition of DNA synthesis takes place. The barrier function of the plasmalemma and its ability to retain calcium are adversely affected. Ca^{2+} concentration in the cytoplasm is increased; Ca^{2+}—-ATPase activity is inhibited. Mitochondria with swollen cristae and packed matrix become noticeable, the plastids are electron-dense and enlarged.

- Prolonged action of contaminants leads to a widening of the cisternae of the endoplasmic reticulum and Golgi apparatus, and vacuolization of the cytoplasm. The size of the cytoplasm is thereby decreased and the periplasmic space concomitantly enlarged. In some cortical cells of the rootspices, the number of ribosomes in the hyaloplasm is increased, and the formation of polysomes is observed. Lysis of mitochondria and depletion of ribosomes from the endoplasmic reticulum take place. Multiple contacts between the endoplasmic reticulum and the plasmalemma, vacuoles, nucleus, and membranes of the mitochondria are detectable. The enhancement of the size of the nucleus and chromatin coagulation, indicating a disturbance of the DNA synthesis process, is observed. Nuclei acquire deviant shapes because of the development of many protuberances of the nuclear membrane. In leaf cells, chloroplast shape and composition become ill defined, the external membrane is not visible, the orientation of the system is disturbed, and the matrix is characterized by large osmophilic inclusions. In the cytoplasm, accumulation of the differentiated cells of the root caps that secrete mucus is visible. Some of these hypertrophied vesicles fuse, forming a large deposit of mucus. Inhibition of the process of maturing secretory vesicle translocation towards the cell periphery is often correlated not only with the swelling of vesicles, but also with the disappearance of the normal dictyosomes.

Prolonged exposure to environmental contaminants causes extensive destruction of the cell and plant death.

16.6 Plants as remediators

Plants are able to act most effectively as remediators at low concentrations of contamination of soil and air, when no significant changes in the cell ultrastructure might be detected. Nevertheless, plants subjected to high concentrations for relatively short periods in most cases are able to recover from slight deviations in cell ultrastructure and thus maintain their vital activities.

Planting of almost any kind of vegetation, including agricultural flora, is beneficial for the environment. However, in order to ensure the maximum possible ecological benefit, plants should be selected according to their potential to assimilate/accumulate toxic compounds of different structures.

Phytoremediation is a unique cleanup strategy. The effective realization of phytoremediation technologies implies the planting on a contaminated area with one or more specific, previously selected plant species with known obility to extract contaminants from the soil. A precise survey of the existing vegetation on site should be undertaken to determine what species of plants would have the best growth at the contaminated site. Based on extensive experimental results, including the use of labelled xenobiotics and electron microscopic observations, the deep degradation of anthropogenic contaminants in higher plants should be considered as a capacity-limited but permanently working pathway, having much less capacity than the conjugate formation process (especially in the case of contaminant saturation). Nevertheless plants, greatly depending on the variety, are able to completely eliminate toxicants by metabolic degradation.

During the last decade phytoremediation, starting from a conceptual methodology, has become an ecologically important commercial technology for the cleaning of the environment. In order to increase this ecological potential of plants, definite progress has already been achieved by the cloning of genes of the enzymes participating in contaminant transformation/accumulation. A number of modified plants, having especially high accumulation abilities and corresponding by large intracellular volumes to deposit metabolite xenobiotic conjugates, have been created. Some recent publications (Li, 1996; Lu, 1997; Song, 2003) are devoted to the discussion of these and other problems concerning the uptake of inorganic contaminants. In these publications, transgenic plants with hyper-accumulation potential and characterized by enhanced tolerance to cadmium and lead (70–75 mM), their are described. Doubling of the lead content in transgenic plants has also been demonstrated (Peuke, 2004). Note also that the successful realization of phytoremediation technologies greatly depends on the synergetic action of micro-organisms and plants.

Among the large diversity of plants with potential for phytoremediation, the poplar family attracts special interest. Owing to its strong root system it is characterized by a high absorption ability. Multiple gene-engineering modifications have yielded convincing evidence for its expediency in practical use. Cloning of glutathione S-transferase was successful in the creation of several transgenic plants. The transfer of cytochrome P-450 genes to different plants has been a widespread activity for the last decade (Ohkawa, 1999). Some of the transgenic plants created are generally characterized by high resistance to herbicides of different structure and have clearly observable high detoxification potential (Morant, 2003). Transgenic plants have also been studied in connextion with the degradation of particular contaminants such as the widely distributed explosive TNT. In order to increase the degradability of TNT and similar compounds, the transgenic plants contained the gene of the bacterial enzyme pentaeritrole tetranitrate reductase (EC 1.6.99.7) (French, 1999). Transgenic tobacco has been analysed for its ability to assimilate the residues of TNT and trinitroglycerine. Seedlings of the transgenic plants, able to de-nitrate the compounds, extracted explosives from the liquid area much faster than the seedlings of common forms of the same plants, in which growth was inhibited by the contaminant (Hannink, 2001). Transgenic tobacco thus differs substantially from the common plant by its tolerance, fast uptake and assimilation of significant amounts of TNT. Analogous experimental results have been obtained with other plant species (Hannink, 2002).

There are many publications concerning the successful improvement of plant detoxification abilities by cloning the genes of transferases and oxidases, which intensively participate in contaminant transformation processes (Ohkawa, 1999; Morant, 2003).

Obviously, attempts to improve artificially the ecological potential of higher plants will be continued, and the results will enhance their eventual practical realization. The positive effect of these investigations would doubtless be much more impressive if all aspects of the complicated and multistage detoxification process were better elucidated with regard to plant physiology and biochemistry. Such information would allow the creation of a more rational and effective strategy for the genetic engineering of plants for potential application.

Until recently plants were considered as organisms having a naturally limited potential for contaminant conjugation and accumulation. Now, depending on the nature of the organic xenobiotic and the type of plant, typically 1 kg of green biomass takes from the air daily amounts ranging from micrograms to fractions of a milligram of pollutant (Ugrekhelidze, 1980; Ugrekhelidze, 1997; Kvesitadze, 2006). Plants possessing universal (i.e. applicable to soil, groundwater and air) cleaning capabilities are the only agents carrying out the process of remediation by transporting metals to the aboveground parts of the plants. Some plants are indeed known as hyperaccumulators of metals. Transgenic plants of Indian mustard, poplar, tobacco, thlaspi, arabidopsis, etc. possess especially high potential for metal accumulation and transportation (Peuke, 2004; Macek, 2002).

Elimination of contaminants located deeper than two metres in the soil faces limitations in time, since the mass transfer processes at that depth and deeper proceeds much more slowly. The extraction by roots and the subsequent transport may become the rate-limiting factor of the whole process. Therefore, plant-microbial action-based technologies would need excessive time to achieve a satisfactory remediation. In the case of high contaminant concentration, phytoremediation as a final "polishing step" must follow other technologies such as excavation, treatment and disposal, etc. Other cases where phytoremediation may not be successfully applied are when high concentrations of soil contaminants such as polychlorinated biphenyls and dioxins are present. At high concentrations of these compounds no plants can grow. Plants are very promising detoxifiers and constitute an ecologically safe technology ("green filter") responding to the problem of anthropogenic contamination (Kvesitadze, 2006). The universality of phytoremediation consists in the uptake of nearly of all types of organic contaminants and heavy metals and their accumulation in intracellular structures or their oxidative degradation to carbon dioxide. Plants represent a complex system that has evolved to deal with life-threatening toxicants formed in the Earth and released naturally and now, increasingly, by mankind. It is indeed fortunate that nature has provided us with such a robust system, and to counteract the excesses of civilization and production we must utilize plant remediation to the full through understanding all its complexities.

16.7 References

Adamia G, Ghoghoberidze M, Graves D, Khatisashvili G, Kvesitadze G, Lomidze E, Ugrekhelidze D, Zaalishvili G (2006) Absorption, distribution and transformation of TNT in higher plants. Ecotoxicol Environ Safety 64: 136–145

Buadze O, Sadunishvili T, Kvesitadze G (1998) The effect of 1,2-benzanthracene and 3,4-benzpyrene on the ultrastructure on maize cells. Int Biodeterior Biodegrad 41: 119–125

Cassagne C, Lessire R (1975) Studies on alkane biosynthesis in epidermis of *Allium porrum* L. leaves. 4. Wax movement into and out of the epidermal cells. Plant Sci Lett S5: 261–266

Chrikishvili D, Sadunishvili T, Zaalishvili G (2005) Benzoic acid transformation via conjugation with peptides and final fate of conjugates in higher plants.

Ecotoxicol Environ Safety 2006: 390–399

Coleman JOD, Mechteld MA, Kalff B, Davies TGE (1997) Detoxification of xenobiotics in plants: chemical modification and vacuolar compartmentalization. Trends Plant Sci 2: 144–151

Collins PJ, Dobson ADW (1997) Regulation of laccase gene transcription in *Trametes versicolor*. Appl Environ Microbiol 63: 3444–3450

Colombo JC, Cabello MN, Arambarri AM (1996) Biodegradation of aliphatic and aromatic hydrocarbons by natural soil microflora and pure culture of imperfect and ligninolytic fungi. Environ Pollut 94: 355–362

Didierjean L, Gondet L, Perkins R, Lau S-MC, Schaller H, O'Keefe DP, Werck-Reichhart D (2002) Engineering herbicide metabolism in tobacco and *Arabidopsis* with CYP76B1, a cytochrome P450 enzyme from Jerusalem artichoke. Plant Physiol 130: 179–189

Durmishidze S, Ugrekhelidze D (1968) Absorption and conversion of butane by higher plants (in Russian). Dokl Akad Nauk SSSR 182: 214–216

Durmishidze S, Ugrekhelidze D (1968). Oxidation of ethane, propane and pentane by higher plants (in Russian). Bull Georgian Acad Sci 50: 661–666

Durmishidze S, Ugrekhelidze D, Djikiya A, Tsevelidze D (1969) The intermediate products of enzymatic oxidation of benzene and phenol (in Russian). Dokl Akad Nauk SSSR 184: 466–469

Durmishidze S, Ugrekhelidze D, Djikiya A (1974) Absorption and transformation of benzene by higher plants (in Russian). Fiziologiya i Biochimiya Kulturnikh Rastenii 6: 217–221

Durmishidze S, Ugrekhelidze D, Djikiya A (1974) Absorption and transformation of toluene by higher plants (in Russian). Appl Biochem Microbiol 10: 673–676

Durmishidze S, Djikiya A, Lomidze E (1979) Uptake and transformation of benzidine by plants in sterile conditions (in Russian). Dokladi Akademii Nauk SSSR 247: 244–247

Eckardt NA (2001) Move it on out with MATEs. Plant Cell 13: 1477–1480

Fonné-Pfister R, Kreuz K (1990) Ring-methyl hydroxylation of chlortoluron by an inducible cytochrome P450-dependent enzyme from maize. Phytochemistry 9: 2793–2804

French CE, Hosser SJ, Davies GJ, Nicklin S, Bruce NC (1999) Biodegradation of explosives by transgenic plants expressing pentaerythritol tetranitrate reductase. Nature Biotechnol 17: 491–494

Galante C, Esposito S., Stile MR, Monti L, Filoppone E (2003) *In planta* expression of a fungal laccase gene useful for phytoremediation of PAHs contaminated soils. Proceedings of the XLVII Italian Society of Agricultural Genetics, SIGA Annual Congress, Verona, Italy, 24–27 September 2003.

Guillén F, Martýnez MJ, Muñoz C, Martýnez AT (1997) Quinone redox cycling in the ligninolytic fungus Pleurotus eryngii leading to extracellular production of superoxide anion radical. Arch Biochem Biophys 339: 190–199

Guillén F, Gómez-Toribio V, Martýnez MJ, Martýnez AT (2000) Production of hydroxyl radical by the synergistic action of fungal laccase and aryl alcohol oxidase. Arch Biochem Biophys 382: 142–147

Hannink N, Rosser SJ, French CE, Basran A, Murray JA, Nicklin S, Bruce NC (2001) Phytodetoxification of TNT by transgenic plants expressing a bacterial nitroreductase. Nature Biotechnol 19: 1168–1172

Hannink N, Rosser SJ, Bruce NC (2002) Phytoremediaition of explosives. Crit Rev Plant Sci 21: 511–538

Hansikova H, Frei E, Anzenbacher P, Stiborova M (1994) Isolation of plant cytochrome P450 and NADPH:cytochrome P450-reductase from tulip bulbs (*Tulipa fosteriana*). Gen Physiol Biophys, 13: 149–169

Hawf LR, Behrens R (1974) Selectivity factors in the response of plants to 2,4-D. Weed Sci 22: 245–249

Helman C, Dirty appetite, Forbes Magazine, 2001

Jansen EF, Olson AC (1969) Metabolism of carbon-14-labelled benzene and toluene in avocado fruit. Plant Physiol 44: 786–791

Johannes C, Majcherczyk A (2000) Natural mediators in the oxidation of polycyclic aromatic hydrocarbons by laccase mediator systems. Appl Environ Microbiol 66: 524–528.

Korte F, Kvesitadze G, Ugrekhelidze D, Gordeziani M, Khatisashvili G, Buadze O, Zaalishvili G, Coulston F (2000) Review: Organic toxicants and plants. Ecotoxicol Environ Safety 47: 1–26

Kvesitadze G, Khatisashvili G, Sadunishvili T, Ramsden JJ (2006) Biochemical mechanisms of detoxification in higher plants: the basis of phytoremediation. Springer, Heidelberg.

Laurent FMG (1994) Chloroaniline peroxidation by soybean peroxidases. Pestic Sci 40: 25–30

Li ZS, Szczypka M, Lu YP, Thiele DJ, Rea PA (1996) The yeast cadmium factor protein (YCF1) is a vacuolar glutathione S-conjugate pump. J Biol Chem 271: 6509–6517

Lu YP, Li ZS, Rea PA (1997) AtMPR1 gene of *Arabidopsis* encodes a glutathione S-conjugate pump: isolation and functional definition of a plant ATP-binding cassette transporter gene. Proc Natl Acad Sci USA 94: 8243–8248

Macek T, Macková M, Pavlíková D, Száková EJ, Truska M, Singh-Cundy A, Kotraba P, Yancey N, Scouten WH (2002) Accumulation of cadmium by transgenic tobacco. Acta Biotechnologica 22: 101–106

Macek T, Sura M, Francova K, Chrastilova Z, Pavlíková D, Sylvestre M, Szekeres M, Scouten WH, Kotraba P, Macková M (2004) Approaches using GM plants for the removal of xenobiotics (Cd, Ni, PCB) including experiments in real contaminated soils. In: Phytoremediation: environmental and molecular biological aspects. OECD workshop, Hungary, Abstr, p 27

Marrs KA (1996) The function and regulation of glutathione S-transferases in plants. In: Ann Rev Plant Physiol Plant Mol Biol 47: 127–158

Martinova E (1993) An ATP-dependent glutathione-S-conjugate "export" pump in the vacuolar membrane of plants. Nature 364: 247–249

McComb AJ, McComb JA (1978) Differences between plant species in their ability to utilize substituted phenoxybutyric acids as a source of auxin for tissue culture growth. Plant Sci Lett 11: 227–331

Mithaishvili T, Scalla R, Ugrekhelidze D, Tsereteli B, Sadunishvili T, Kvesitadze G (2005) Transformation of aromatic compounds in plants grown in aseptic conditions. Z Naturforsch, 60, 97–102

Morant M, Bak S, Moller BL, Werck-Reichhart D (2003) Plant cytochromes P450: tools for pharmacology, plant protection and phytoremediation. Current Opinion Biotechnol 2: 151–162

Mougin C, Cabanne F, Canivenc M-C, Scalla R (1990) Hydroxylation and N-demethylation of chlortoluron by wheat microsomal enzymes. Plant Sci 66:

195–203

Niku-Paavola ML, Viikari L (2000) Enzymatic oxidation of alkenes. J Mol Cat 10: 435–444

Ohkawa H, Tsujii H, Ohkawa Y (1999) The use of cytochrome P450 genes to introduce herbicide tolerance in crops: a review. Pestic Sci 55: 867–874

Peuke AD, Kopriva S, Rennenberg H (2004) Phytoremediation with the help of transgenic trees. In: Phytoremediation: environmental and molecular biological aspects. OECD workshop, Hungary, Abstr, p 33

Robineau T, Batard Y, Nedelkina S, Cabello-Hurtado F, LeRet M, Sorokine O, Didierjean L, Werck-Reichhart D (1998) The chemically inducible plant cytochrome P450 CYP76B1 actively metabolizes phenylureas and other xenobiotics. Plant Physiol 118: 1049–1056

Rodríguez-López JN, Tudela J, Varón R, Fenoll LG, García-Carmona F, Garcí-Cánovas F (1992) Analysis of a kinetic model for melanin biosynthesis pathway. J Biol Chem 267: 3801–3810

Rompel A, Fischer H, Meiwes D, Buldt-Karentzopoulos K, Magrini A., Eicken C, Gerdemann C, Krebs B (1999) Substrate specificity of catechol oxidase from *Lycopus europaeus* and characterization of the bioproducts of enzymic caffeic acid oxidation. FEBS Lett 445: 103–110

Sánches-Ferrer A, Rodríguez-López JN, García-Cánovas F, García-Carmona F (1994) Tyrosinase: a comprehensive review of its mechanism. Biochim Biophys Acta 1247: 1–11

Sandermann H (1987) Pestizid-Rückstände in Nahrungspflanzen. Die Rolle des pflanzlichen Metabolismus. Naturwissenschaften 74: 573–578

Sandermann H (1994) Higher plant metabolism of xenobiotics: the "green liver" concept. Pharmacogenetics 4: 225–241

Sargent JA, Blackman GE (1972) Studies on foliar penetration. 9. Patterns of penetration of 2,4-dichlorophenoxyacetic acid into the leaves of different species. J Exp Bot 23: 830–839

Schönherr J, Bukovac MJ (1972) Penetration of stomata by liquids. Dependence on surface tension, wettability, and stomatal morphology. Plant Physiol 49: 813–823

Schuler MA (1996) Plant cytochrome P450 monooxygenases. Crit Rev Plant Sci 15: 235–284

Sharma MP, Vanden Born WH (1970) Foliar penetration of picloram and 2,4-D in aspen and balsam poplar. Weed Sci 18: 57–65

Shida T, Homma Y, Misaio T (1975) Absorption, translocation and degradation of N-lauryl-L-valine in plants. 6. Studies on the control of plant diseases by amino acid derivatives. J Agric Chem Soc Jap 49: 409–418

Shinohara A, Kamataki T, Ichimura Y, Opochi H, Okuda K, Kato R (1984) Drug oxidation activities of horse-radish peroxidase, myoglobin and cytochrome P-450cam reconstituted with synthetic hemes. Jap J Pharmacol 45: 107–114

Song WY, Sohn EJ, Martinoia E, Lee YJ, Yang Y-Y, Jasinski M, Forestier C, Hwang I, Lee Y (2003) Engineering tolerance and accumulation of lead and cadmium in transgenic plants. Nature Biotechnol 21: 914–919

Stiborova M, Anzenbacher P (1991) What are the principal enzymes oxidizing the xenobiotics in plants: cytochrome P-450 or peroxidase? Gen Physiol 10: 209–216

Tateoka TN (1970) Studies on the catabolic pathway of protocatechuic acid in mung bean seedlings. Bot Mag (Tokyo) 83: 49–54

Taylor H, Wain R (1978) Studies of plant growth-regulating substances. 52. Growth retardation by 3,5-dichlorophenoxyethylamine and 3,5-dichlorophenoxybutyric acid arising from their conversion to 3,5- dichlorophenoxyacetic acid in tomato plants. Ann Appl Biol 89: 271–277

Tkhelidze P (1969) Oxidative transformation of benzene and toluene in vine grapes (in Russian). Bull Georgian Acad Sci 56: 697–700

Tsao DT (2003) Phytoremediation. Advances in biochemical engineering and biotechnology. Springer, Heidelberg

Ugrekhelidze D, Kavtaradze L (1970) The question of metabolism of a-naphthol in higher plants (in Russian). Bull Georg Acad Sci 57: 465–469

Ugrekhelidze D (1976) Metabolism of exogenous alkanes and aromatic hydrocarbons in plants (in Russian). Metsnieraba, Tbilisi

Ugrekhelidze D, Durmishidze S (1980) The biosphere chemical pollution and plant (in Georgian). Metsniereba, Tbilisi

Ugrekhelidze D, Phiriashvili V, Mithaishvili T (1986) Uptake of salicylic acid and aniline by pea roots (in Russian). Fiziol Rast (Moscow) 33: 165–170

Ugrekhelidze D, Korte F, Kvesitadze G (1997) Uptake and transformation of benzene and toluene by plant leaves. Ecotoxicol Environ Saf 37: 24–28

Wilson L. Williamson T. Gronowski J, Gentile GI, Gentile JM (1994) Characterization of 4-nitro-o-phenylenediamine activities by plant systems. Mutation Res 307: 185–193

Zaalishvili G, Lomidze E, Buadze O, Sadunishvili T, Tkhelidze P, Kvesitadze G (2000) Electron microscopic investigation of benzidine effect on maize root tip cell ultrastructure, DNA synthesis and calcium homeostasis. Int Biodeterior Biodegrad 46: 133–140

Complexity and Security
J.J. Ramsden and P.J. Kervalishvili (Eds.)
IOS Press, 2008

Chapter 17

Modern trends in integrated information systems development

Karine Kotoyants

Institute of Telecommunications, Almaty, Kazakhstan

17.1 Introduction

The computer world has undergone great technological changes during the last two decades. Information technologies for business applications have been transformed beyond recognition. But these changes here called forth new specific problems. One of them can be defined as the lack of system thinking in the information technology (IT) industry. A main consequence of this drawback is the disability to produce systems complicated enough to correspond to the demands of consumers.

In accordance with the definition of complexity in general systems theory, complexity in infosystems has a number of characteristic inherent features, including:

- Unpredictability, as system behaviour is a result of interconnexions and relations between components;
- Brief interactions between the components of the system;
- The relationships between the system components are not linear;
- Relationships between the system components may include feedback coupling;
- A complex system by definition is open; its boundaries in accordance with the system's nature must be transparent either for information or for

energy. None of the elements possesses full information about the system as a whole;

- The characteristic feature of complex systems is nesting. For example, economics as a system may consist of a number of enterprises that are systems. Each enterprise comprises separate employees, which are in their turn also systems, etc.

As a result of globalization, the development of new technologies, the increase in consumer demands and the appearance of new social exigencies, IT systems cover more and more functions and become more and more complex. This requires a more comprehensive approach to the designing and construction of the systems, corporate IT systems in particular. The development of web services, and then architectures oriented to these services, represent not only the appearance of new engineering solutions, enabling the construction of loose information systems; but also, connexions between the components of the system via the services is a possible engineering solution that enables the systems to be provided with new qualities.

17.2 Service-oriented architecture

The evolution of IT consists of two global stages—the *technology-led era* and the *business-led era* (Figure 17.1). Over the years, as the price of technology spiralled downwards at an ever-increasing rate and new—technology-led–infrastructure models promised increased efficiencies, it became consistently easy to just add more hardware to solve immediate, tactical issues and to support the implementation of new business systems. IT infrastructures were built to be stable, not to react quickly to changing business strategies and processes—which is why one survey found that the business environment is changing seven times as fast as the underlying IT applications. So, IT departments typically spend nearly half their time finding stop-gap fixes to outmoded solutions that can't keep up with yesterday's demands, let alone today's. In essence, IT departments are solving the same problem over and over again—how to manage change brought about by changing business needs.

The *technology-led era* resulted in two major problems:

- There has been enterprise-wide over-deployment of over provisioned servers and storage, and over-redundant network infrastructures. Gartner estimates that the typical utilization rate of the infrastructure is only about 30 %.

- The 'islands of computing' serving individual application needs have become inflexible silos that resist accommodating change (and are exceedingly difficult to manage).

The business-led era

The current economic and business environment is regarding as creating an inflexion point for the computing industry, which is leading to a new era of computing that will be driven by business imperatives more than by technology capabilities. Whereas the last decade was fueled by the possibilities of

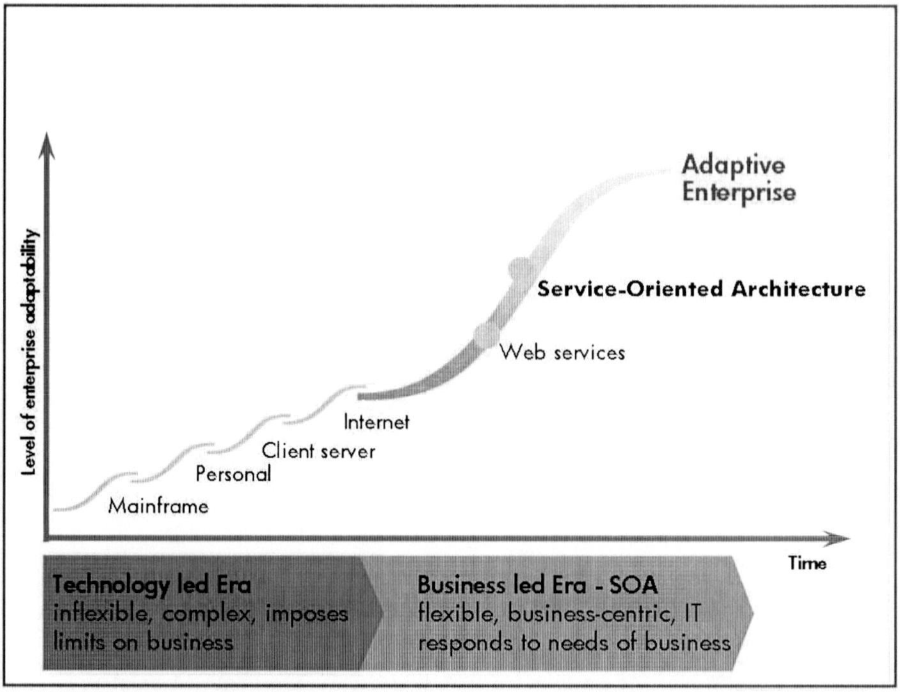

Figure 17.1: The technology-led and the business-led eras.

technology—personal computing, client server computing, and the Internet—this new era will be driven by a business focus on increasing the value and return from information technology. Businesses will be responding to market and business demands in real time; balancing a dynamic supply of owned and on-demand assets and services. Individuals will be empowered by immediate knowledge available where and when needed.

Computing infrastructure will be:

- tightly and dynamically tied to business requirements;
- managed as a single globally distributed resource;
- enabled by enterprise grids;
- powered by modular, standards-enabled components.

In this new adaptive computing environment, the total enterprise technology infrastructure will increasingly be viewed as a business asset rather than an expense line. This will shift the job of the Chief Information Officer (CIO) and the role of IT from being a cost centre to being a service provider for the enterprise, and that in turn will place new demands on the providers of technology.

This era, in which the IT foundation for a corporation becomes a flexible, utility-like business service increasingly vital to the operation, and increasingly an enabler of new opportunities and competitiveness, is named the "adaptive era"

We believe that in the future business will be real-time and agile, informed by an immediate view of all core business operations. The service and technology infrastructure will be adaptive. It will be a shared resource, local and distributed, that can be dynamically tapped to provide services and computing resources, as the business requires. The link between business and IT will be instinctive and responsive, based on service level agreements, automated rules, and intelligent management.

Service-oriented architecture (SOA) is an architectural approach centered on the concept of services. It is all about decomposing applications into services that can be combined and recomposed into new, flexible business processes. SOA enables processes to be composed of reusable services and exposed as services. The main strength of SOA is synchronizing the link between business and IT.

Service oriented architecture has made itself widely known as a logical extension and development of web-service technologies. The idea of services in information systems is not new in itself. The following approaches to its realization are well known:

- Java RMI (Java remote method invocation) from Sun Microsystems;
- CORBA (common object request broker architecture) from the Open Management Group;
- DCE (distributed computing environment) introduced by the Open Group consortium;
- DCOM (distributed component object model) from Microsoft.

Each architecture that manifests itself by these means can be characterized as service-oriented. But at the same time each one defines its own formats and protocols, call mechanisms, and interfaces for applied programs. The drawback of universalism restrains their wide diffusion (i.e. it is less widely diffused than it might be). The modern interpretation of SOA treat services as web services, the basis of which is formed by conventional Internet technologies and already-developed infrastructure.

In general, entities (people and organizations) create capabilities to solve or support a solution for the problems they face in the course of their business. It is natural to think of one person's needs being met by capabilities offered by someone else; or, in the world of distributed computing, one computer agent's requirements being met by a computer agent belonging to a different owner.

There is not necessarily a one-to-one correlation between needs and capabilities; their granularity of needs and capabilities varies from fundamental to complex, and any given need may require the combining of numerous capabilities, while any single capability may address more than one need. The perceived value of SOA is that it provides a powerful framework for matching needs and capabilities and for combining capabilities to address those needs.

In the remainder of this article the approach of the Hewlett-Packard company to SOA architecture development is described.

What are the basic benefits of SOA?

SOA:

- Decreases cost by driving reuse and vendor agnosticism;
- Increases adaptability by allowing easy modification of IT solutions to meet changing business demands;
- Speeds up development by moving developer focus from software functions to business functions;
- Helps visualize business processes by creating a tighter link between business needs and IT capabilities;
- Focuses on simplicity and modularity to radically reduce development costs;
- Enables services to be developed ahead of business process definition.

SOA consists of switching on working systems and processes; and common standards of creation and integration of IT organization systems. It implies modernization of existing systems if it is necessary; and the existence of a flexible architecture model; it comprises an integral solution to build service-oriented infrastructure; and there are several variants for its realization.

The functional view of the SOA solutions reference model (Figure 17.2) envisages the functional components required in an enterprise's SOA environment. It consists of services and a service infrastructure base upon which the services sit. The SOA environment is shown on the left, in relation to an enterprise's existing pre-SOA environment on the right.

Pre-SOA environment

Typically, an enterprise's pre-SOA environment consists of applications making use of an application infrastructure, layered over a common technology infrastructure. Applications make use of the application infrastructure by running on application servers that are interconnected by a middleware component. Examples of applications are packaged applications (such as SAP, Siebel, Oracle applications), legacy applications, and custom applications (such as custom-built telecommunications billing application, health care application). The applications often are accessed via application clients with some sort of user interface (GUI, web browser) that are within the presentation tier. These application clients may be run within a portal framework.

The SOA environment

Services make use of the service infrastructure to run. The service infrastructure should have the following basic components:

- An enterprise service bus,[1] which provides a SOAP engine, transports, routing, and transformation of messages;

[1] This term was coined by the Gartner Group: "An enterprise service bus (ESB) is a new architecture that exploits web services, messaging middleware, intelligent routing, and transformation." (Source: Roy Schulte, Gartner Group).

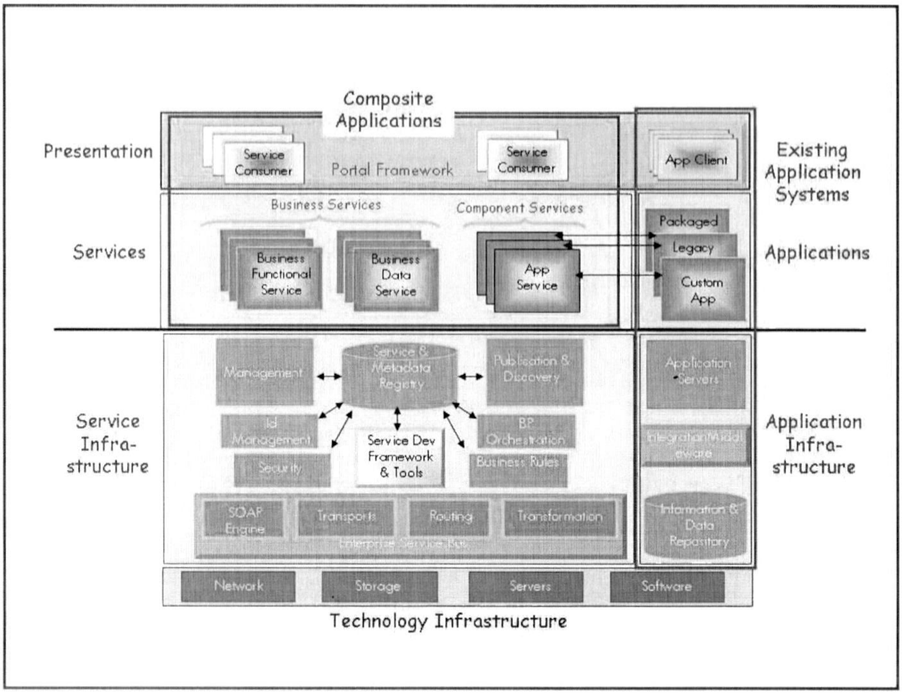

Figure 17.2: The functional components required in an enterprise's SOA environment.

- A service and metadata registry, used to locate all an available information about the services;
- A publication & discovery component, which should be standards-based (e.g. UDDI, which was designed specifically for the publication of, and searching for, services;
- A business process orchestration component to automate business processes;
- A business rules component to manage and execute business rules;
- A management component to provide various aspects of web services management, including policy management;
- An identity management component to manage identities, including attributes and properties of identities such as contact details, personal preferences, roles and memberships, access permissions etc;
- A security component to provide various aspects of web services security.

The service infrastructure components mentioned are all used mainly at run-time, although publication & discovery may also be used at design-time. For design-time, additional components of the service infrastructure, depicted by a different shape, are a service development framework and service development tools.

Services, broadly speaking, can be classified into two main types—business services and component services. Business services imply IT automation of the capabilities that are provided by the business unit as services to its business customers. These are the customer-facing services. Component services imply IT automation of the supporting capabilities that are required in order to provide the Business services. Component services are generally not intended to be customer-facing, but provide the back office capabilities. The Component services are often implemented by creating web services out of back end applications. Services are invoked or used by software components known as service consumers, whereas applications are accessed by Application clients.

The technical view of the SOA. Solutions reference model (Figure 17.3) shows how the functional components of an SOA environment relate to each other in terms of message flow through the environment. Messages normally arrive through the delivery channels of portals, web UIs, or web service routers. They then go through security checks for various security-related controls. At the same time, the messages, in flowing through the ESB, are also monitored and checked by the service management components that are often integrated with the ESB.

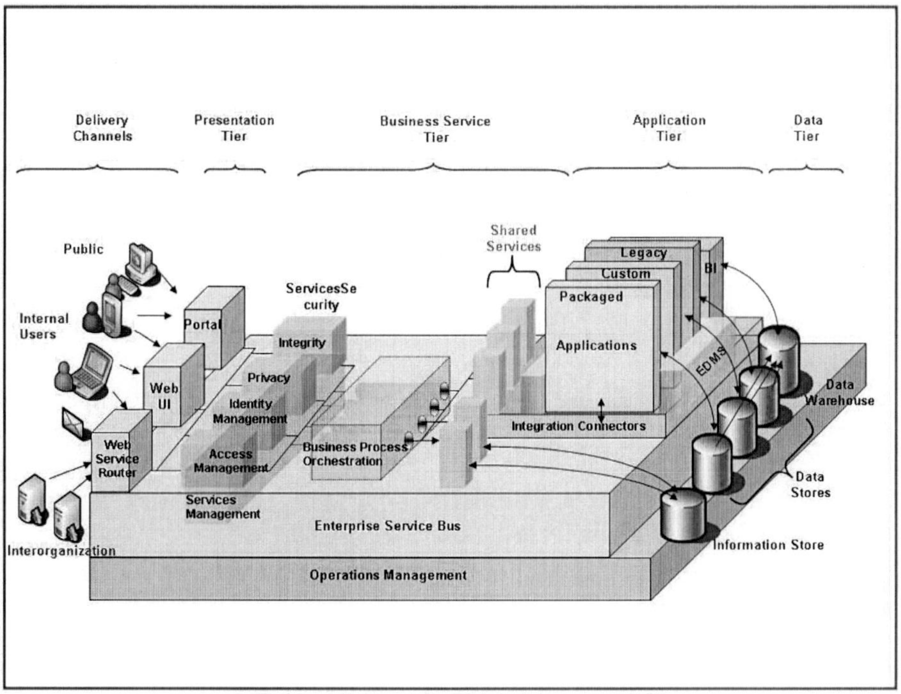

Figure 17.3: Functional components of an SOA environment.

If business process management (BPM) or orchestration is used, the messages are then passed onto this component for orchestration of the business processes. The orchestrations define where messages are destined, and which shared services are to be called upon to support the business process. The services registry and metadata repository may hold information for the control of security (e.g. what security policies to apply), monitoring (e.g. what parts of the messages to monitor), transformation (e.g. format change, fragment extraction) and destinations, of messages. Where the service is implemented by back-end applications (packaged, custom, legacy), the messages may go through some transformation in the integration connectors before being sent onto these applications. Applications and services make use of various information stores and data stores to support their functionality. Finally, the whole environment sits on top of an operations management base that provides the basic technology infrastructure and lower-level management.

As the technical view shows how messages flow through the various SOA functional components at run time, the portfolio management, governance, development and Testing components are shown in a different dimension from the SOA service infrastructure.

Bibliography

Chernyak, L. Open systems and problems of complexity. Otkrytye systemy 8 (2004).

Chernyak, L. SOA—steps beyond the horizon. Otkrytye systemy 9 (2003).

Service Oriented Architecture. Reference Models. Hewlett-Packard Development Company, L.P. (2005).

Complexity and Security
J.J. Ramsden and P.J. Kervalishvili (Eds.)
IOS Press, 2008

Chapter 18

The synthesis of information protection systems with optimal properties

Alexei Novikov and Andrii Rodionov

National Technical University of Ukraine
(Kiev Polytechnic Institute)

Abstract. This article describes the general problems, classified by types of threats, of information security in modern information and communication systems. Classification of the typical problems in the synthesis of information security systems are suggested with optimal properties. Moreover, a methodology for the problem of setting the allocation of the protection tools, and the protection strength parameter synthesis, is described.

18.1 The problems of information security in modern information and communication systems

The volume of information that is circulating and accumulating in modern information systems is increasing almost exponentially. Currently the technologies for building information systems with an open architecture (open systems) are being actively developed, and the open systems belong to a class of complex systems (Gasser, 1988] An example of an open system is a typical corporate Intranet—a network, in which the exchange of information between a basic structure and users is realised through the Internet (Figure 18.1).

In this case, exchange of information is formalized by use of the seven levels

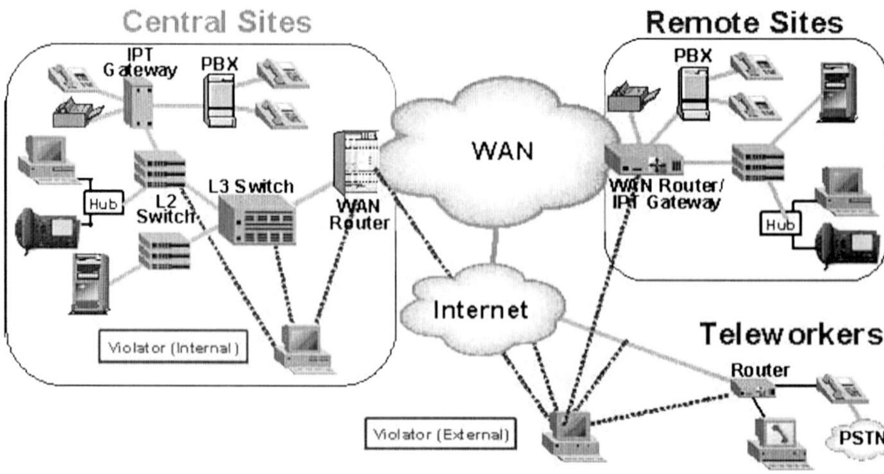

Figure 18.1: Network reference model.

of the ISO-OSI open systems interconnexion model (Figure 18.2). Open systems have advantages and serious weaknesses. The advantages of the open systems are: interoperability—the possibility of complex coöperation between heterogeneous computer systems; portability—the ability of transferring application programs between different platforms; and some other characteristics.

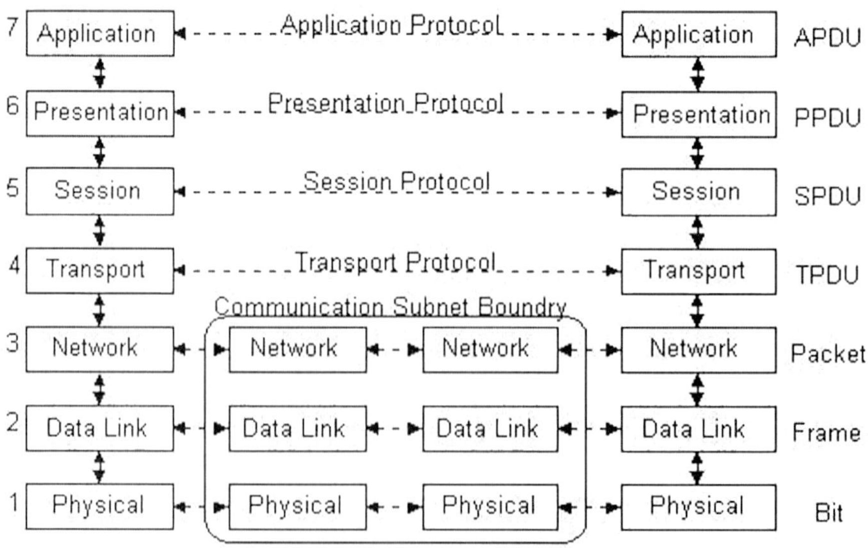

Figure 18.2: ISO-OSI open systems interconnexion model.

A serious weakness of the open system is the high level of threats capable of compromising the safety of information. The common ones are: the threats of violating confidentiality, integrity and accessibility of the information properties (CSTB, 2002). A possible violation of confidentiality of information may be through unauthorized listening to the traffic ('sniffing') by the violator;

although unauthorized penetration into the data warehouse, where an authorized object of telecommunication exchange is substituted by an unauthorized object ('spoofing') (Figure 18.3). The threat to information integrity is realized through attacks, such as interception of information and its substitution, modification, or destruction during transmission and penetration into the data warehouse (Figure 18.4). The threat to information accessibility is realized through DoS (denial of service) attacks and others (Figure 18.5). The problem of the synthesis of information security systems in modern information and communication systems is being actively developed at present.

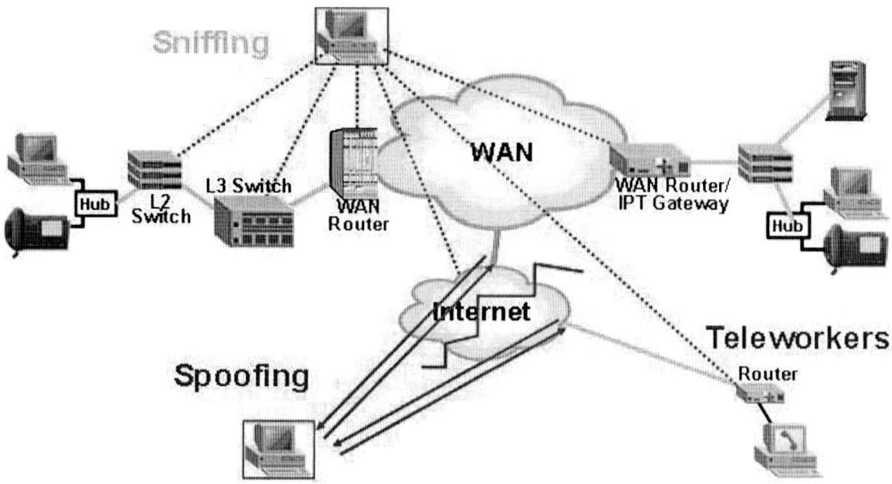

Figure 18.3: Threats of confidentiality violation.

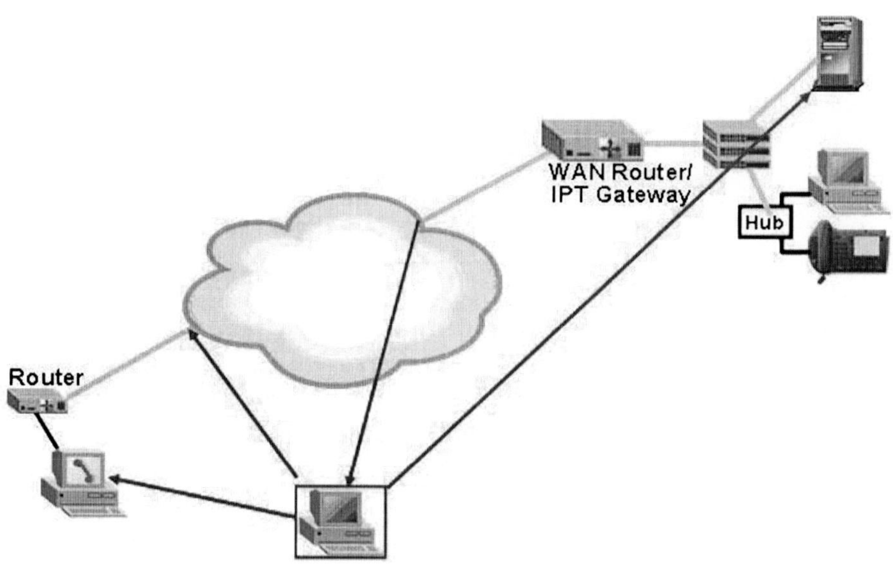

Figure 18.4: Threats of integrity violation.

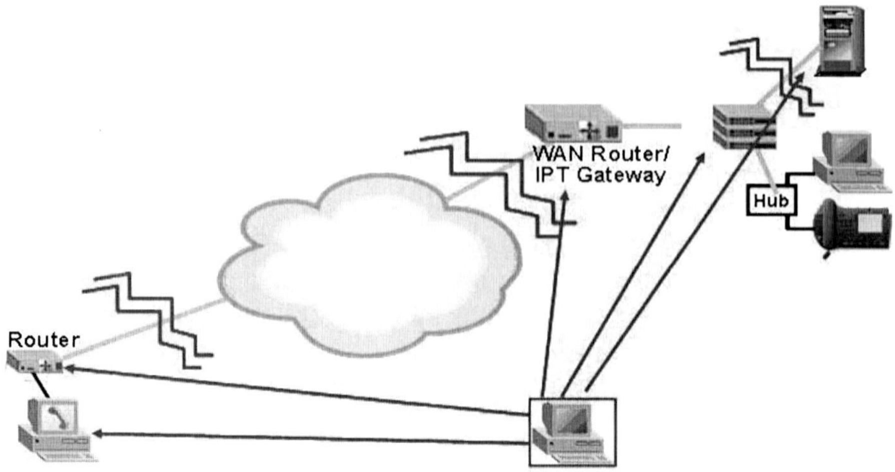

Figure 18.5: Threats of accessibility violation.

18.2 The methodology of synthesis of information security systems in information and communication systems

In the field of synthesis of an information security system there are some methodological problems, among which the most important are: the *problem of complexity in analysing the system* and the *problem of synthesis of the "guaranteed secure information of security systems"* (Harrison et al., 1976). The problem of complexity in analysing information security systems can be solved using the method of organizing the security systems into a hierarchy. According to this method the whole complex system can be described by several hierarchical levels: the level of security policy, the level of supporting security policy systems (access control—mandatory or discretionary control, auditing and other methods); the effectiveness of the security tools (cryptography protocols and algorithms); and the level of realization of the security tools (virtual memory, protected modes of the processor).

For each level, the analysis problems are solved by using specific methods: the level of security policy (foresight of the system's behaviour—modelling the reliability of the security systems, analysis of risks through the threat realization capacity), the level of systems supporting security policy (access control—modelling of mandatory or discretionary access policies); the effectiveness of the security tools (modelling of cipher strength); and the level of realization of the security tools (modelling the reliability of the devices or facilities—memory, processors).

The solution to the methodological problem of synthesizing the "guaranteed safety system" is achieved in three stages: the first stage lies in defining the 'proper' level of security, which depends on the level of information value. The second stage involves quantifying the safety systems in terms of standards of security. The third stage identifies particular architectures with the 'proper' level of security by using security standards in the form of profiles of functional security and the security tools.

Usually a 'proper' level of security is not published. Hiding this information makes the threat of violation more complicated when estimating their resources (capacity of the devices, the amount of time available) required for breaking the information security system.

The problems of effective usage of the security tools must take into account the cost, risk, and reliability and are left to the system developer. At this stage of the synthesis computer aided design (CAD) tools may be used.

18.3 Typical problems of the synthesis of information security systems with optimal properties

The classification of typical problems are shown in Table 18.1.

Table 18.1: Typical problems of synthesis of information security systems.

Class of the problem of optimal synthesis	**Criterion**	**Restriction**
Structural synthesis of ISS[a] with an optimal level of information protection	Maximizing reliability	A fixed cost for the ISS
Parametric synthesis of the ISS with an optimal level of information protection	Maximizing reliability	A fixed cost for the ISS
Structural synthesis of the ISS with optimal economics	Minimizing cost	A fixed level of information security
Parametric synthesis of the ISS with optimal economics	Minimizing cost	A fixed level of information security
Combined problems of structural and parametric synthesis	Combined	Combined

[a] Information security system.

The synthesis structure of the information protection system with the optimal level of information safety (Timoshenko and Novikov, 2002) synthesis of parameters of information protection with the optimal information safety level (Novikov et al., 2007) are critical, and technological indicators are used when determining quality and economical indicators are used with regard to limitations (Timoshenko and Novikov, 2002), (Bonya and Novikov, 2007). A contrary set of problems may arise if only the most economical solution is sought.

18.4 The problem of structural synthesis of the information security system with an optimal level of information protection

As an example of structural synthesis consider the problem of protection tool allocation (Timoshenko and Novikov, 2002). In order to formalize the decision, consider the widespread protocol stack TCP/IP. The seven levels of the ISO-OSI open systems interconnexion model may be concatenated to four: the level of network accessibility (or lower), the network level (or Internet level), and the transport or application level (Figure 18.6).

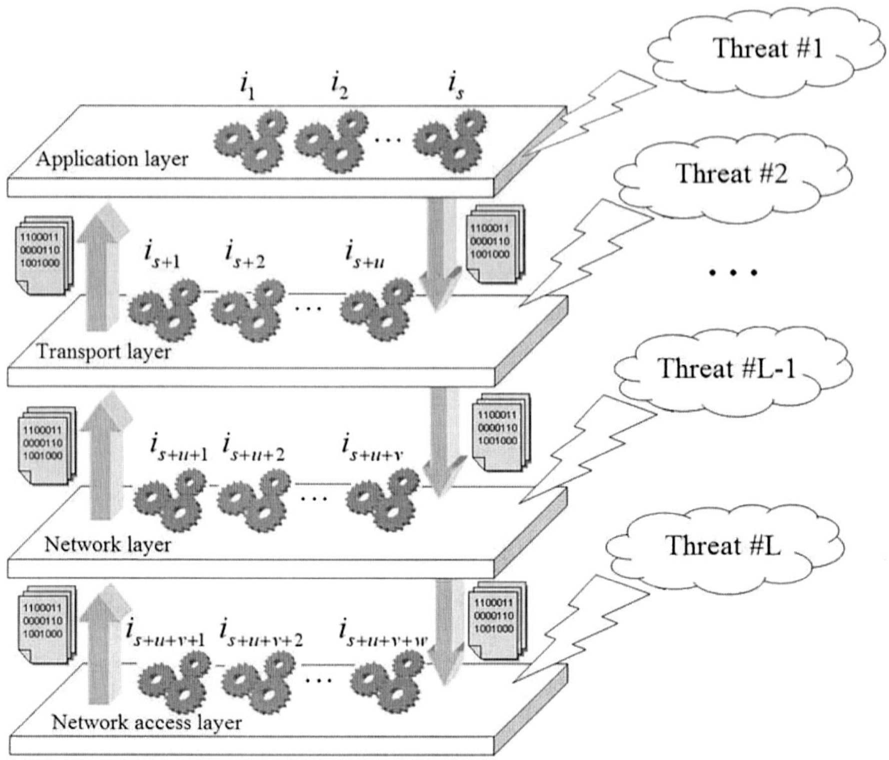

Figure 18.6: Four level stack TCP/IP protocol.

Let us take $P(M)$, the probability function of saving information for protection under conditions of limited cost ($C_{\max}$) of the protection tools:

$$P(M) = \begin{cases} \prod_{i=1}^{L}\left[1 - \sum_{j=1}^{N}\left(E_{ij}\prod_{k=1}^{j-1}(1-E_{ik})\prod_{k=1}^{j}(1-\alpha_{ik}M_{ik})\right)\right] \to \mathcal{M}\ , \\ \sum_{i=1}^{L}\sum_{j=1}^{N} M_{ik}C_{ij} \le C_{\max}\ , \end{cases} \tag{18.1}$$

where $\mathcal{M}$ represents $\max_{M_{ik}\in\{0,1\}}$, a_{ik} is the coefficient of strength of the ith protection tool, realized on the kth level, α_{ik} takes values in the interval [0,1]; N is the number of levels of the protocol stack; L is the number of threats; E_{ik} is the index of efficiency of realization of the ith threat to the information on the kth level; M_{ik} (taking a value of 0 or 1) determines the presence of the ith protection tool operating on the kth level; C_{ik} is the cost of using protection tool M_{ik}. The coefficients were developed using probability methods when testing the reliability of the structure. This class of models forms a risk dependency which is defined as a function of the probability of preserving the protection system from probable characteristic threats. The methods of linear Boolean programming and separable programming are used to optimize the equations. Results are shown in Figure 18.7 (Novikov et al., 2007).

where $M = [m_{1,1}$ (the encoding of data); $m_{1,2}$ (filtration); $m_{1,3}$ (authenti-

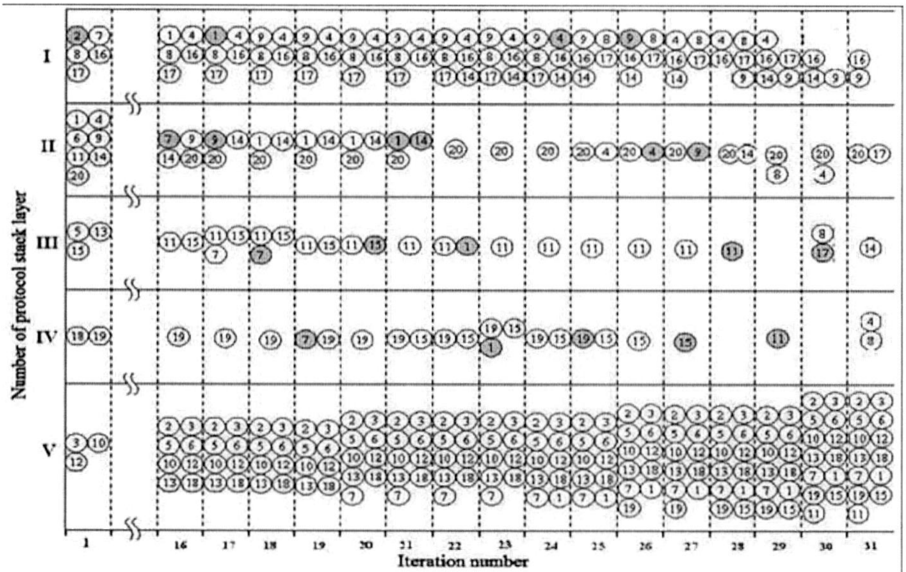

Figure 18.7: Allocation of security tools.

cation); $m_{1,4}$ (digital signature); $m_{2,1}$ (routing); $m_{2,2}$ (checksum); $m_{2,4}$ (access control)].

18.5 The problem of parametric synthesis of an information security system with an optimal level of information protection

An example of parametric synthesis follows Novikov et al. (2007). The aim is to calculate the strength characteristics of the information protection tools at a sufficient level. The level of safety may be taken from the optimal safety level with defined security structure either predefined or as a result of optimization (calculated above).

Let us take $P(\alpha)$, the probability function of saving information for protection in the information system under the conditions of an existing security tool structure M^*, and try to find strength tool parameters, providing the probability P_{thr}, a sufficient level (threshold) of protection.

$$\begin{cases} I(\alpha) & = [P(\alpha)|_{M=M^*} - P_{\text{thr}}]^2 \to \min_\alpha ; \\ \alpha & = \{\alpha_{ij}, 0 \le \alpha_{ij} \le 1 . \end{cases} \quad (18.2)$$

The P_{thr} can be taken as a equal to βP_{opt}, where β lies the interval [0,1] and P_{opt} is the probability function of saving information for protection with this security tool structure:

$$P_{\text{opt}} = P(M^*)|_{\alpha=1} . \quad (18.3)$$

The solution of this problem can be found using a gradient method:

$$\alpha_{ij}^{\xi+1} = \Pr\{\alpha_{ij}^{\xi} + \lambda \frac{\partial I(\alpha)}{\partial \alpha_{ij}^{\xi}}\} . \tag{18.4}$$

Here ξ is the number of the gradient iteration, λ is the step of the gradient procedure and Pr is the projection on an acceptable definition interval. Some manipulation gives:

$$\frac{\partial I(\alpha)}{\partial \alpha_{ij}} = 2[P(\alpha) - P_{\text{thr}}] \frac{\partial P(\alpha)}{\partial \alpha_{ij}} ; \tag{18.5}$$

$$\frac{\partial P(\alpha)}{\partial \alpha_{ij}} = \frac{\partial P_i(\alpha)}{\partial \alpha_{ij}} \prod_{k=1}^{i-1} P_k(\alpha) \prod_{k=i+1}^{N} P_k(\alpha) ; \tag{18.6}$$

$$\frac{\partial P_i(\alpha)}{\partial \alpha_{ij}} = M_{ij} \sum_{k=j}^{N} E_{ik} \prod_{l=1}^{k-1} (1 - E_{il}) \prod_{l=1}^{j-1} (1 - \alpha_{il} M_{il}) \prod_{l=j+1}^{k} (1 - \alpha_{il} M_{il}) ; \tag{18.7}$$

and

$$P_k(\alpha) = \prod_{l=1}^{N} (1 - E_{kl}) + \sum_{l=1}^{N} \left[E_{kl} \prod_{m=1}^{l-1} (1 - E_{km}) \left(\sum_{n=1} \alpha_{km} M_{km} \prod_{n=m+1}^{l} (1 - \alpha_{kn} M_{kn}) \right) \right] . \tag{18.8}$$

The gradient procedure begins with initial parameters a_{ij}^0 and finishes when the termination criteria are fulfilled:

$$|I\xi + 1(\alpha) - I^{\xi}(\alpha) \leq \varepsilon \tag{18.9}$$

where ε is a solution error. With four levels of protocol stack, we have

$$P_{\text{thr}} = 1. \tag{18.10}$$

For the threshold level of protection take 80% of the optimal security level,

$$P_{\text{thr}} = \beta P_{\text{opt}} = 0.8 . \tag{18.11}$$

The efficiency of threat realization on information according to Bonya and Novikov (2006) is contained in: $E = [e_{1,1} = 0.3$ (corresponds to sniffing of information traffic disclosure of information); $e_{1,2} = 0.05$ (corresponds to the denial of service (DoS)); $e_{1,3} = 0.15$ (corresponds to unauthorized access); $e_{1,4} = 0.4$ (corresponds to violation of information integrity); $e_{2,1} = 0.05$ (corresponds to spoofing of information traffic); $e_{2,2} = 0.35$ (corresponds to violation of information integrity); $e_{2,4} = 0.6$ (corresponds to unauthorized access)]. The iterative procedure for finding the solution begins from the starting parameter $a_{ij}^0 = 0$ with $i = 1, 2$; $j = 1, 4$ and ends in 140 steps (Figure 18.8).

The resulting coefficient of strength of any protection tool i, realized on level k for the optimum protection of information, is: $\alpha = [0, 1] : \alpha = [\alpha_{1,1} = 0.78$ (encoding of data); $\alpha_{1,2} = 0.26$ (filtration); $\alpha_{1,3} = 0.22$ (authentication); $\alpha_{1,4} = 0.14$ (digital signature); $\alpha_{2,1} = 0.68$ (routing); $\alpha_{2,2} = 0.56$ (checksum); $\alpha_{2,4} = 0.23$ (access control)]. Perhaps, not surprisingly, data encryption is the most effective protection at the application layer.

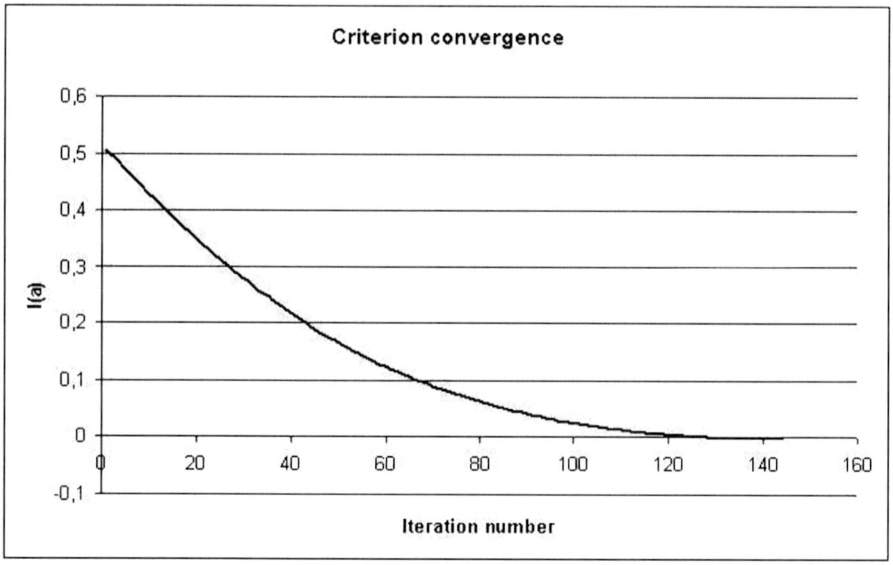

Figure 18.8: Criterion convergence.

18.6 Summary

In this paper, we have described the complex problems of information security system synthesis with optimal properties. Information security system synthesis is only small part of this complex system. Two further areas concern:

1. The analysis (or reverse engineering) of an existing information security system, where the level of information safety provided requires to be quantified and compared to the demands of a security policy;

2. The inverse problem (or penetration testing), where the information available to an intruder is estimated in order to assess the likely damage of a potential attack.

18.7 References

Bonya, Y.Y. and Novikov, A.N. (2006). Synthesis of information protection systems with a minimum cost of protection mechanisms. J. Automation Information Sci. 38, 72–82.

Bonya, Y.Y. and Novikov, A.N. (2007). Synthesis of information protection systems with optimal risk level. Pravove, normativne ta metrologichne zabezpechennya sistem zahistu informacії v Ukraïni 3, 50–54.

Gasser, M. (1988). Building a Secure Computer System. New York: Van Nostrand Reinhold.

Harrison, M., Ruzzo, W. and Ullman, J. (1976). Protection in operating systems. Comm. ACM, 19, 461–471.

Novikov, A.N., Rodionov, A.N. and Timoshenko, A.A. (2007). Optimal parameters synthesis for an information security system. Naukovi visti NTUU

KPI 4, 146–151.

Timoshenko, A.A. and Novikov, A.N. (2002). Defining security tools structure for optimal information system protection. Pravove, normativne ta metrologichne zabezpechennya sistem zahistu informaciï v Ukraïni 4, 98–105.

CSTB (Computer Science and Telecommunications Board) (2002). Cybersecurity Today and Tomorrow: Pay Now or Pay Later. Washington D.C.: National Academy Press.

Complexity and Security
J.J. Ramsden and P.J. Kervalishvili (Eds.)
IOS Press, 2008

Chapter 19

Complexity and security of coupled critical infrastructures

O. Udovyk

National Institute for Strategic Studies, Kyiv, Ukraine

Abstract. An overview is provided for policy-makers and opinion leaders of the physical structure and of the governance structures and processes for electricity, gas and water supply, transport and systems for general information and communication services. Their vulnerabilities and the main drivers of these vulnerabilities are also summarized, as well as possible political and institutional shortcomings. Based on these findings, a number of technical, management and organizational strategies and policy options are outlined, which may help to reduce the probability of disruption to these systems and consequent interruptions to the vital services they supply. Additionally, some suggestions are offered for areas in which further study may be needed before definitive policy recommendations can be made.

19.1 Introduction

Modern civilization needs a set of systems that supply energy and information. In at least limited ways, these systems have always been dependant on each other. Recent decades have witnessed a much greater and tighter integration and interdependence between them—effectively the creation of a 'system of systems', which has no single owner or operator.

While this has often yielded improved service and convenience and promoted greater efficiency, it has also led to increased social vulnerabilities in the face of accidental or intentional disruption. Today, a disruption or malfunction often has much greater impacts than was typically the case in the past, and can also propagate to other systems, resulting in further additional disruptions.

This paper focuses on risk assessment for next infrastructures: the electric power network; gas and water supply systems; transport; and, general information and communication services particularly as provided by the Internet as well as ICT as used to monitor and control other infrastructures.

These infrastructures are highly complex and interconnected, challenging our abilities and willingness to assess and understand their vulnerabilities and to take appropriate actions to reduce these vulnerabilities.

Typically, these actions involve increased costs, which must be paid for through increased service prices or from other sources such as government subsidy. The systems are all subject to increased stress, to different degrees, and are also dependent on the different market environments and operational contexts. All of these factors raise questions concerning conflicting objectives and the adequacy of risk governance.

Therefore further efforts are needed to understand these complex issues, to share that understanding with decision makers and the public, and to increase cooperation among the parties responsible for risk management of these systems. These parties include the system owners and operators, and governmental departments, agencies and regulators at levels extending from local to regional, national, and international.

19.2 Characteristics of critical infrastructures

Much is being written about critical infrastructures at the present time (see, for example, the U.S. Patriot Act, 2001) but just what is meant by this term? Different authors adopt slightly different meanings, with a recent EU communication document (EC, 2004) listing many examples.

A more academic explanation of the term is given by (IRGC, 2006): critical infrastructures are a network of large-scale, man-made systems (set of hard and soft structures) that function collaboratively and synergistically to produce a continuous flow of essential goods and services and are, finally, essential for economic development and social well-being.

19.2.1 Risk-shaping factors

The infrastructures are coupled or interconnected to different degrees and finally must be regarded as a 'system of systems'. Their operating strategies and end-user behaviors are subject to significant contextual changes and an increasing number of risk-shaping factors (Table 19.1).

19.2.2 Assessment matrix

Focusing on society as a whole at a higher level, the criticality of the system can be described (EC, 2004) in terms of scope (extent of geographic area affected), magnitude (degree of impact or loss) and effects of time. Based on the summary of each infrastructure given in the References, Figure 19.1 provides a template (assessment matrix) for an initial assessment of the characteristics of the infrastructures.

This figure addresses certain of the different dependencies between the infrastructures. For example, transport relies on continuous electricity supply and

Table 19.1: Risk-shaping factors.

Market organization (e.g. competition, oligopoly, monopoly, hybrids)
- Transition from one market system to another and the speed of transition
- Control structure (unbundling, ownership patterns, legally operational rules)
- Investment incentives and financial risks (maintenance and new facilities)
- Business principles (redundancy versus cost of service trade-off, profit max)
- Price and price regulation as paradigms: how price of service is based on cost
- Behavioral issues (e.g. of corporate and political leaders, service end-users)
Government policy-making (e.g. renewable, nuclear energies)
Legislation/regulation (responsibilities, institutional complexity, differences within integrated networks, e.g. between EU and non-EU member States)
Technology-related
- Potential for storage; inherent inertia
- Localized versus pan-state and multi-state vulnerabilities
- Customized versus off-the-shelf systems
- Susceptibility to failures and/or accidents
- Speed of developments and/or innovations
Infrastructure-related
- Degree of 'criticality', potential for choice
- Technical design and operating principles (e.g. the N-1 criterion,[1] maintenance)
- Space extension and exposure
Degree of interconnectedness, complexity
- Interdependences within single infrastructures
- Interdependences across infrastructures and regions
Availability of resources
- Shortage, depletion of scarce resources
- Contamination or degradation of supply
Natural conditions (weather) and hazards
Context of risk and threats, openness of society
- Attractiveness for, and vulnerability to, malicious attacks (cyber, terrorism)
- Public acceptance and risk awareness
- Strategic issues

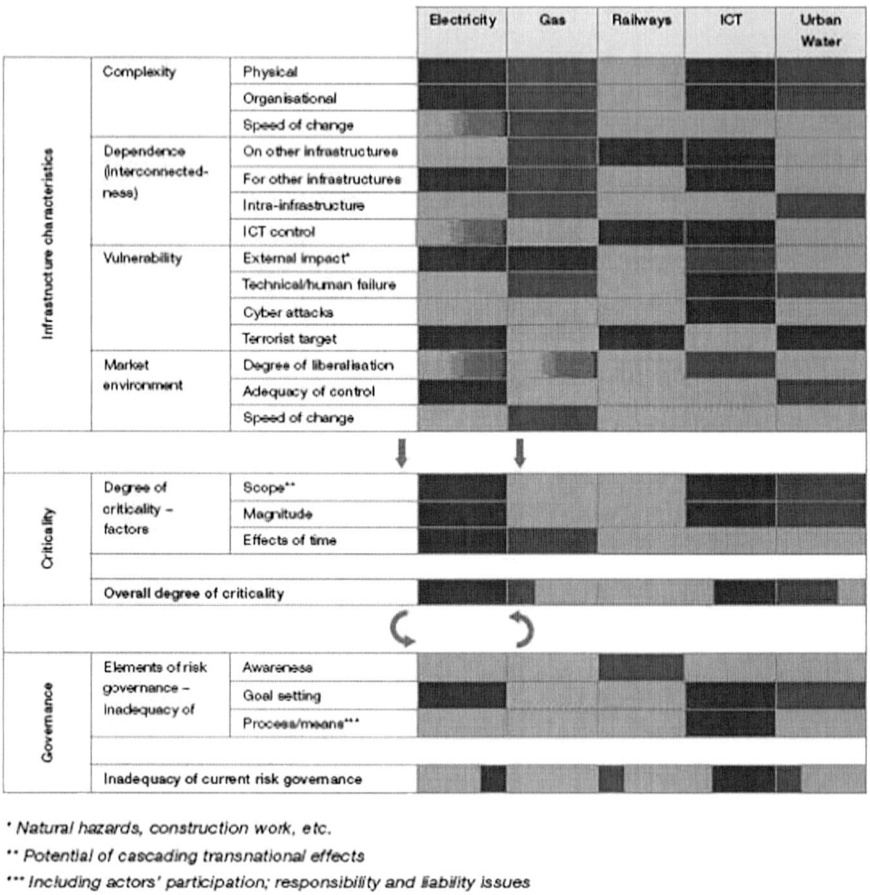

Figure 19.1: Assessment matrix for critical infrastructures. Tones are used for gaining a rapid initial judgment of the interdependencies between the infrastructures. Dark grey corresponds to strong interdependency: for example, transport relies on continuous electricity supply and ICT support, hence the corresponding cells are shaded dark grey. More moderate interdependencies are shaded grey; transitions from one tone to another within a rectangle indicate changes/trends.

ICT support, so these cells are marked dark grey. The importance of electricity to other infrastructures and the associated dependencies are more moderate and thus are marked grey. This assessment matrix may provide initial guidance (in the absence of more detailed assessment and analysis) on where to put emphasis on risk governance strategies, and how to tailor the measures that are outlined below.

19.3 Risk governance strategies

Risk governance in the context of critical infrastructures includes the totality of players, rules, conventions, processes, and mechanisms concerned with how relevant risk information is collected, analysed and communicated and management decisions are taken.

Strategies to reduce the probability of disruption to services provided by infrastructures, as well as the social vulnerabilities associated with them, should encompass technical, management, and organizational measures. Adequate strategies must consider the different characteristics of the various infrastructures such as their complexity, dependencies and interconnectedness, as well as such important contextual factors as the market environment (Figure 19.2).

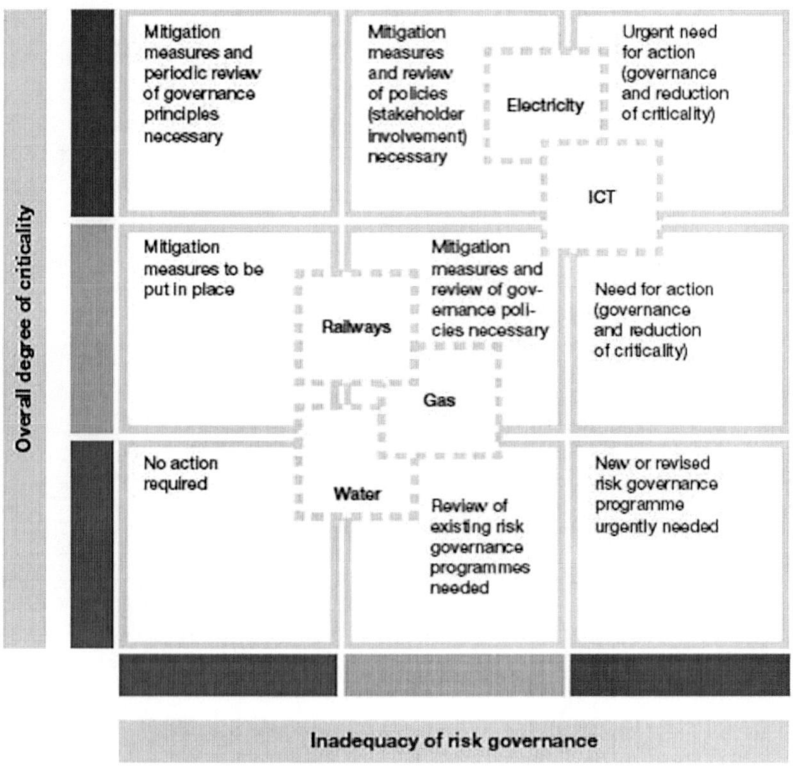

Figure 19.2: Inadequacy of risk governance.

19.3.1 Step by step

Infrastructures can be vulnerable to a variety of events including failures of system components, human errors, natural hazards such as extreme weather conditions or earthquakes, and malicious attacks. Critical infrastructures are vulnerable to a variety of disruptions.

The first step in guarding against such events is an adequate assessment of the range of possible accidental and intentional disruption scenarios as well as of possible weaknesses, including 'bottlenecks'.

Having analysed the events that might give rise to system failure, the next step is to perform contingency and failure analysis appropriate to meeting pre-agreed societal needs and objectives (e.g. appropriate levels of security; balanced degree of redundancy; alignment of the criteria for automatic protective devices with those needs; etc.).

However, in many important areas, there is as yet no agreement on such needs and objectives, especially in an international context. Simple safety criteria (N-1, N-2)[1] and failure consequence methods are widely used in assessing and in shaping the design of many infrastructures. In many of today's complex systems, more sophisticated approaches are needed.

In the creating or modifying of such rules, consideration must be given to balancing conflicting social objectives. Market mechanisms may play a role in this process, but much of the need is for improved and more explicit political objectives and enabling frameworks, especially at the international level.

Of course, rules alone are not sufficient. System operators must also know what is happening so that they can take informed actions. This means that, for safe and reliable system operation, one must have real-time situational awareness and emergency preparedness along with adequate system-wide scope based on improved instrumentation and communications. The need is perhaps greatest in the case of electric power and transport. However, in a world in which terrorism is a growing threat, improvements are also needed in a number of other settings, such as urban water distribution systems, gas and oil supply.

---

[1] The N-1 security criterion specifies that any probable signal event leading to a loss of a power system element should not endanger the security of the interconnected operation, that is, trigger a cascade of tripping or the loss of a significant amount of consumption. The remaining network elements, which are still in operation, should be able to accommodate the additional load or change of generation, voltage deviation or transient stability régime caused by the initial failure. It is acceptable that in some cases, transmission system operators (TSOs) show a loss of consumption in their own areas on condition that this amount is compatible with a secure operation, which usually implies predictable and locally limited. A TSO monitors the N-1 criterion for its own system through observation of the interconnected system (its own system and some defined parts of adjacent systems) and carries out security computations for risk analysis. After an unexpected event occurs, each TSO works to rapidly restore its power system to an N-1 compliant condition and, in case of any delay, immediately informs other TSOs affected. The approach is deterministic and does not address the possibility of occurrence of more than one failure. Its methodological deficits and inappropriate application of the N-1 criterion have clearly contributed to major blackouts. While the N-1 criterion deals with the ability of the transmission system to lose one linkage without causing an overload failure elsewhere, the N-2 criterion is a higher level of system security, dealing with the ability of the system to withstand any two linkages going down. See also the NATO Energy Security Forum, 24 February 2006, Prague.

19.3.2 Identification and prediction

Before one can address the risks posed by potential common-cause or causal failures, they must first be identified. That is often not easy to do and requires careful and extended data collection and analysis informed by real-world experience. One solution is to add independence, redundancy or spatial separation, but these can also add unintended complications.

The performance of large complex interconnected systems is not easy to predict. In some cases, such as the electric power system and many ICT systems, the complexity can be so great that complete analysis is simply not possible. Nevertheless, more comprehensive and holistic approaches need to be undertaken and, for many areas, more sophisticated methods developed.

ICT systems present a range of challenges for all of the infrastructures. Many key systems for situational awareness and control are still highly vulnerable to accidental or intentional disruption or spoofing. Such systems should not make use of, or be interconnected to, the public Internet, which is inherently insecure and will remain so for the foreseeable future. However, at present, a number of such systems are connected to the Internet and are thus vulnerable to accidental disruption or intentional cyber attack. Further investigation and actions to reduce such vulnerabilities are urgently needed. Adequate physical system maintenance and support are also vitally important.

A number of critical infrastructures suffer from the fact that they have grown in a rather unplanned and unstructured way, sometimes without basic changes in operation and control. Often, decentralized control areas are maintained while the system has expanded spatially, hence requiring better coordination and data exchange. Coherent expansion planning and associated capacity expansion is critically important if these systems are to evolve in ways that are consistent with the interests and needs of all affected parties.

However, such systematic planning can run counter to market competition objectives and privatization. Gradually, strategies are being evolved to reconcile these tensions but, in the case of several infrastructure systems, much additional attention is needed.

19.3.3 New technologies

New technology, such as more capable SCADA systems, can sometimes play an important role in relieving previous technical or institutional constraints, as well as providing new functionality. But this may also introduce new vulnerabilities.

Even the best-designed systems will fail occasionally. When this happens, operators may never have experienced such circumstances before and may not know how to react. Several actions can reduce the risks in such circumstances:

- Designs that support 'graceful' degradation of capabilities ('island solutions' in power systems' control and grid structure and reduced bandwidth and traffic priority in ICT, to give two examples);

- Demand management, including priority setting;

- The incorporation of rapid-acting, cooperating, distributed autonomous computer control agents;

- Careful contingency preparation, including operator training conducted in realistic simulators.

Attention should also be directed at enabling critical social services to continue to operate in the face of primary system failure. Thus, for example, if the natural gas system fails or is degraded, storage of fuel near the gas turbine may be needed to assure that the pumps can continue to run in the event that both the electrical grid and the natural gas systems are unavailable.

Similarly, if traffic lights are converted to low power Light Emitting Diodes (LEDs) and backed up with solid state controls and trickle-charged batteries, traffic can continue to flow, even when the power goes out.

Since occasional service outages are unavoidable in a world with storms, floods, earthquakes, and terrorism, it is important that system operators maintain equipment and prepare effective plans for the rapid restoration of services. This deceptively simple observation carries some very profound implications in terms of stockpiling critical components and sharing resources among different system operators, as well as training and preparing work crews.

19.3.4 Standards

Here a number of strategies to promote the growth of effective system design standards without resorting to inflexible government regulation are suggested, including:

- Best professional practice
- Certification
- Acquisition specification
- Legal frameworks
- Tort and liability
- Insurance
- Taxes or fees on uncertified systems

The combined effects of such actions could prove far reaching and widespread adoption of best professional practice and certification standards should, over time, help to create a culture in which system designers routinely think about issues of anonymity and security as they develop systems.

One way to assure continuity of the services that critical infrastructures provide is to find ways to allow, or perhaps even promote, multiple service routes and providers. This is most easily achieved in telecommunications. It is also possible in electric power through the use of distributed controls, distributed generation, micro-grids, and intelligent distribution system management. But what is possible is not always allowed.

Some critical infrastructure systems (or elements of them) are owned and operated by private parties, some by local or national governments. There is no single owner of coupled infrastructures. Clearly, governance options differ in these two cases. Yet, even if in private ownership, if the system is truly critical, other parties who depend upon the services it provides (end-users) must be given a role in developing the policies and practices that govern its operation and in overseeing their effective implementation. Classical decision-making and risk management processes should be revisited and, where necessary, supplemented or even replaced by more participative governance strategies.

Information technology is evolving so rapidly that in many cases mandatory standards can be counter-productive, seriously impeding innovation or otherwise causing problems. While a few basic rules, such as no use of the Internet for system-critical control functions, make sense, in general a more flexible approach will be more appropriate.

Finally, research can often create new options which can better meet and balance private and social interests. While R&D investments in ICT are substantial (8–10% of sales), too few are focused on addressing issues of security and reliability. In electric power R&D investments are much too low to meet societal needs (< 0.5% of sales). The industry has not had a strong research tradition and restructuring has complicated matters, focusing many players on short-term bottom line issues and creating a 'free rider' problem.

Unless R&D investments are mandated for all players as a 'cost of doing business' it is difficult to see how this situation can be expected to change. This could be done by specifying that some proportion of value added (e.g. 1%) must be invested in R&D. Firms that do not want to bother to manage such research could be required to support a government R&D programmer.

19.4 Towards an integrative approach

This initial study examined critical infrastructures, issues of interdependencies between them, and a number of socio-economic, contextual and physical factors which impact on them; it is acknowledged that there are other important infrastructures that have not been considered.

An infrastructure-by-infrastructure approach has been followed. Further study, involving a region-by-region approach that looks across several infrastructures simultaneously, could provide additional insights, especially if it is expanded to more regions and explores the influence of different cultures, regulatory environments and legal frameworks.

The main focus has been on reducing social vulnerabilities by increasing the reliability and robustness of the systems. There is a need for additional work which focuses on identifying social vulnerabilities and developing strategies to maintain critical services when the main infrastructures on which they depend fail or malfunction.

There is a need to develop and refine appropriate risk and vulnerability assessment methods. This should facilitate more effective assessment of the relative criticality of different infrastructures and related services.

This study has given some consideration to the duration of disruptions, although the principal focus has been on short-term impacts. Some of our conclusions could be different if, for example, long-term impacts have been looked at—a long-term loss of water supply would certainly have enormous criticality.

Future studies should give greater consideration both to more extended disruptions and delayed effects arising from initial disruption, which may persist even after the original service has been restored, and to input supply issues, particularly the security of their supply.

Although a broad spectrum of threats have been addressed, including natural events, human failures and malicious attacks, more work needs to be done on:

- Natural disasters of large spatial extent and duration such as strong earthquakes, hurricanes, ice storms and floods;
- Occurrence of multiple failures or attacks on a system, or simultaneous attacks on several systems, which may amplify total impacts;
- Strikes and other labour actions;
- Epidemics, pandemics, mass evacuation, etc.;
- Longer-term developments such as migration or the impacts of climate change (see Part Energy and Climate).

The importance of stable social and political conditions has not been emphasized, although their importance has been clearly demonstrated by instances of political sabotage and destabilizing activities affecting key industries and infrastructures. Besides direct consequences such as loss of production, the unavailability of ICT support may seriously worsen the situation. More investigations are needed to better understand such complex situations and to propose clear, adequate governance strategies.

19.5 References

[EC 2004] EU Document COM (2004) 702 final, concerning critical infrastructure protection in the fight against terrorism.

[EC 2005a] The Future of ICT for Power Systems: Emerging Security Challenges. Report on a workshop held on 3–4 February 2005 in Brussels.

[EC 2005b] Green Paper on a European Programme for Critical Infrastructure Protection, COM (2005) 576 final. 17 November 2005.

[EC 2006] Green Paper on a European Strategy for Sustainable, Competitive and Secure Energy, COM (2006) 105 final. 8 March 2006.

[Gheorghe 2006] Gheorghe, A.V., Masera, M., Weijnen, M. and de Vries, L. Critical Infrastructure at Risk, Securing the European Electric Power System. Springer (2006).

[IRGC 2005] IRGC White Paper No 1, Risk Governance—Towards an Integrative Approach, Geneva (2005).

[IRGC 2006] IRGC White Paper No 3, Managing and Reducing Social Vulnerabilities from Coupled Critical Infrastructures. Geneva (2006).

[Kirschen 2005] Kirschen D. Why do we get blackouts? Presentation given at the EC Workshop on The Future of ICT for Power Systems: Emerging Security Challenges, February 2005, Brussels.

[Moteff 2003] Moteff, J., Copeland, C. and Fischer, J. Critical Infrastructure: What Makes an Infrastructure Critical? Report for Congress, Order Code RL31556, Congressional Research Service, Library of Congress, January 2003.

[NATO 2006] NATO Energy Security Forum, February 24, 2006, Prague, Westby, J., CEO Global Cyber Risk LLC, Cyber Security, 2006

[OECD 2005] OECD Futures Project on Global Infrastructure Needs: Prospects and Implications for Public and Private Actors, Second Meeting of the Steering Group, Discussion Paper, December 2005.

[Shea 2003] Shea, D. Critical Infrastructure: Control Systems and the Terrorist Threat, Report for Congress, Order Code RL31534, Congressional Research Service, Library of Congress, 2003.

[US Patriot Act 2001] Uniting and Strengthening America by Providing Appropriate Tools Required to Intercept and Obstruct Terrorism (U.S. Patriot Act), Act of 2001.

[WFS 2003] Toward a Universal Order of Cyberspace: Managing Threats from Cybercrime to Cyberwar, Document WSIS-03/GENEVA/CONTR/6-E, World Federation of Scientists, Permanent Monitoring Panel on Information Security, August 2003.

Complexity and Security
J.J. Ramsden and P.J. Kervalishvili (Eds.)
IOS Press, 2008

Chapter 20

The formation of a global information society, digital divide and trends in the Georgian telecommunications market

Otar Zumburidze and Guram Lezhava

Telecom Georgia, Tbilisi

Historically, telecommunications and informatics—two independent fields of science—were formed in different periods of time and passed different stages of development (technological perfection). At the same time, these fields always represented the strategic directions of progress and prosperity.

Recently, we have been experiencing the convergence of telecommunications and information technologies into one, stable and fast developing *Infocommunication industry*. It is considered that in the 21st century this moving force will further the formation of the so called global information society (GIS), the elements of which are already visible to us in the forms of the Internet, mobile and satellite communication systems, fibre-optic telecommunication backbones, multimedia, SDH, WiMax, TriplePlay and many other technologies.

The formation of the GIS is taking place against a background of general moving forces of globalization and trends of the post-industrial development of society in the 21st century. The end of the 20th century was marked by the completion of the age of industrial development and the beginning of the evolutionary transition to the GIS, which, in the opinion of experts, will take 50–60 years, i.e. will last approximately until the middle of the 21st century.

Special summits, conferences and workshops have been devoted to the issues of the stepwise formation and criteria of the GIS, and the leading role of the infocommunication industry and the services within it. The most recent significant events were the World Summit on the Information Society (WSIS) held in Geneva and Tunis in 2003 and 2005 respectively, the "Green Paper on the Convergence of Telecommunications" issued by the European Commission, and "Five Challenges to the Telecom World" by Gore et al. [4,5]. These trends, processes and some solvable problems are shown in Figure 20.1.

The first definition of the term *postindustrial society* as the society of services was given by Bell [2] as far back as 1973. Having used the three-sector model of an economy (manufacturing (industrial, I), agrarian (A) and service sectors (S)) he identified the *postindustrial society* at the beginning of 21st century as the *society of services*. This forecast is apparently being completely fulfilled. According to World Bank data, in 2001 the share of the service sector constituted 66% of the total GDP of 31.1×10^{12} USD of the world community, and for the EU countries the share constituted 69% of their total GDP of 6.1×10^{12} USD.

Figure 20.2 shows the evolution of the shares of these three sectors of the world economy of the total GDP per capita worldwide on average, over an interval of 17 years. Based on these charts we can see that for this period A and I have been decreasing and S increasing along with the growth of GDP. Besides, during the 17 years the GDP per capita and the share of the service sector in almost all countries have increased by 10% [9]. It is therefore quite reasonable to refer to the *postindustrial society* as the *society of services*. The source [9] provides the following summary of evidence for the existence of this global society of services:[1]

- the average share of services in the world economy is equal to 66%, and this value much exceeds the 'majority' value of 50%;
- out of 152 countries the service sector prevails in 112 (i.e. in 74%);
- the economy of postindustrial countries makes up 92% of the world economy;
- out of 63 large countries 58 (i.e. 92% of them) are postindustrial countries where 60% of the total world population lives.

GIS is based on the *global infocommunication complex*, i.e. the *global infocom* (GIC), which can be represented as a kind of the sandwich-pyramid structure, whose foundation represents the integrity of terminals of subscribers (fixed and mobile telephones, faxes, PCs and other), and whose second layer is the different networks of access: PSTN, Mobile, Internet (Intranet, Extranet), Broadband. Then comes the layer of local transport communication networks (RRL, fibre-optic and others), and last comes the layer of global information infrastructure consisting of the global satellite communication system and the global network of transcontinental fibre-optic backbones (the so called global digital communication loop).

[1]The evidence is incontrovertible regarding the quantity of services; but the picture looks very different if quality is taken into account. For example, nowadays many small railway stations in Europe that twenty years ago were fully staffed, offering a full range of real services to the traveller (such as the possibility of leaving luggage), even in remote rural areas, are nowadays unstaffed, and the only available 'service' might be a ticket machine. "Society of services" implies, therefore, some distortion of meaning.

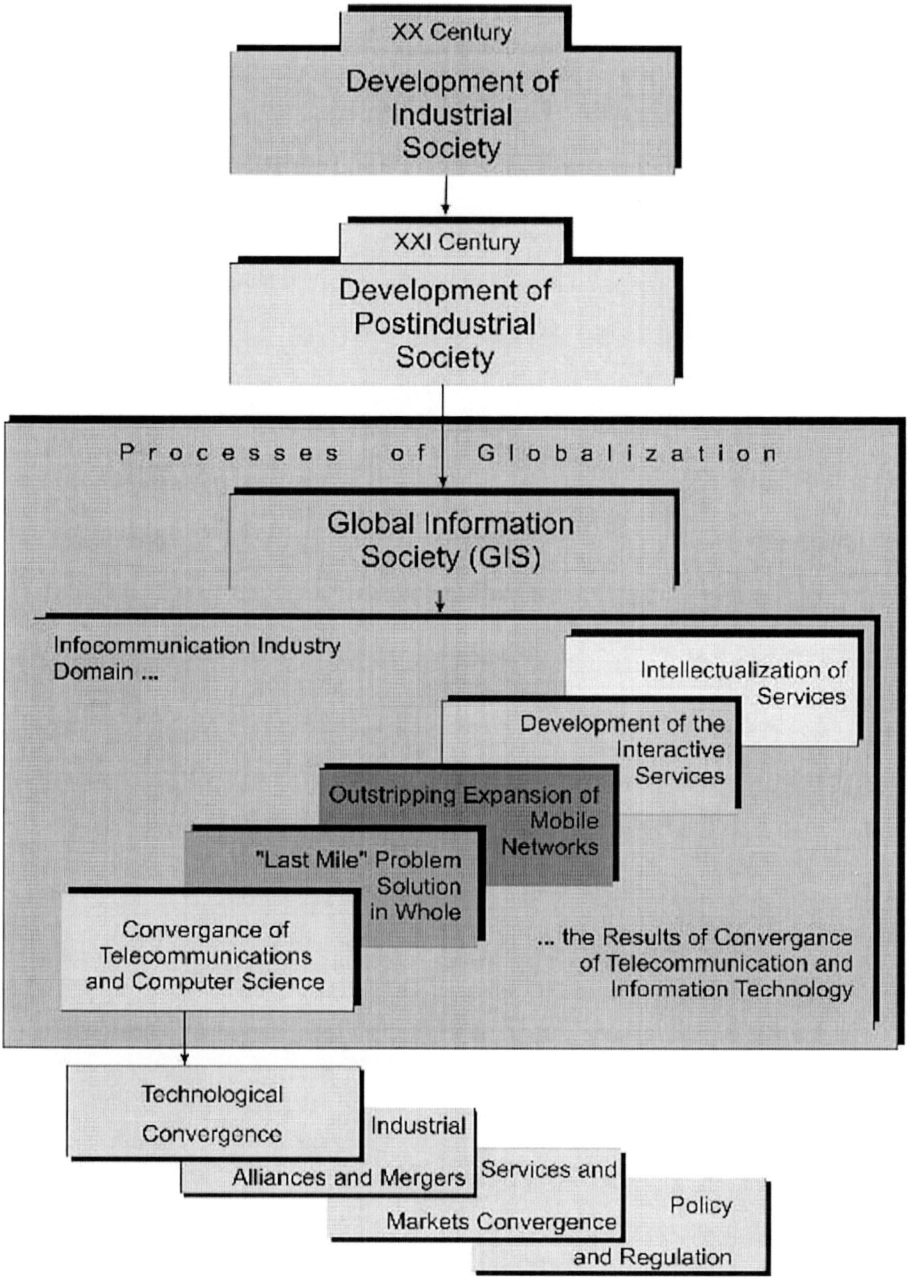

Figure 20.1: Trends, processes and some solvable problems.

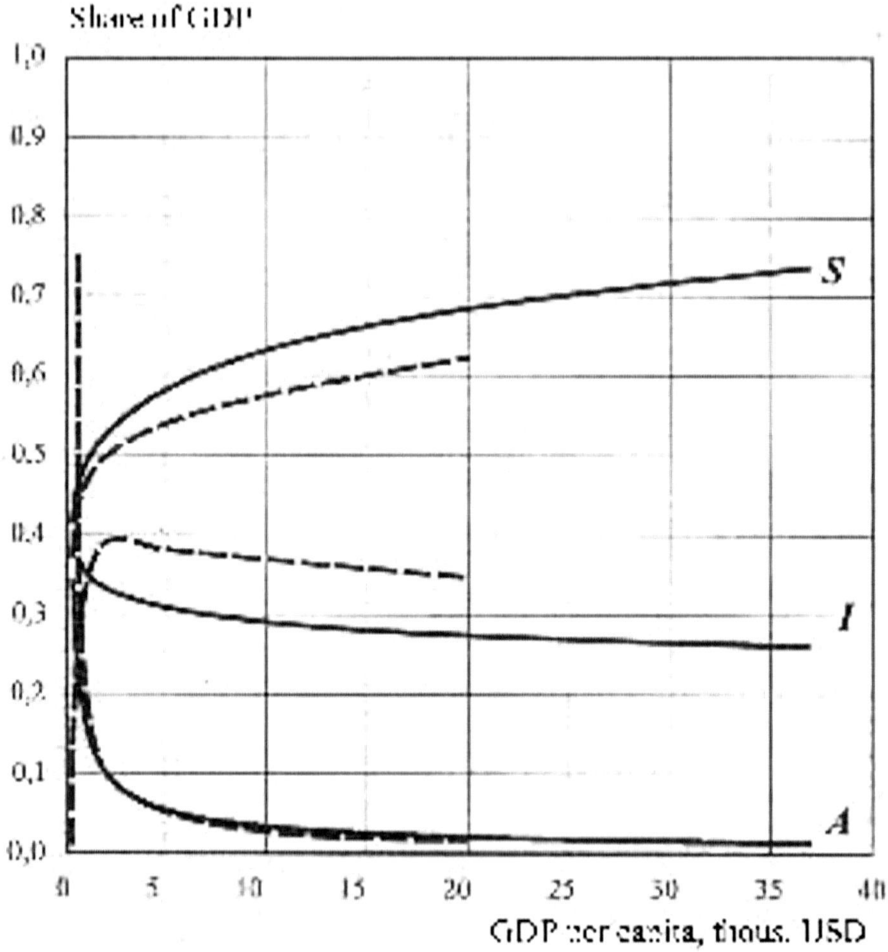

Figure 20.2: Sketch of the performance of the three sectors of the economy: Agrarian (A); Industrial (I); and Services (S) over a 17 year interval from 1987 to 2004. Abscissa: GDP per capita in thousands of USD, scale from 0 to 40; ordinate: share of GDP, scale from 0.0 to 1.0.

It is evident that the basis of this infrastructure is the subscriber's terminals as in the long run their availability determines accessibility to the world's information resources. For this reason the basic indicator of development of infocommunication is taken as the quantity of these terminals per 100 inhabitants (i.e. the *density*), viz.: TD and MD, fixed and mobile telephone density; PCD, personal computer density; and IHD, Internet host density. Of course, the higher the GDP of a country, the better developed is its telecommunication infrastructure and the services provided by this sector are more affordable and available for the general population [7,10].

Such correlation can be represented as a Jipp diagram, which has been used to determine the relation between GDP per capita and telephone density (TD) since 1963. On Figure 20.3 are given modern Jipp diagrams made by the International Telecommunications Academy (ITA) on the basis of statistical data from the ITU [9]. These correlation dependencies are in kinds of straight lines (in logarithmic coordinates) increasing with GDP per capita. To gain a better insight into GIS, the multiparameter task was reduced to a single parameter one in [9], where the author introduces the so called *multidimensional infocommunication vector* (ICV) of some country (region) as

$$\|A\| = \sqrt{\frac{1}{n}\sum_{i=1}^{n} a_i^2}\ , \tag{20.1}$$

where $\|A\|$ is its magnitude (length, norm), a_i the ith coordinate of this vector, i.e. components of the above-mentioned density (TD, MD etc.), and n the number of parameters of the correlation dependence.

Figure 20.4 shows the ICV correlation dependencies and the share of services S plotted against GDP per capita for 63 large countries, adapted from [9], and Figure 20.5presents the situation in ICV-S coordinates (from the same data source). These diagrams also show the differences in these parameters between developed and developing countries, and in particular between big cities and rural regions of those countries. The obligatory prerequisite for the formation of GIS is to overcome the so called digital divide, which implies the provision of up-to-date infocommunication services to the population.

Of course, overcoming the digital divide is directly related to the dynamics of the economic development of a country and its diversification, the so called *country/economy profile*. Based on the Global Competitiveness Report [3], countries pass through three stages of economic development, as illustrated in Figure 20.6.

The great differences between the profiles of different countries evinces the large competitiveness gaps between countries, also comprising the digital divide . Switzerland, which ranks as number one according to the global competitiveness index (GCI) [3], and most other developed countries are at the innovation-driven stage,[2] while NIS countries are mostly either at the factor-driven (e.g., Armenia, Azerbaijan, Georgia) or the efficiency-driven (e.g., Estonia, Latvia, Lithuania) stages.[3] However, it may be mentioned that the dynamics of the economic development of Georgia are quite favourable and promising.

[2]This roughly corresponds to the K-selection régime *(Ed.)*.

[3]This is despite their lavish scientific and technical legacy. Other indices, for example the Civilization Index (CI) of the Collegium Basilea, or the Human Development Index (HDI) of the United Nations Development Programme (UNDP), to an extent take some of these other aspects into account *(Ed.)*.

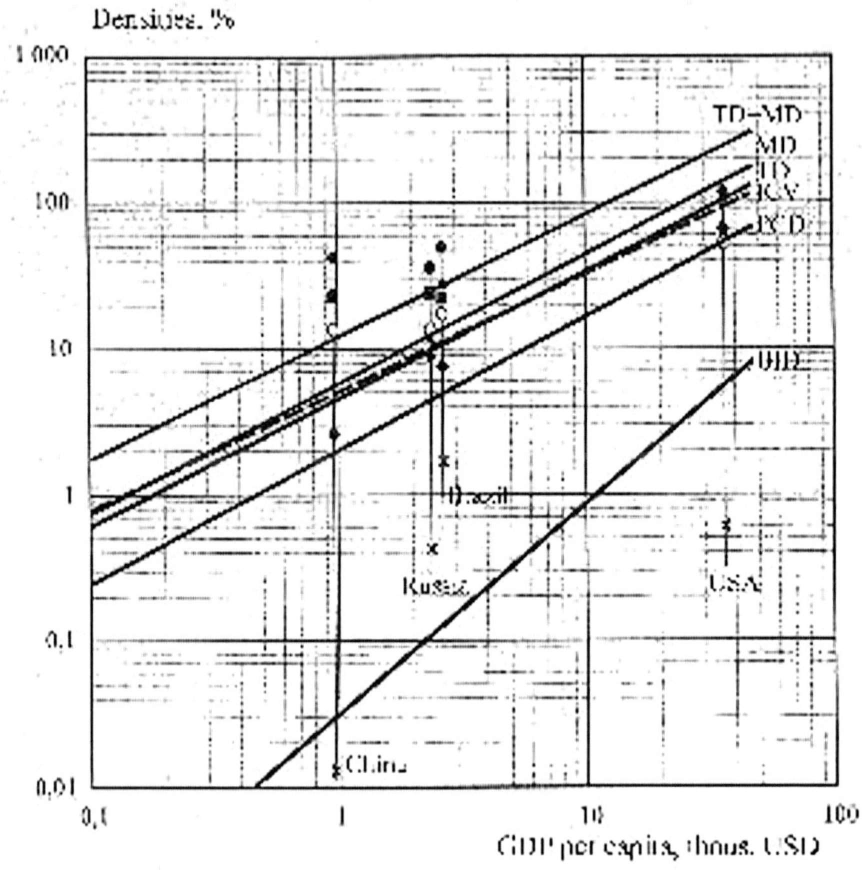

Figure 20.3: Sketch of correlation dependencies for infocommunication (2003). •, TD+MD (total density of fixed and mobile communication); ▲, MD (density of mobile communication); ■, TD (density of fixed communication); ○, ICV (density of infocommunication vector); ♦, PCD (density of personal computers); ×, IHD (density of Internet hosts). Abscissa: GDP per capita in thousands of USD, logarithmic scale from 0.1 to 100; ordinate: densities in percentage, logarithmic scale from 0.01 to 1000.

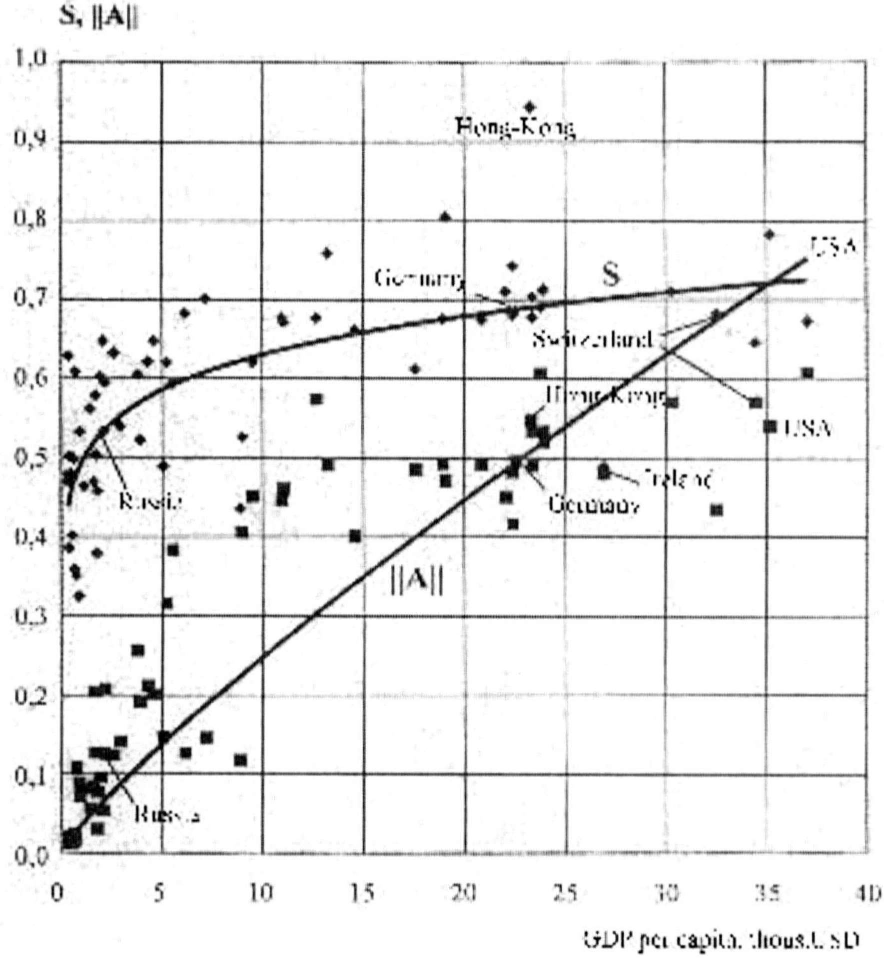

Figure 20.4: Sketch of ICV correlation dependencies (share of services S and infocommunication vector versus GDP). Abscissa: GDP per capita in thousands of USD, scale from 0 to 40; ordinate: S and $\|A\|$, scales from 0.0 to 1.0.

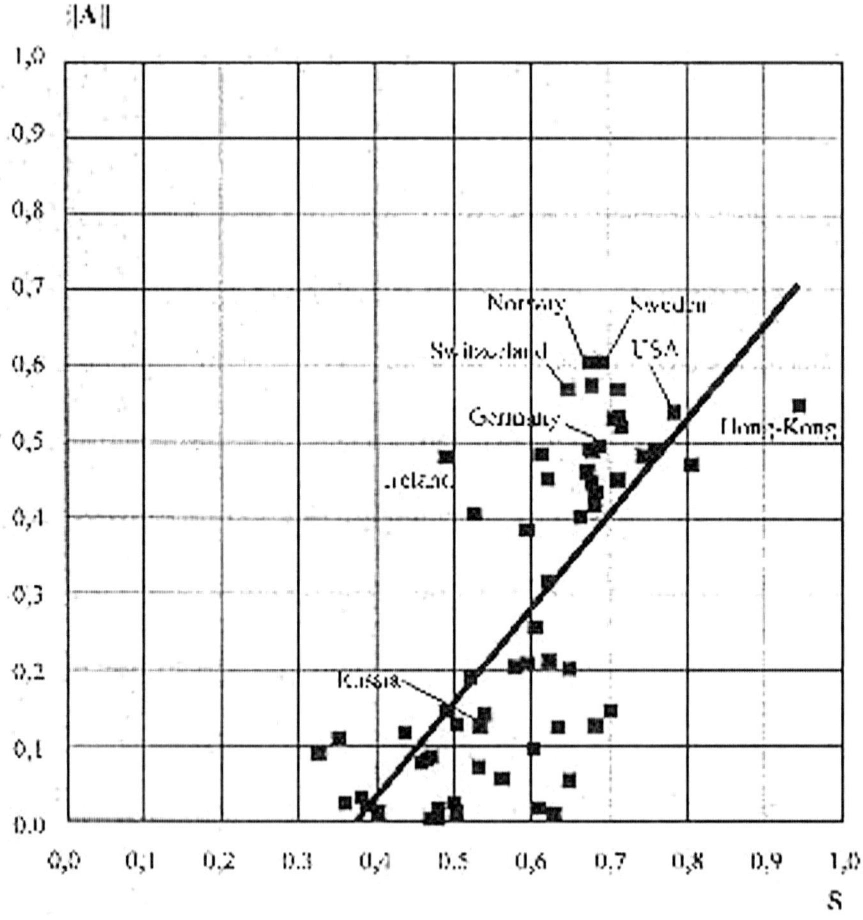

Figure 20.5: Sketch of the ICV-share of services relation for different countries. Abscissa: S, scale from 0.0 to 1.0; ordinate: $\|A\|$, scale from 0.0 to 1.0.

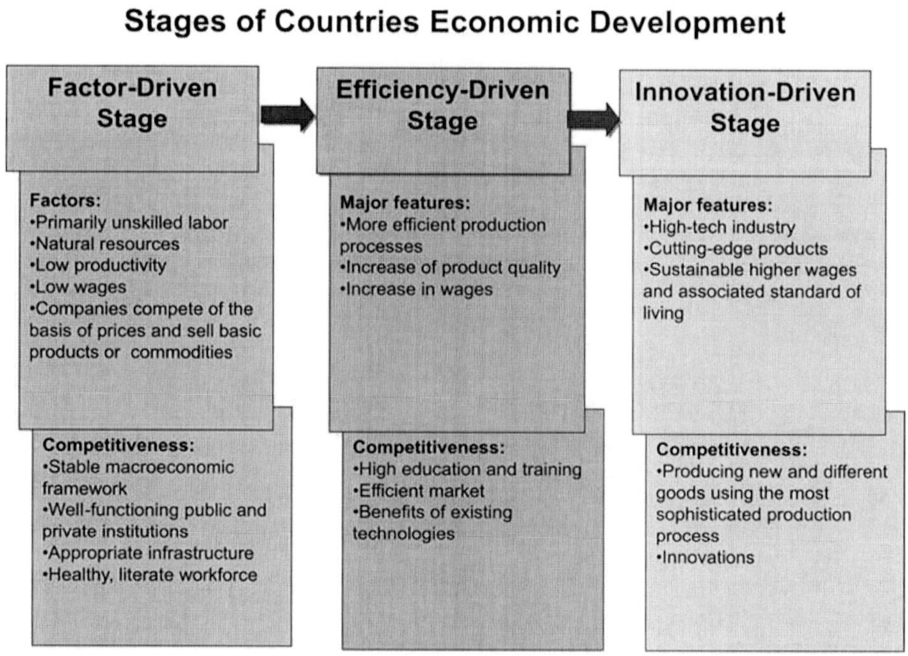

Figure 20.6: Explanation of the country/economy profile.

The existence of the digital divide is the particular manifestation of a more general problem, namely the unequal distribution of income, technology and services. This problem is very actual for the worldwide community as it raises the problem of *security* and *integrity* as global problems of contemporary society, its progress and the evolutionary transition to GIS. Indeed, the existence of significant economic inequality (and, consequently, of infocommunication, education and other features of society) between the richest and poorest parts of the world population is far-reaching in its implications. The negative consequences of such gaps (poverty growth, striving for redistribution of property and income, widespread resentment) can sometimes result in the violence, conflicts and even in hostilities at a local or larger scale. These trends are of great concern to the United Nations and other international organizations, leaders of various countries and NGOs. The matter of infocommunication inequality (the "digital divide") are actively treated by the ITU. A part of this activity is the programme of achieving *universal access* to the basic infocommunication services (the so-called *Universal Service Obligation*).

For an impartial assessment of distribution of income, technology and services, various indicators are introduced (some of them have been mentioned above). Data for these indicators are published according to countries and regions in reports of the World Bank (World Development Reports), UNDP (Human Development Reports), ITU (World Telecommunication Indicators), the Economist (Pocket World in Figures), Siemens (International Telecom Statistics) and others. All countries including Georgia are supposed to participate in the above-mentioned global process by developing telecommunication infrastructures, contributing to the proper functioning of the telecom markets and carrying out necessary regulatory procedures.

External factors including the geopolitical situation are also very important in the integration process of the global 'infocommunication' environment. Georgia and the South Caucasus are good examples of successful realization of the advantageous geopolitical locations. Historically, very important trade routes went through Georgia (Figure 20.7). Noteworthy are the well-known Great Silk Road (GSR), and the telegraph backbone built in the second half of the 19th century (1865–1870) by Siemens. At the time it was the largest project of its kind, connected London and Calcutta and passed through Tbilisi. This telegraph line remained operational even during World War II.

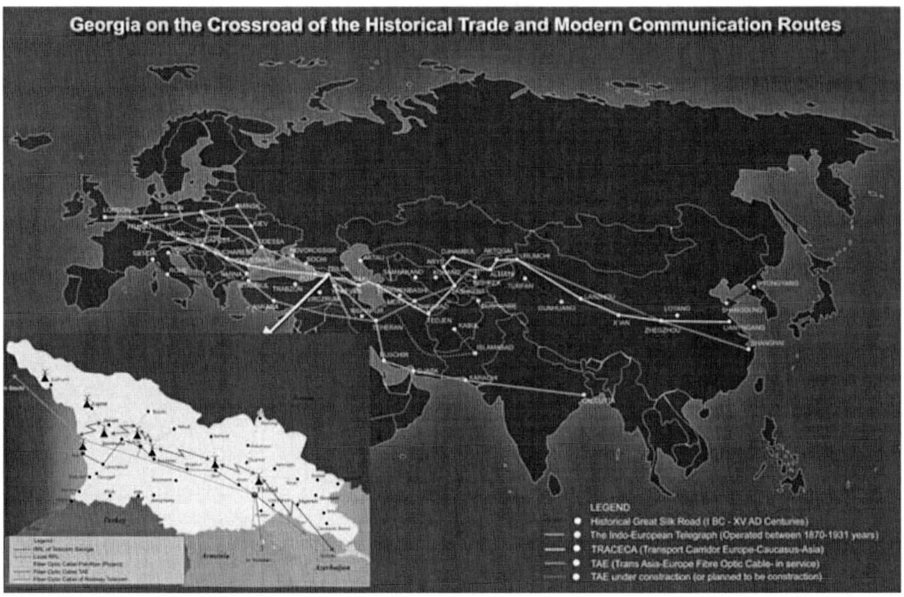

Figure 20.7: Eurasian trade routes.

At present, the geopolitical factor is also a widely applied, strategically important issue. Today several modern telecommunication, energy and transport highways (the Trans-Asia-Europe (TAE) fibre-optic highway, the Baku-Tbilisi-Ceyhan and other oil pipelines, the TRACECA—Europe-Caucasus-Asia transport corridor) pass through Georgia. Other telecommunication highways passing through Georgia are shown on Figure 20.7, including fibre-optic links operated by Georgia Railways and the company Foptnet, the digital microwave link of Telecom Georgia, and some others. It is also worth mentioning that the formation of the Georgian telecommunications market could be characterized by the main trends taking place worldwide, notably:

- the market, like in the most developed countries, is fully liberalized. For example, in 2005 the privatization process of the largest state-owned PSTN company with significant market power—the Georgian United Telecommunications Company—was completed;
- the most up-to-date mobile technologies and networks are developing very

fast. For example, the Georgian mobile operators have already offered third generation (3G) mobile communication services to their customers. At the same time, in 2003 the number of mobile customers exceeded the number of PSTN customers, as has happened in many countries worldwide. The mobile communications sector generates about 65% of the total telecom revenues, which amounted in 2006 to 650 million GEL;[4] In 2000, Georgia was one of the first countries in the post-Soviet area where a regulatory body—the "Georgian National Communications Commission" (GNCC) was established. GNCC played an important positive role in the development of the telecom industry by undertaking flexible and adequate regulatory activities, increasing trust and accountability between telecommunications companies and customers, contributing to the establishment of a highly attractive and stable business environment;

- in 2005 the Law on Electrical Communications, harmonized with European Union legislation and ITU recommendations, was adopted. Particularly, the law includes the following principles:

    1. authorizations in the sphere of electrical communications;
    2. establishment of a competitive environment and definition of 'authorized player' with significant market power;
    3. technical neutrality;
    4. definitions of "network operator" and "service provider";
    5. creation of a secondary market for scarce resources (radio frequencies and numbering schemes).

20.1 Development dynamics of the Georgian telecommunications market, 2000–2006

For the years 2000–2006 the revenues of the telecom industry increased almost 5 times from 211 million GEL to 1001.4 million GEL [1,8]. This is a clear manifestation of the successful reforms undertaken in this field of the economy. It is worth mentioning that year 2006 was characterized by trends of significant growth. The overall revenues increased by 28.6% compared with the previous year and amounted to 285.9 million GEL (Figure 20.8). For the last six years, telecom revenues as a share of overall GDP were also characterized by an increasing trend. In 2000 this share made up 3.52% of GDP, in 2006 the industry's overall revenues more than doubled and made up 7.49% of GDP (Figure 20.9). According to the data of the International Telecommunication Union (ITU) in developed countries the share of telecom revenues typically makes up 2–3% of national GDP.

In 2005 the Parliament of Georgia adopted the Law on Electrical Communications. Under this law a person (legal or natural) who wishes to commence his activity in the telecommunication sphere shall be subject to authorization by the GNCC. During 2005–2006 194 legal and natural persons achieved authorization in 9 various activities, with a total of 372 authorizations (see Table

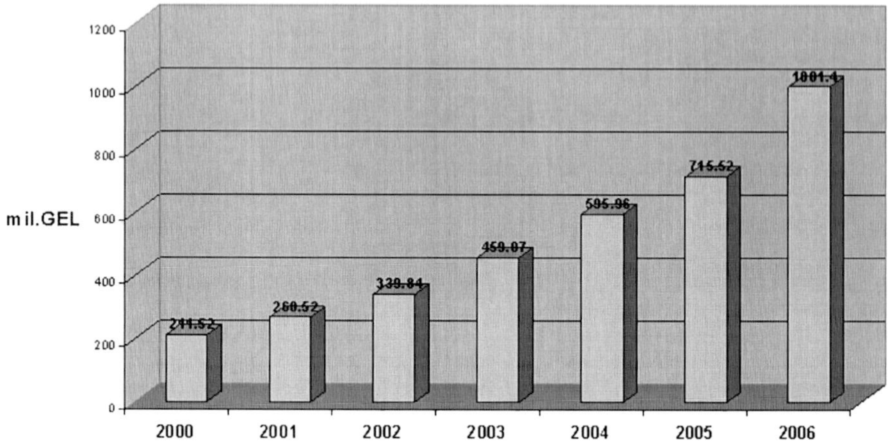

Figure 20.8: Telecom revenues.

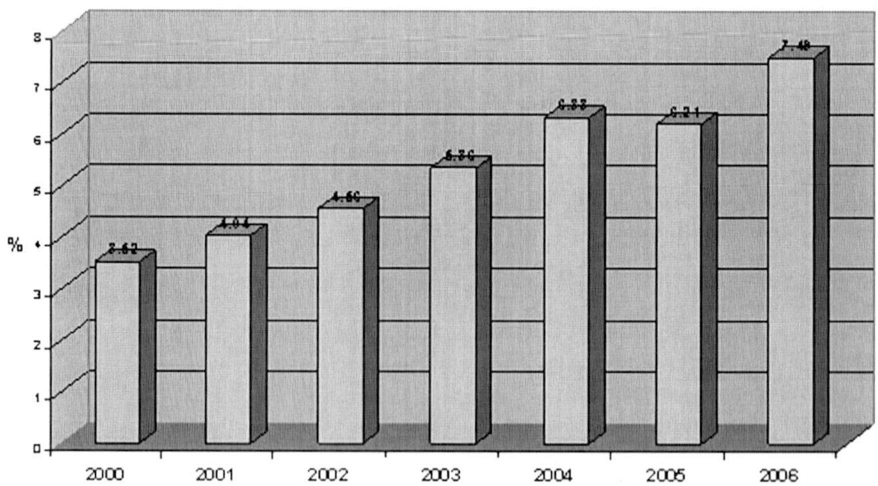

Figure 20.9: Share of telecom revenues of GDP.

Table 20.1: Telecommunications authorizations in Georgia.

No	Activity	2005	2006	Total
1	PSTN	14	49	63
2	Internet service	13	51	64
3	Cable transit TV broadcasting	2	83	85
4	International long-distance communication	15	46	61
5	Mobile communication	3	11	14
6	Local long-distance communication	11	44	55
7	Broadcasting	2	1	3
8	Cable transit radio broadcasting	12	12	24
9	Other communication service	12	3	15

20.1).

During the past few years various investors (including foreigners) paid millions of US dollars for participation in the tenders for different licences (mainly 3G) organized by the GNCC (Table 20.2).

Table 20.2: Recent telecommunications licence payments in Georgia.

Year	frequency	amount paid/(USD $\times 10^6$)	licensee
2005	800 MHz	11.7	Magticom
2005	2.1 GHz	15.6	Magticom
2006	800 MHz	40	Bloomfield Ltd
2006	2.1 GHz	11	Argotex
2006	2.1 GHz	10.4	Telecominvest[a]

[a] Subsequently transferred its rights to Geocell.

Figure 20.10 shows the structure of Georgian telecommunications sector at the beginning of 2005, with its component segments, interrelations and shares in the total sector revenues. The major operators engaged in each segment are also represented, with their estimated market shares in the respective services. As we see, the shares of the different segments in the total revenue structure of the Georgian telecommunications sector vary greatly: the leading place is held by the mobile telecommunication segment, the second place is held by the fixed line segment, and the third by the international gateway facility operators (IGFO) (Figure 20.11). In year 2006, 65% of total telecommunication sector revenues came from the mobile telecommunications sector. From year 2000 to 2006 the share of fixed line operators in total telecommunication market revenues decreased from 42% to 19%. The above-mentioned facts can be explained by the propensity of telecommunication services customers to choose mobile services and show the beginning of the convergence of fixed and mobile networks.

[4]GEL denotes the Georgian lari, the national Georgian currency. The average rate of exchange of GEL to USD (for 2000–2006) was 1 USD = 1.95 GEL.

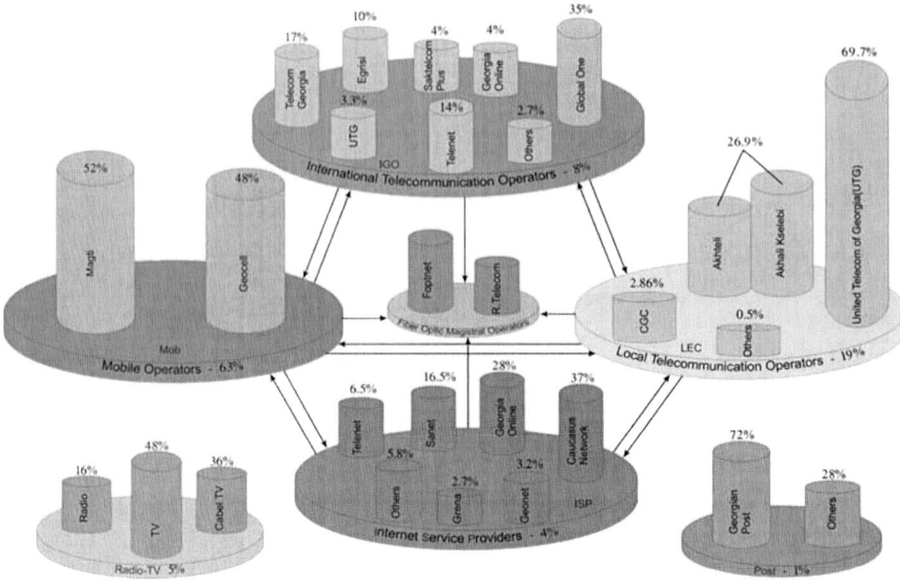

Figure 20.10: Impression of the structure of the Georgian telecommunications sector at the beginning of 2005.

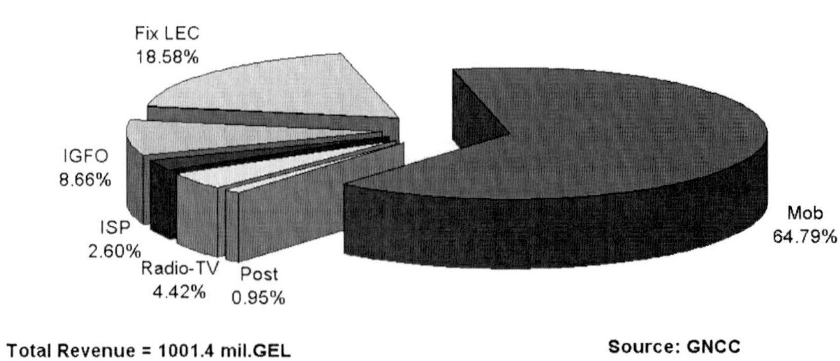

Figure 20.11: Georgian telecom market shares.

20.1.1 Mobile telecommunication services

The increasing tendency of the mobile telecommunication services market fixed in the previous years was maintained in year 2006. Two mobile operators, Magti GSM and Geocell, currently function on this segment. (Lately, a third mobile operator, Mobitel, commenced activity in Georgia.) As compared to the previous year, the number of active subscribers of mobile telecommunication services during the year 2006 increased by almost 45% and amounted to 1.7 million. During the five-year period (2000–2005) the number of active subscribers of mobile telecommunication services increased eight times. It must be noted that in 2003 the number of mobile subscribers already exceeded the number of fixed line subscribers and in 2006 the number of mobile telecommunication services subscribers thrice exceeded the number of fixed line telecommunication services subscribers.

As compared to the previous year in year (2006) the revenues of mobile telecommunication services increased by 53% or by 226 million GEL and amounted to 650 million GEL. As compared to year 2000, the revenues increased by 543 million GEL, and the mobile telecommunication services still represent the fastest growing and most profitable segment of the telecommunication market (Figure 20.12).

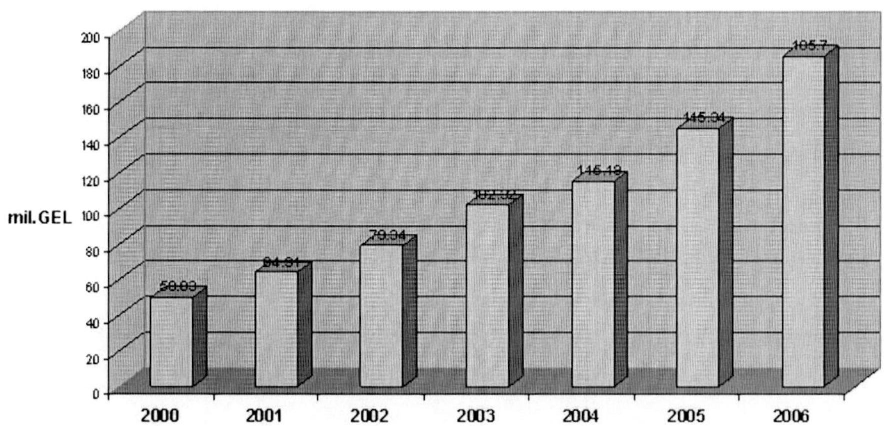

Figure 20.12: Total annual revenue from fixed line services.

20.1.2 Fixed line telecommunication services

During the period 2000–2006, the number of fixed line subscribers increased by 11% and in 2006 they amounted to 553 thousand. The major share of revenues from this segment comes from JSC Georgian United Telecommunications Company, New Net Ltd and Akhteli Ltd, which hold approximately 90% of the Georgian fixed line network. The total annual revenue from fixed line service providers increased from 50 million GEL in 2000 to 186 million GEL in 2006 (Figure 20.12).

20.1.3 International/long-distance telecommunication services

Also the increasing trend can be seen in the international gateway facility operators' market. In the year 2000, the total revenue of operators functioning in this segment amounted to 35 million GEL, while in 2006 this figure was 87 million GEL (Figure 20.13). It should be noted that the growth of total revenues was due to: increasing competition in the international telecommunication operators market (the main rivals in this market are: Telecom Georgia, Georgia Online and Global One); abolishment of the tariff cap and floor by GNCC; and an increase in the amount of international inbound and outbound traffic. In 2006, compared with 2000, outbound international traffic from the Georgian telecommunication networks increased by 78% to a total of 105 million minutes, while inbound international traffic into the country during the same period increased by 360% and amounted to 324 million minutes. Figure 20.14 shows the dynamics of the amount of the outbound international traffic (the same as the MOU indicator) in 1996–2006. However, in the period from 2005 until 2006 the amount of outbound international traffic even decreased slightly although in 2005 GNCC abolished the tariff floor for all directions. Due to the intense competition among the operators the tariffs in some directions decreased two to three times but this has not resulted in the expected increase in the amount of traffic. The reason of this could be the limited demands of the local population and saturation of the local telecommunication market.

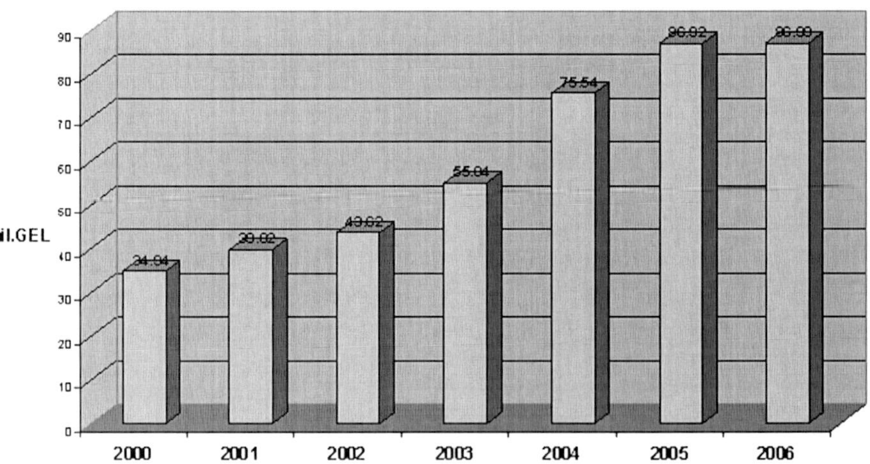

Figure 20.13: Total annual revenue from the international gateway facility.

20.1.4 Internet services

During the period 2000–2006 the total revenue of internet service providers increased from 3.7 million GEL to 26 million GEL (Figure 20.15). It should be noted that internet services represent the fastest growing segment on telecommunication market: from 2005 to 2006 annual revenue increased by 53%. There were 27.7 thousand users of ADSL internet technology at the end of year 2006.

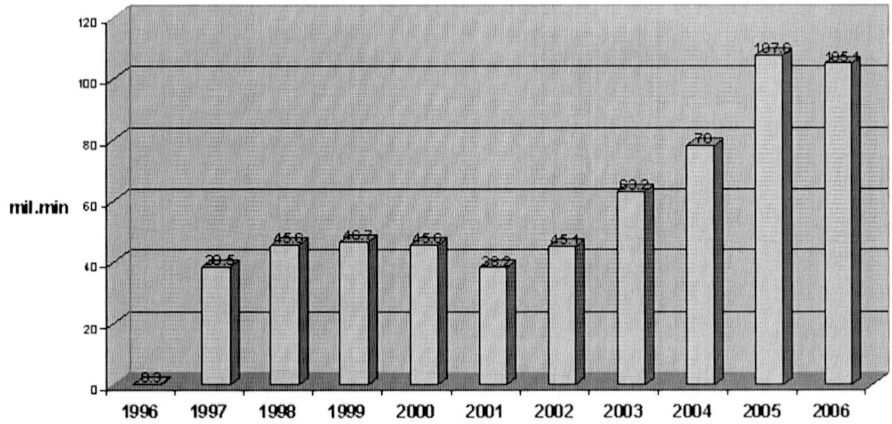

Figure 20.14: Dynamics of the amount of the outbound international traffic.

Compared to the previous year (2005), the number of users almost doubled. The number of dial-up technology users in 2006 increased by 40% compared to the previous year.

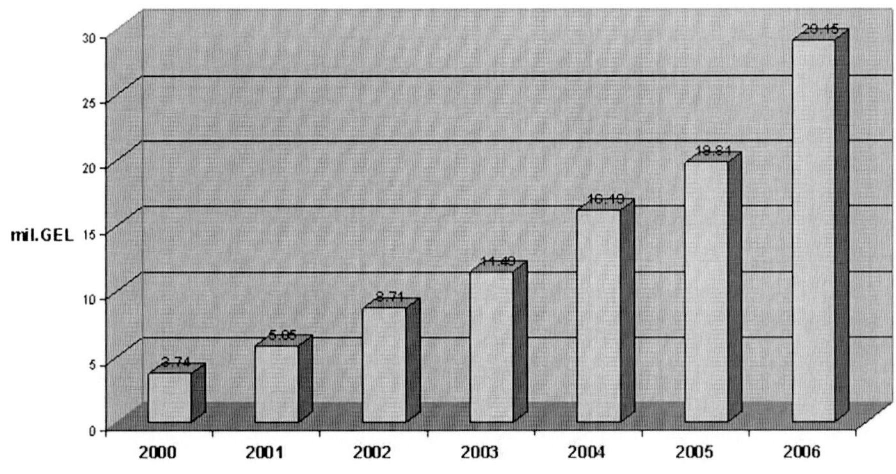

Figure 20.15: Total revenue of internet service providers.

Below is shown the so called Herfindahl index (HI) that measures the degree of competition in the various telecommunication market segments. It is computed by the following formula:

$$\mathrm{HI} = \sum_{i=1}^{m} x_i^2 \tag{20.2}$$

where m is the number of operators and x_i is the fractional market share of the ith operator. Table 20.3 shows the competition in the various Georgian telecommunication market segments using this index.

Table 20.3: The Herfindahl Index (HI) for various telecommunications sectors in Georgia in 2005 and 2006.

Market segment	2005	2006	Change
Mobile operators	5338	5008	−7%
Local (PSTN) operators	5595	5357	−4.3%
IGF operators	2594	2497	−3.75%
ISP	2519	5808	229%

It can be seen from Table 20.3 that in all segments other than ISP the competition level is increasing. As for the ISP segment, because of the merger of three major operators, Caucasus Network, Sanet and Georgia Online, a dominant operator (Caucasus Network) with significant market power (SMP) appeared, and the level of competition has drastically dropped.

It should be noted that in general, in spite of some drawbacks, the Georgian telecommunications sector is developing in the right direction. The forecasts made two years ago by GNCC have proved to be correct and the total telecom revenues exceeded 1 milliard GEL as early as in 2006. Hence the gradual integration of Georgia into the GIS is a quite realistic prospect.

20.2 Conclusion

We emphasize that the gradual formation of the GIS is an accomplished fact and this process has involved all countries worldwide (depending on their financial-economic potential), including Georgia and its neighbours. However, the differences in standards between countries in the field of infocommunication result in the so-called digital divide, which is a component of an extended range of problems such as the unequal distribution of income, technologies and services that impedes the further process of enhancing the integrity and security of the world community. Therefore, elimination of such gaps is one of the highest priority challenges facing contemporary post-industrial society.

20.3 References

1. Annual Report of the Georgian National Communication Commission. Tbilisi (2006).

2. Bell D. The Coming of Post-Industrial Society: A Venture in Social Forecasting. Basic Books (1973).

3. Global Competitiveness Report, 2006–2007. Geneva: World Economic Forum (2006).

4. Green Paper on the Convergence of the Telecommunications, Media and Information Technology Sector and the Implications of Regulation. Brussels: European Commission Records (December 1997).

5. ITU releases its year 2000 guide for telecommunications operators. ITU News No 3 (1999).

6. Jipp, A. Wealth of nations and telephone density. Telecommunications Journal, July 1963, pp. 199–201.

7. Moskvitin, V.D. Two criteria of the information society. ITA Proceedings (1998).

8. Statistical Yearbook of Georgia. Tbilisi: State Department of Statistics (2005).

9. Varakin, L.E. The Digital Divide in the Global Information Society. Moscow: ITA (2004).

10. World Telecommunications Development Report. Geneva: ITU (2004).

Part V

Psychological, Social, Political, Economic and Ethical Aspects of Security

Complexity and Security
J.J. Ramsden and P.J. Kervalishvili (Eds.)
IOS Press, 2008

Chapter 21

Psychological, social, economic and political aspects of security

Jeremy J. Ramsden

Cranfield University, Bedfordshire, MK43 0AL, UK

We have established in Chapter 2 that security is above all a *feeling* of safety. While safety may (or may not—cf. Chapter 14) be assured by technology, and also depends on factors that we may call exogenous, such as climate (leaving aside for now the question of possible anthropogenic influence—cf. Chapter 10) and the exhaustion of finite resources (the timescale of which is obviously anthropogenically determined), security, as a feeling, must depend essentially on the individual psyche and how it processes its mental inputs. Simultaneously, an important role must surely be assigned to structures collectively organized by society, such as the police force, the fire and rescue service, the ambulance brigade, and so on. Their presence and observable efficacy of course feed into the individual psyche. A rather similar process at work was surely responsible for the great desirability of, and emphasis on, the erection of solid, imposing buildings—often "with columns", *pace* Gogol—by institutions that needed to gain the public trust, such as banks, insurance companies, and (especially in Russia) scientific research institutes. The fact that today banks operating entirely via the telephone or the Internet have notwithstanding gained the public's confidence presumably represents a profound shift of psyche.

The link between security and psychology and sociology seems then to be clear enough; complexity is then immediately implicated, the human brain being the most complex object known to man.[1]

The matter of individual and collective psyche is a great current preoccupation, certainly in most Western European countries. Repeated surveys of public

[1]See for example J.J. Ramsden, Computational aspects of consciousness. Psyche: Problems, Perspectives 1 (2001) 93–100.

opinion place concern about insecurity as a major source of dissatisfaction with government The type of insecurity that is so strongly abhorred is not the risk of war, or even of a terrorist attack, but what has been called chronic insecurity (Chapter 2)—the risk of harm to life or limb while simply walking along the street.[2] The response on the part of many governments and government agencies to this chronic insecurity seems to be converging on the provision of mass surveillance. Great Britain has so far gone furthest in this regard than most other countries, with the provision of video cameras in public places being the highest (per capita) in the world, but other countries such as France are following suit—despite some doubts about their efficacy.[3] The private security industry—the provision of watchmen and patrollers—is now a major growth area, and the general trend is to expand already large police forces. These are significant sources of employment on a national scale. Furthermore, new ways of combating crime by technology are being actively sought (e.g., in the UK the state Engineering and Physical Sciences Research Council (EPSRC) has repeatedly called for research proposals in this area during the past few years). Relatively less effort seems to have gone into establishing the fundamental cause of crime. One idea, so widely held, especially in political circles, that it can safely be put into the category of "received ideas", seems to be that crime is due to poverty, although the actual evidence for it is not especially convincing. It is better to start at a more basic level, for example to simply examine how crimes increase with population. I propose

$$C \sim N^{\chi} \tag{21.1}$$

where C is the total number of crimes committed in a country (considered as a more or less culturally homogeneous bloc, within which people can freely exchange ideas and desires) in one year, N is the population of that country, and χ is a characteristic exponent. Figure 21.1 is a log-log plot of annual numbers of crimes versus population for different countries. The slope of the regression line is, within statistical uncertainty, equal to one.[4] From the fact that $\chi = 1$ one can infer that crime is essentially based on individual motivation.[5] There

[2]For example, "Reducing the Fear of Crime" was declared a Best Value Performance Indicator for Bedfordshire County Council in 2006. In May 2007, the Council decided to distribute £1 million (to be called the "Confident Communities Fund") among the county's 131 town and parish councils (the population of the county is about 560 000) to spend on things to make residents less fearful of crime.

[3]The main problem seems to be one of information overload: the variety of possible criminal incidents is too great to be monitored automatically (unlike the now very successful automatic monitoring of the police registration marks ("number plates") of motor-cars contravening regulations such as those prohibiting driving in bus lanes). For such a general monitoring system to be effective, it would require an unrealistically high proportion of the population to be engaged in monitoring the recordings. Even in the German Democratic Republic, which probably had a much higher than average proportion of the population engaged in surveillance, although foreign telephone calls for example were routinely recorded, the recordings were merely stored for possible future reference and analysis. The main protection against overweening surveillance is thus actually an economic one. The collapse of the Yugoslav economy, which preceded the political collapse of that country, was largely driven by the trend to more and more surveillance—towards the end it has been estimated that only about 10% of the active population was actually producing something, the remaining 90% being engaged in surveillance of one kind or another (including quality and safety checks in industry).

[4]A similar plot of the annual numbers of crimes (in 2006) versus the populations of major British cities also yields a slope of 1.0.

[5]This kind of approach is called "evidence-based sociology".

is clearly no justification for arguing (as some have tried to do) that crime is based on the "rottenness of society" (the assertion of which might itself be a better indication of rottenness than the crime rate), for that would imply $\chi \geq 2$. Interestingly, a plot of crimes per capita versus population density (Figure 21.2) shows no correlation, indicating that physical crowding is not per se a contributory factor.[6]

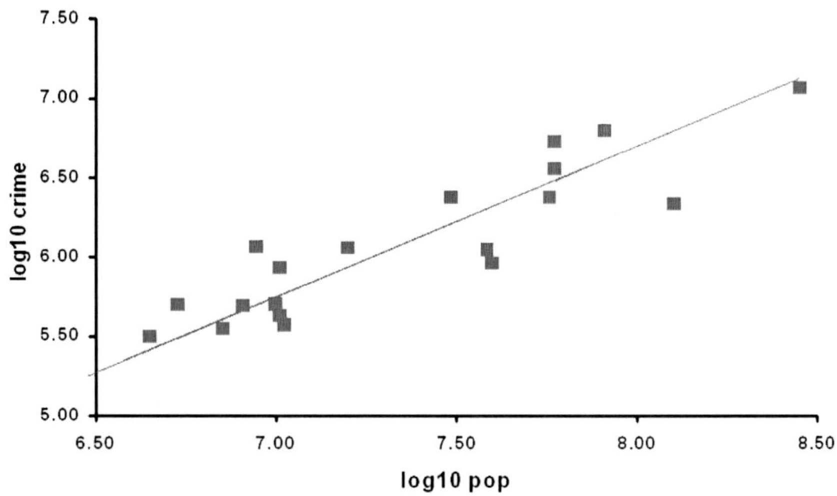

Figure 21.1: Annual numbers of crimes versus population for Australia, Canada, China, Denmark, Finland, France, Germany, India, Indonesia, Ireland, Italy, Japan, Korea, New Zealand, Norway, Spain, Sweden, Switzerland, the UK and the USA. Solid line: least squares linear regression line (slope = 1.0).

Does the establishment of this fact ($\chi = 1$) mean that the subject is thereby closed to further scientific investigation?—since individual motivation is subject to "unseen feelings".[7] When one thinks of the sheer variety of those unseen feelings, merely by reflecting on one's own, and who knows how imperfectly they reflect those of others, one might well conclude that it is hopeless task to attempt to achieve some understanding of motivation. This is a situation that indeed appears to lie well beyond the complexity ceiling (cf. Chapter 5).

[6] At least in the past, there appears to have been some evidence for a correlation. For example, J.T. Bunce has written "Crime and overcrowding always go together, like overcrowding and disease. One cause of the comparative freedom of Birmingham from crime, especially serious crime, is that—for its population—it covers more ground than any other town in the kingdom." (The social and economic aspects of Birmingham. In: S. Timmins (ed.), The Resources, Products and Industrial History of Birmingham. London: Robert Hardwick, 1866), a view reinforced by the results from perhaps rather naïve experiments with rats crowded in cages.

[7] Bain, A., The Senses and Intellect, p. 3. London: Parker (1855). The full quote reads "We cannot trace any uniformity in the operations of a human being by merely looking at the actions themselves, as we can in the fall of a stone or the course of a planet. It is the unseen feelings that furnish the key to the vast complication of man's works and ways."

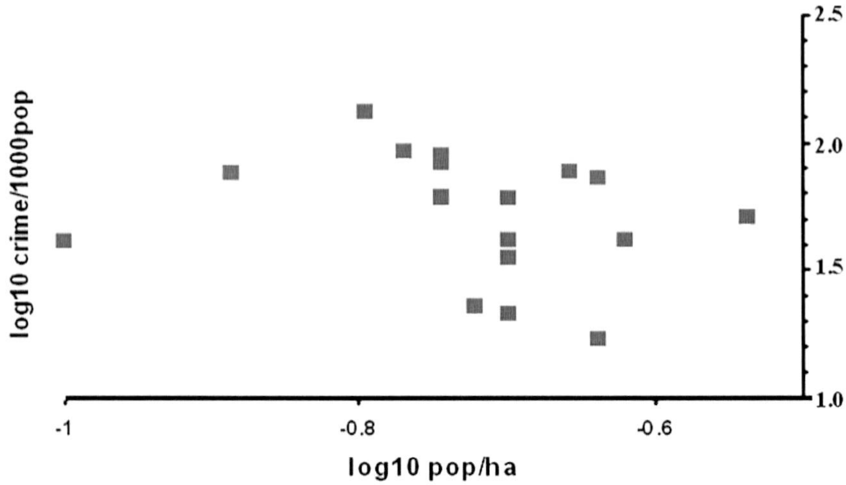

Figure 21.2: Annual numbers of crimes per capita versus population density for the countries plotted in Figure 21.1.

Greed

Nevertheless, there is one motivation that appears to have a peculiar significance when it comes to threatening security, namely greed. Especially when discussing what we have called chronic insecurity, it would appear to be the obvious, and principle, motivation.[8] Why else does the pickpocket steal a purse? Or the mugger a mobile telephone? Is it not the dominant motive behind crime such as bank robbery? At the same time there seems to be little evidence in favour of the proposition that greed is simply innate among some people (and absent in others). It appears to be something inculcated in our individual psyches.[9] There is clearly a strong economic motivation for this inculcation. Greed drives consumption, or rather, to be more precise, it drives the unnecessary consumption that is such a necessary adjunct for the continuation (i.e. growth) of the economy. We are constantly surrounded by advertisements trying to persuade us to buy goods or services that we do not require. It is rather ironical that the only countries where this bombardment was absent were the Soviet Union and some of its socialist satellites—countries built on an overtly materialistic philosophy,

[8]This inference is but a simple application of the maxim "is fecit, cui prodest."

[9]It may be that mankind can be divided into two categories, those that are susceptible to this inculcation, and those that are not. The former appear to constitute the overwhelming majority; the latter, whom we might conveniently label as 'brahmin' after India's ascetic and intellectual élite, seem to be very much in the minority (does anyone doubt that a book entitled "How To Get Rich" will sell much better than one entitled "Deepening Spirituality", for example?), but in practice the minority depends very much on the majority in order to maintain a comfortable lifestyle. Both groups in fact require each other: the former requires the latter to provide direction and ultimate motivation; the latter requires the former in order to provide the aeroplanes and computers (etc.) that seem to be so indispensable nowadays to daily life.

but in actual daily life a good deal less materialistic than our Western so-called "Christian democracies". I suppose that those of us who are immune to being hypnotized are more or less impervious to persuasion by advertisements—I do not know what fraction of the population is immune, but that fraction must be fairly small, otherwise the returns on advertising expenditure would be too small to make it worthwhile—yet worthwhile it obviously is (as well as being an important source of employment for those of an artistic bent!).[10]

Returning now to the postulated relation between greed and poverty, we can use a result of Daniel Bernoulli, who very convincingly showed that the marginal utility of money decreases with the amount already possessed,[11] giving precise expression to what is an easily observable empirical fact, namely that the more one has, the more one wishes to possess.[12] This law of Bernoulli clearly implies that greed increases with wealth. Can we obtain any evidence of a link between greed and crime at this level of quantitation? According to the "received idea", crime should decrease with growing prosperity, but the actual data (Figure 21.3) do not support it. A more subtle approach does, however, reveal a correlation. A powerful driver for greed is envy, which is (by definition) engendered by the perception of disparity. A useful way of quantifying the unevenness of the distribution of wealth in the country is to determine the range of annual incomes in a country, divide that range into N equispaced bins, and determine the proportion p_i of total income, or the proportion b_i of the total population earning income in that category, in each bin. The Shannon entropy ("weighted variety") H is then

$$H = -\sum_{i=1}^{N} p_i \log_2 p_i \; . \tag{21.2}$$

According to this measure of variety, the lowest possible value (zero) occurs when the entire wealth of a country falls in one bin, and the highest value ($\log_2 N$) occurs when the wealth is evenly spread among all bins. In Figure 21.4, this is used as the independent variable against which per capita crime is plotted for the same countries as previously. If the three labelled countries in the lower left hand corner are taken out of account, there is a rather clear

[10]Further evidence in favour of the proposition that greed is artificially built up comes from a comparison with the Soviet Union: it was almost always a foreign visitor or journalist who pointed out the absence of an abundance of material goods—in other words, that absence was perceived relative to the usual situation in the West. But looking at the situation more objectively (as I did myself while living in Moscow as a visiting scientist in 1989 and 1990), a more sober assessment is simply that what was necessary was available (e.g. in the "Gastronom" supermarket near my lodgings where I often did my shopping), and rather more conveniently because the shelves were not cluttered up with all the superfluous merchandise that one finds in a typical Western supermarket. The luxury of variety was anyway always available in abundance (albeit expensively) at the Danilovska market, one of many in the city, where excellent produce from all over the Union could be found.

[11]$\mathrm{d}U \sim \mathrm{d}M/M$, where U is the utility (value) and M the amount of money, which integrates to $U \sim \log M$.

[12]A further example, were any needed, of the irrelevance of the actual state of wealth to motivation by greed, is given by recent events in Hungary, where since the end of the socialist era there has been a conspicuously high number of people in high office, mostly connected with the former communist government, amassing vast riches at the expense of the state through a variety of corrupt actions. Although some of these actions were uncovered and widely publicized in the media, the reaction of the general public was not the outrage that had been anticipated by the investigating journalists, but mostly indifference. A typical comment was "we would have done the same, had we had the opportunity."

inverse correlation, suggesting that income disparity does fuel crime, with greed presumed to be the intermediary.

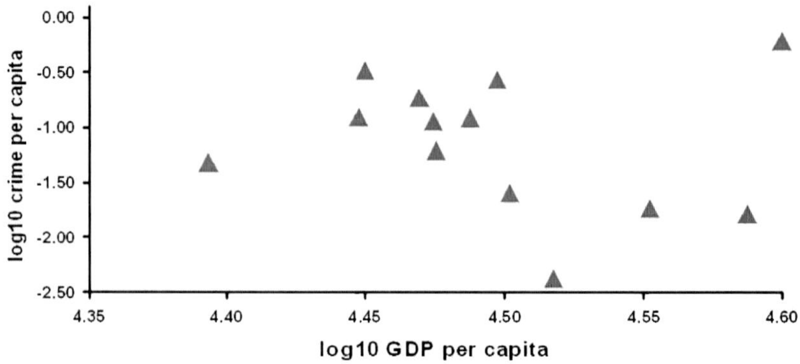

Figure 21.3: Annual numbers of crimes versus GDP, both per capita, for most of the countries plotted in Figure 21.1.

Greed is certainly not an elevating emotional state—there will hopefully be near-universal consensus on that point—and for that reason it would appear to be dangerous to deliberately cultivate it mainly for the sake of boosting consumption. The danger—real, concrete danger to life and limb as well as a general feeling of insecurity—comes not only from the robbery as such, which by itself would be annoying but hardly threatening, but because it is so often attended by violence against the person, even to the extent of shooting the victim, or the guard in the case of a bank robbery, to death. Incidents of this kind, involving mainly a few individuals, are however only the tip of the iceberg. What else, other than greed, lies behind the motivation of the building contractor to substitute inferior materials for those specified? Such actions have lead to the collapse of structures with dozens of fatalities. What else lies behind the substitution of offal unfit for human consumption for meat for the table?[13] Or the adulteration of wine with methyl alcohol or ethylene glycol? If undiscovered, such acts can also lead to fatalities. In some cases, vast and costly epidemics can be unleashed. A fairly recent example is bovine spongiform encephalopathy (BSE) or "mad cow disease", probably initiated by the unlawful sale of material deemed to be unfit for animal consumption for that purpose. Recalling the health threats to which the World Health Organization and others have called attention (Chapter 14), it often appears that these epidemics are considered to be initiated endogenously, whereas in reality many of them have fairly direct human origins. As well as the degradation of safety implied by these crimes, there is a long-lasting impact on security. After the emergence of BSE, sales of beef to households declined precipitously, and beef is still viewed with a certain

[13]Many examples of such corruption are to be found documented in House of Lords Select Committee on the European Communities, Session 1993–94, 12th Report, Financial Control and Fraud in the Community (HL paper 75). London: HMSO, 1994.

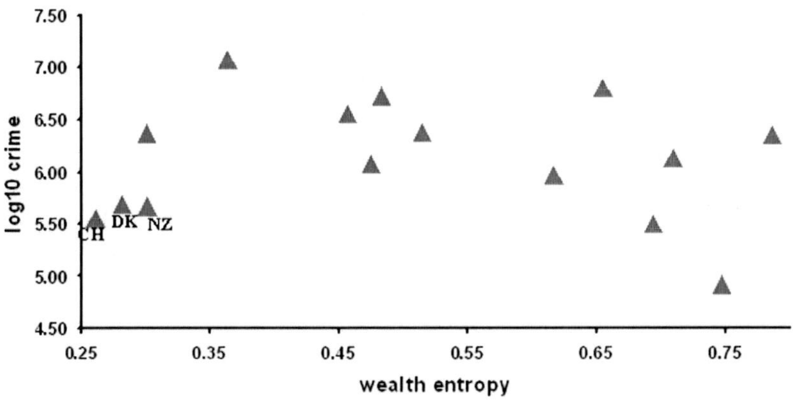

Figure 21.4: Annual numbers of crimes per capita versus wealth entropy (equation 21.2) for most of the countries plotted in Figure 21.1. Note that Denmark, New Zealand and Switzerland have "anomalously" low crime rates for their wealth entropy. Strictly speaking, however, it is the numbers of individuals with incomes in the various brackets that should be evenly spread to generate a high wealth entropy, i.e., equation (21.2) with b rather than p; were this done, possibly the anomalies would disappear. For now, the present figure can only be considered as indicative.

suspicion. Similarly, sales of Austrian and Italian wine declined after the discovery of the contamination with ethylene glycol and methyl alcohol respectively.[14] From this we can learn that impacts on safety are generally greatly amplified when they turn into impacts on security, which possibly justifies what might otherwise seem to be disproportionately onerous measures taken to maintain safety.

In comparison with the above, the so-called "white collar" crime that consists purely of embezzlement of funds (including a raft of qualitatively similar offences such as insider share trading) seems to be relatively harmless. It is paradoxical that nowadays such crimes are punished extremely, and indeed inappropriately, harshly. After all, expropriated funds can be restituted, but similar restitution is not available to the victim of violent crime. Outrage is frequently expressed at fraudulent schemes to gather money, but it seems to be forgotten that the gullible 'victims' who have lost their funds are themselves victims of the same emotion, namely greed, that drove the scheme to be set up in the first place—what else prompted them to send away their money in the hope of extravagantly high returns? Most of this white collar crime simply involves reshuffling paper money, usually without any noticeable—or at any rate demonstrable—impact on the productive sectors of the economy.[15] In contrast, and ironically enough, the other kind of crime can result in actual increases in productivity: not only does inculcating greed encourage people to buy things that they do not really want, but greed-driven robbery means that people must replace the stolen goods, repair the broken windows after a burglary, and so on, creating much employment. The situation appears to be complex in the extreme. Yet it would be unwise to be sanguine about the apparent harmlessness of financial unruliness, despite the evidence for considerable resilience in the overall system. The complex financial instruments, whose basis may be unduly shaky,[16] and the motivation for which is greed for ever higher returns on investment, may cause such widespread financial chaos when the instruments collapse that firms may no longer be able to pay wages to their employees, factories may not be able to invest in essential repairs to machinery, thus making the firm technically less competitive and perhaps ultimately leading to its demise, and so on: in other words "paper money", and even the slight feeling of insecurity that leads people to prematurely sell their bonds, driving down their value, can thus be transformed into concrete form that impacts directly onto safety and people's livelihoods, in turn further influencing security, often in a vicious circle.

The need for penal reform

We shall return to economic matters later on in this chapter, but for now let us explore what appears to be a secondary (in comparison with greed) crime-

[14] The difference between Austrian and Italian wine, it was joked in Germany and Switzerland, is that the former is "frostsicher", whereas the latter is "todsicher".

[15] This assertion of course needs expansion into a fuller justification. But the reader might at any rate be persuaded in advance that it is plausible, looking at, for example, the very significant losses periodically announced by banks (since the Workshop took place, we have had the collapse of Northern Rock in the UK, and heavy losses at UBS due to exposure to the U.S. subprime mortgage market), yet such events do not in themselves appear to engender economic recession.

[16] See, for example, the 77th Annual Report, especially Chapter 8. Basel: Bank for International Settlements (2007).

driver, but one that is still significant, above all as a mechanism for amplifying criminality. I refer to recidivism among released prisoners. According to official statistics, the majority of those released from prison after serving a sentence following a conviction for a crime reoffend, often within a year of release. Prisons seem to have become veritable incubators for crime,[17] rather than reformatories. The concept of reforming character appears to be considered unfashionable, and is therefore not attempted. We cannot at present say we know that a tendency to delinquency is innate (or not). The most conservative (and hence least biased) interpretation of what data is available would be to say that perhaps the tendency is innate in some, and that the presence of this innate tendency is inevitable in human society, part of its mechanism for exploring unknown territory, and comparable in its role in human development to the presence of disease. Hence, just as we accept a certain proportion of disease, we could accept a certain proportion of criminality;[18] at the same time this does not mean that one allows the disease to spread rapidly, and by the same token we should make every effort to confine criminality as much as possible, knowing that it cannot be, and perhaps should not be, completely stamped out.

An immediate difficulty (compared with disease, most of which have clear biochemical markers) is how to identify the inveterate criminal, who might only be distinguished by some special pattern of neural firings in the brain.[19] The solution surely is to uphold a régime of reform applicable to all, and if the attempt is futile in the case of the inveterate criminal, then so be it—one has wasted some effort but one has gained a reform of character among the majority.[20]

There seems then to be no conceivable justification for what appears to be the philosophy underlying prisons at present—to degrade the personality of the criminal at minimal cost. It is astonishing, for example, the prisoners are allowed access to television, which mostly broadcasts programmes featuring violence.[21] All inputs to the brain, whether mental (ideas) or physical (e.g.

---

[17]Here, tribute should be paid to those inspectors of prisons such as Sir Albert Ramsbotham, who have fearlessly spoken out on these issues—and made themselves deeply unpopular among both prison governors and the Home Office (in the UK) as a result.

[18]The analogy can be made more precise by reference to allergy, an overreaction of the immune system: the presence of allergies among the population pool might be a very necessary consequence of the immune system maintaining the utmost flexibility of response to new, unknown pathogens.

[19]The monitoring of people suspected of intending to commit crimes and the identification of definite intent when it occurs have been proposed as a means of pre-empting crime (J. Baumberg et al., Where is nano taking us? Nanotechnology Perceptions 3 (2007) 3–14).

[20]And the majority it must surely be, for as H.T. Buckle has pointed out (History of Civilization in England, vol. 1. London: Longmans, Green & Co., 1869) if evil were to preponderate over good, it would long ago have driven out good completely.

[21]Most television programmes are, at best, in doubtful taste, but it is above all their screening of violence that is very damaging. Given that the primary purpose of programmes is as vehicles for advertising (i.e., as we have seen, for inculcating greed), it is in a spirit of perfect complementarity that they also promulgate violence, as an instrument for satisfying the impulses of greed. Violence is actually innate to this form of entertainment: since television leaves relatively little room for the imagination, the only way that the interest of viewers can be maintained is to expose them to a constant stream of rapidly changing scenes. The same defect applies of course to the cinema, as well as its modern reincarnation, the video. When it was still new, the cinema did of course attract some of the best and most creative spirits of the age, such as Fritz Lang, but already eighty years ago one commentator could write, "the cinema is the other great national amusement in United States. It is said that not less than 20,000,000 people [the population of the USA was then about 110 million—*Ed.*] attend these theatres every day. The cinema has the doubtful advantage of giving a maximum of

drugs) are able to influence it.[22] However slight the effect of a single viewing, however slight the effect of being exposed to such ideas day after day, week after week, it is nevertheless a bias in the wrong direction. This would appear to be incontrovertible. Therefore, it must be considered a bad influence that needlessly hinders any attempt to reform the criminal.

One cannot propose a costly solution, even though there would appear to be a wide margin of worthwhile expenditure that could be undertaken.[23] Common sense and humanity would suggest a solution that is both genuinely austere and elevating. The régime should be physically harsh but not gratuitously vindictive; above all it should be morally elevating and maintain human dignity. Some measures may be adopted from the monastery, such as strictly silent meals, taken while listening to one of the prisoners, appointed for the day or for the week, reading some elevating or inspiring book. An important duty of the prison governor would be to select the book to be read at those meals—books of the exploits of the heroes of the past, including the classics of Homer and Virgil, translated into the vernacular, but also books concerned with exciting and elevating contemporary developments, such as those in science, technology and engineering. Of course, the chosen books must not only be suitable regarding content, but must also be written in a superior style, able to command the attention of their diverse audience. However small, even imperceptible, the effect of one individual reading, when repeated day after day, week after week, year after year, there cannot but result some beneficial effect from such a bias in the right direction.

Another very important ingredient of a truly reforming prison régime is the provision of meaningful and challenging work, including any requisite training. Although something in this regard is attempted in most prisons, it would appear to be far too little to be really effective. There is unfortunately often pressure from local businesses opposing such developments, because they see it as 'unfair' competition. The argument is, however, spurious. Everyone in society stands to benefit from a humane, dignified and reforming prison régime, because the result should be a gradual and general diminution of crime.

Tribute should be paid to past attempts to institute a more rational and effective penal system, such as that of Captain A. Maconochie RN, Superintendent of the Norfolk Island Penal Colony from 1840 onwards. He instituted

---

recreation at a minimum of mental effort. Certainly it is in no respect intellectually stimulating, nor does the average American picture seemed to be particularly elevating. There are, of course, numerous censorial bodies; but, as is the way of censors the world over, they are chiefly distinguished by the oddity of their actions." (Anon., Where Freedom Falters, p. 283; London: Charles Scribner's Sons, 1927.) During the last 50 years it was generally only in the Soviet Union and other communist states that the cinema was interesting—one thinks, for example, of the outstanding work of Andrei Tarkovsky—because it was an important medium for satirizing or criticizing the government, yet it had to be done in a manner subtle enough not to attract the attention of the censor.

[22] For more discussion on this point, see for example J.J. Ramsden, Computational aspects of consciousness. Psyche: Problems, Perspectives 1 (2001) 93–100.

[23] In their 2001 document, *Criminal Justice: The Way Ahead* the UK Home Office reckoned that there were about 100 000 persistent offenders who carried out about half of all crime (and that about 20 000 were in jail at any one time; there are currently around 80 000 inmates in UK prisons). The average annual running cost of a prison place is now about £40 000. Hence the incarceration of all persistent offenders would cost approximately £4000 million per annum (the cost of building one prison place amounts to around £100 000). These figures might be compared with the total annual cost of crime, estimated by the Home Office to be £60 000 million.

a system whereby convicts were punished for past crimes, and trained for the future in progressive, attainable stages. Following his experience, he was able to provide choice in personal and group responsibility, and believed that "the fate of every man should be placed unreservedly in his own hands."[24] Moreover, the system was strictly equitable, eschewing the granting of favours with partiality.

Nor to be forgotten in this vision of reform are other possibilities, such as those offered by music—but here it must be remembered that tastes and appreciation vary very widely among individuals, which limits the practical applicability of those possibilities—and architecture: the prison building itself should elevate, not oppress, the human spirit, but in an austere manner that nevertheless does not attempt to deny the gravity of the reasons for imprisonment.

In this secular and pluralist age one probably cannot demand that the prison régime be tied to any particular religion, although that would certainly make it easier for those charged with administering the system. In Britain at least there would be no constitutional bar against doing so, but presumably the reason preventing the Church of England from attempting to play a more proactive role in advocating reform of the present system is the fact that Britain, thanks to its liberal post-Empire immigration policy, has become a land of many faiths and religions. Yet the chance of developing a secular approach has not been taken up either. Given the veritable army of sociologists presently graduating from British universities, the lack of any intellectually forceful movement of reform emerging from that army is not merely a disappointment, but either an indictment of the calibre of the graduates or an indication of the effeteness of the discipline.

At any rate, Christianity[25] does have one advantage, namely that its "turn the other cheek"[26] morality allows the spiral of revenge to be broken. How wrong it is to insist that prison is first and foremost an instrument of punishment, for that cannot but breed a deep spirit of resentment within the mind of the criminal, which almost guarantees recidivism upon release.

Even the best managed prison system will have difficulty in effacing the social stigma associated with a custodial sentence; reform can only be consummated by forgiveness, and society as a whole needs to work out a way of achieving that, in parallel with a reformed penal system aimed at achieving genuine reform of the individual criminal. A very important element of that effacement, that actually serves the whole world, is the French Foreign Legion. Certainly in former times—I do not know what the current situation is—it was possible to enlist with total effacement of one's individual past.

In the spirit of one of the major conclusions of this Workshop on Complexity and Security, here I would at least emphasize that it would be highly important to put in place a *variety* of prison régimes, so that results can be compared objectively, one with another, before deciding to follow whichever proves to be the best in results.

---

[24]Cf. P. Hunter, Creating sustained performance improvement. In: J.J. Ramsden, S. Aida and A. Kakabadse (eds), Spiritual Motivation: New Thinking for Business and Management. Basingstoke: Palgrave (2007).

[25]L. Denny, Christianity. In: J.J. Ramsden, S. Aida and A. Kakabadse (eds), Spiritual Motivation: New Thinking for Business and Management. Basingstoke: Palgrave (2007).

[26]Matthew 5, 38–42; Luke 6, 27–31.

Boredom

In the zeal to reform the character of the malefactor, another significant cause of crime should not however be overlooked, namely that of boredom. Daily survival 500 years ago, let alone 5000, required the constant and lively application of a whole range of human faculties, many of which are scarcely used today. Boredom is almost certainly the major cause of much youthful unruliness and what is called delinquency. In a certain sense, in Britain at any rate there is little excuse for having neglected this cause, since in the fairly recent history of colonial administration the problem has already arisen (although not perhaps dealt with wholly effectively). In East Africa, a major occupation of the menfolk of many essentially pastoral tribes was raiding the cattle of neighbouring tribes, essentially as a pastime (although of course there was a concrete economic benefit to the successful raider). Wise administrators realized that although this practice was incompatible with progress towards a modern, orderly state, nevertheless its abolition would create a huge problem of under-engaged menfolk. Hence to some extent the practice was officially tolerated.

Even though one continues to be amazed by reports of biological evolution (among, for example, certain birds) taking place very rapidly, that is, over far fewer generations than have been imagined to be necessary, and cultural evolution anyway seems to take place even more rapidly than biological, the current pace of the reorganization of our society is so swift that it cannot be supposed that the biological evolution of our species keeps up with the relentless general diminution of the incidence of situations of danger and excitement requiring the essential exercise of all our faculties. The fact that this problem has not been solved in any contemporary society points to its complexity and concomitant difficulty of solution. In a nutshell, the problem is that on the one hand exploratory behaviour is very necessary to ensure the survival of our species, and it cannot take place without mistakes, i.e. some damage and destruction, but on the other hand it becomes counterproductive if this exploratory behaviour ends up destroying a major part of its environment. At the same time, real exploratory behaviour cannot meaningfully be allowed to take place under artificial constraint (as was attempted in some East African districts, in which raiding could take place with prior authorization from the Commissioner), since then it largely ceases to be truly exploratory, for its boundaries are then necessarily delineated. A similar criticism attends the so-called "extreme sports" currently gaining popularity, or exploits such as crossing deserts or Antarctica with motorized transport, or ascending a mountain by helicopter.

Vested interests

At any rate, we can be sure about what is *not* the solution to the crisis in custodial imprisonment: privatization of the prison service. A single argument suffices to condemn such a proposal, namely that a private prison service acquires a vested interest in the promotion of crime. If a single principle had to be named as underlying the explanation of the governing social and economic phenomena of our present world, then "vested interests" would make a very strong candidate. Many powerful industries lobby intensively and effectively in order to promote their particular interests. Cement and construction provides a good example. Railways were needlessly run down in Britain in the 1960s, to provide the motivation for a massive motorway construction programme. In that case, lobbying was scarcely needed, since the founder of a major civil engineering company (Marples Ridgeway Ltd) that was awarded many of the contracts conveniently became Minister of Transport. An economist (Richard Beeching) was drafted in to draw up a pseudo-economic justification for the mass closure of the then nationally owned railway network.[27] In this example, the communication channels linking the players involved only a small number of key individuals and are relatively easy to delineate. In other cases, the phenomenon is more delocalized. For example, the pharmaceutical industry has an obvious vested interest in promoting ill-health among the population (e.g. by promoting unhealthy lifestyles and the perpetuation of genetically borne disease), but has no real need to take action because the equally large and powerful food industry quite effectively promotes that simply by pushing its own wares. Even without such actions, Aneurin Bevin's founding vision of the National Health Service (NHS), as an organization that would ultimately wither away as the general health of the population increased, has turned out to be illusory, because the NHS, like any large bureaucracy, acquires a vested interest in its own survival, and only the independent and dedicated professionalism of the "front-line soldiers", that is, the consultants, surgeons in the operating theatres and so forth, ensure that some patients at least do get better.[28] This is a complex situation indeed.

What is perhaps puzzling is that publicly elected bodies often fail to oppose such trends. A national parliament should have every reason to promote a healthy lifestyle among its citizens. Although we now have draconian anti-smoking legislation in most European countries, comparably health-harming substances such as alcohol are tolerated, if not encouraged. The policies of most governments on health, anxiety about which is a major contributor to individual insecurity, are practically devoid of consistency, almost certainly because the issues have not been thought about from a fundamental viewpoint. A complicating feature is of course that considerable revenues are raised for the exchequer from sales of tobacco and alcohol—and, one might add, from motoring, which both indirectly through the promotion of a sedentary lifestyle and directly through the injuries and fatalities arising from traffic accidents is a significant contributor to ill-health.[29]

Alas, most modern parliaments are probably incapable of thinking about

[27] From a technical viewpoint, this was particularly absurd since, as is well known, the British Isles are generally prone to rain and fog, and the railway is the only transportation technology that is essentially unaffected by them, whereas wet roads and poor visibility significantly increase the risk of motoring accidents. One wonders how many tens of thousands of deaths could have been avoided if the railway network had been developed rather than dismantled.

[28] One should probably attribute to C.N. Parkinson the emphasis on *large* bureaucracies; recognition of this particular feature immediately suggests the cure, namely to break up large organizations wherever size is not intrinsically essential to their operation. An example of the essentiality of size is provided by the semiconductor industry: the required concentration of effort is such that breaking it up into small constituents would destroy it. This actually happened in one case, as was very clearly shown by the misguided breakup of AT&T in the USA: only a giant corporation was able to properly finance a highly innovative establishment as was the formerly renowned "Bell Laboratories", which today, supported by the much smaller Lucent Technologies, is but a shadow of its former excellence.

[29] At least drinkers do not demand that alcohol tax revenues are spent on the construction of more breweries and distilleries, whereas the motoring lobby is importunate indeed in its demands that motor vehicle taxes are spent on road construction and improvements.

anything from a fundamental viewpoint, because their members have a vested interest in being parliamentarians. It is well-known that whenever salaries have begun to be paid to members of parliament (as happened in Britain in 1911), there has been a noticeable decline in the calibre of members. In Britain, the transformation of real, independent-minded politicians into self-serving functionaries was consummated when they ceased to be self-employed in 1975, and became in effect simple employees of the state. Of course, traditional and continued service from existing members ensured that there was not an immediate collapse, but it is now noticeable that an ever shrinking minority of the present members of ministerial or shadow ministerial rank have had a real job outside parliament. The worst example of an assembly filled with "inutility men" (and women) is probably the so-called European Parliament, whose members were paid (rather high) salaries from its inception, and the best example of a still rather effective body could be the Swiss Nationalrat (Conseil national, Consiglio nazionale), whose members nearly all have proper, ordinary jobs outside parliament and who only serve part-time, and receive a modest and appropriate remuneration (although even they are not exempt from self-initiated attempts to raise it—fortunately usually thwarted by the higher wisdom of the voting population). Salaries in Britain were introduced to overcome the difficulty that people who were not *rentiers* could not afford to become Members, since (unlike in Switzerland) the occupation is essentially a full-time one. Hence in the interests of both social equality and to ensure that representative experience was present, it was decided to pay members of parliament an ordinary salary. Here is not the place to undertake a comprehensive review of the history of the Westminster parliament, but for some decades the results presumably justified the continuation of the experiment. It is doubtful whether this situation still obtains today. For one thing, the present membership is scarcely representative of the population as a whole: the majority of Members appear to have trained (and a diminishing number also practised) as lawyers. The solution is probably to transform parliaments into gerontocracies, by raising the minimum age for elected public office to 60 (at least). By that age many people (and if 70 were selected, almost everybody) are rentiers, therefore the problem of ensuring social equality is solved, and obviously everybody at that age has plenty of life experience.[30] This solution would also reflect the growing proportion of older people in our society. Doubtless numerous other recognizably rather undesirable trends in current parliamentary life, such as the resurgence of the politics of faction, would be resolved through this solution.

In many cases of vested interests, no real moral issue is directly raised, because principally adults, who should be capable of deciding themselves,[31] are involved. This principle would appear to apply to the examples raised above, and countless others, real or imagined, such as arms manufacturers promoting wars, and even shipbuilders promoting the dissemination of inaccurate charts. Other cases, such as the corruption of children's thinking, must be considered reprehensible. Most people are horrified by seemingly irrepressible attempts to associate children with immorality (e.g. child-centred pornography), and the

[30] Cf. the old proverb "Die Junge zur Tat, die Alte zum Rat".

[31] This could in fact form a definition of adulthood. A typical dictionary definition of an adult is "a person who is fully grown and developed", but this seems to implicitly place too much emphasis on purely biological-physical growth and development, rather than mental and psychological.

transformation of children into targets of commerce includes well organized attempts to encourage children to define themselves according to their material possessions. It is a reflexion on this growing depravity that whereas Diogenes noticed with admiration the boy who drank water from his cupped hands, and immediately then threw away his last possession, a cup, nowadays children are taught to acquire objects with almost as much zeal as their elders. Indeed they seem to be strenuously encouraged to become as independent as they can from their elders, who, they are taught, will only put a brake on their naturally acquisitive inclinations.

One of the fascinating aspects of this ultracomplex web of vested interests is the question whether the system is regulated. Presumably what Adam Smith meant by the "invisible hand" was regulation in the absence of well defined and localized sources and sinks of information connected by equally well defined channels; "vested interests" is actually a precisiation of the invisible hand, which belongs to the class of systems with distributed regulation.

Complex economic challenges

The fundamental problem posed by the world's population growing not only in numbers but also in its aspirations is that more and more goods are needed, while the world's resources are ever more acutely finite as they get used up or irreversibly dissipated. In a nutshell, the engineering challenge is that more has to be produced with less (materials, energy and waste).[32] An empirical look at the past suffices to show that technological solutions are unlikely to be sufficient to solve this seemingly impossible challenge; the way that production (which may be defined as the transformation of natural resources into consumer products) is organized will also play a significant role.

The exhaustion of key natural resources should be a primary candidate for promoting insecurity, yet this does not seem to be the case. Indium, nowadays very widely used in almost every electronic device that displays something, will be exhausted three years hence at present rates of consumption and recovery, although I have yet to meet a member of the general public who is worried about it. Perhaps there is a deep underlying confidence in the ability of technologists to solve such problems. Far more insecurity is created by shaky investments—the recent events associated with the (near) collapse of the English Northern Rock bank: near-iconic pictures of enormous queues of people seeking to withdraw their deposits, beamed around the world, illustrate this point rather well. The recent warnings raised by the Bank for International Settlements (loc. cit.) are extremely timely in this regard.

The proportion of insecurity caused by worry over investments is quite difficult to quantify. It has been pointed out that out of the present world population of about six milliard souls, only about one milliard live in a state that can be

[32]The protagonists of this approach usually add that less labour must also be used for the increased production, which seems to constitute a paradox, because almost by definition if the population is increasing there will be an increasing abundance of the availability of labour. The paradox is usually 'resolved' by a statement to the effect that technological progress in the past was usually associated with cutting down on labour; other technological advances taking place at the same time resulted in completely new types of activity emerging, which absorbed the newly freed labour. Since no fundamental mechanism is associated with this explanation, it seems to be rather too bold an assertion to imply that this state of affairs will continue to operate indefinitely into the future.

reasonably described as "stable prosperity". The remaining five milliard largely live in a state of flux characterized by social and political insecurity. Their perceptions of many of the issues raised above might be expected to be very different from those of the stably prosperous.

But are they? I have tried to argue in this brief introduction that a small number of fundamental concepts underlie matters, and should therefore be able to explain the observed phenomena. These concepts are indeed so fundamental that they can be expected to be very general. To take one of them, greed—it can be supposed that it is possessed by the unstable five sixths to the same degree as by the stable one sixth. Much energy is expended on problems associated with poor farmers in the so-called Third World. A typical scenario is of farmers persuaded (by external agents, perhaps associated with high technology seed production, artificial fertilizers and pesticides) to plant lucrative cash crops for export (that also typically require imported seed, fertilizers and pesticides) instead of food for local use. Probably they will then need to borrow money to pay for the seed, fertilizers and pesticides, but readily receive credit on the strength of the vast anticipated income from the exportable cash crops. By the time they are harvested, however, there is already a glut of the product on the market and the consequently low selling price is insufficient to cover the initial outlay. The result may be the confiscation of the farmer's land and his transformation into a virtual serf. The external agents are automatically blamed, but why did the farmers listen to them in the first place? They were tempted by the vision of huge profits from the cash crops—greed combined with gullibility.

The basis for any kind of human system must be the autonomy and personal responsibility of every adult. Otherwise, the recipients of leniency, cancelled debts, etc. become downgraded into a state of permanent dependence. In this regard, the case of Zimbabwe is very instructive. As Rhodesia, subjected to fairly effective sanctions during the 10 years of Unilateral Declaration of Independence (UDI), from 1965 to 1975, it was nevertheless a very successful country according to many reasonable criteria. When finally it became 'officially' independent, different terms of reference were adopted, the realization of which is not yet consummated, and the transition is proving to be lengthy and painful (although that is not in itself a reason for abandoning them). The ultimate reason for the (external) unpopularity of both Ian Smith (the UDI leader) and Robert Mugabe (the post-UDI leader) was or is their refusal to accept locking their country into a system based on terms other than their own. It is as if world opinion rejects an option of autonomous terms involving a certain degree of material sacrifice.

It is appropriate, at this Workshop held in a country that was for many years part of the Soviet Union and therefore adopted the so-called 'command' or planned economy, to ask whether that style of economy offers more or less security than the Western-style capitalist economy. At the level of the individual, the fact that most citizens were materially on the same plane in the Soviet Union (the highest salaries were perhaps only three times greater than the lowest ones) powerfully depressed material greed. Only near the end, when there was a sudden influx of foreign-manufactured consumer electronics, especially personal computers, did robbery (especially of computers) become a problem. Materially, people were generally well looked after by the State, a situation above all appreciated by scientists, writers, painters and other creative thinkers, who could therefore devote their undistracted energies to their intellectual exertions. Backing that up was an excellent system of education at all levels, so that practically anyone with the ability and inclination to join that intellectual élite could do so. The system as a whole eventually faltered economically, involving what one might call a classic complex systems mode of failure. Just as a modern very large-scale integrated electronic circuit has become almost too complex to design in every explicit detail, and could be more advantageously be designed along evolutionary principles,[33] it is not really possible to plan a complex modern economy in explicit detail: therefore it has to be allowed to evolve autonomously. The downside is of course that one loses direct control of the outcomes. As a rule, human beings seem to dislike losing control, but on the other hand many people feel comfortable enough with our evolutionary capitalist economic system, and most of the former states of the Soviet Union and its satellites have opted for the same, rather than trying to progressively reform the command (planned) system, as was originally envisaged under perestroika.

Environmental security

In this connexion, given that environmental security is seen to playing an increasingly important role in the fate of humankind, it is interesting to compare the environmental effects of the two systems (capitalist and socialist) with each other. The socialist system has certainly produced catastrophic environmental disasters, the largest probably being the destruction of the Aral ecosystem (an area of about 70 000 km^2), but on the other hand the capitalist system also produced the dust bowl in North America (an area of about 200 000 km^2) and is now presiding over the destruction of tropical rainforests. Without going to the trouble of carrying out a comprehensive environmental audit of the two systems, the impression is that there does not appear to be a great deal of difference in the results, although the pathways to reach them may indeed differ. The only real possibility of carrying out an objective comparison is probably between the two Germanies, the Federal Republic (BRD) and the Democratic Republic (DDR). The DDR was notorious for poor air quality, especially in winter, primarily due to the extensive use of brown coal combustion for heating; on the other hand 80% of freight was carried by environmentally-friendly rail in the DDR, whereas the corresponding figure in the BRD was only 20%, the bulk of the balance in both cases being transported by road. And it is well known that the volume of domestic refuse rose tenfold in East Berlin after the influx of West German goods following the currency union (presumably largely due to the excessive use of packaging materials in West Germany)—a highly significant statistic bearing in mind that disposal of waste is currently a major environmental problem in most developed countries of the world.

Conclusions

The purpose of this chapter is not only to introduce those that follow, but also to raise some of the issues pertinent to the topics on which papers are regrettably absent. In conclusion, it seems to be important to above all emphasize

[33]Cf. W. Banzhaf, G. Beslon, S. Christensen, J.A. Foster, F. Képès, V. Lefort, J.F. Miller, M. Radman and J.J. Ramsden, From artificial evolution to computational evolution: a research agenda. Nature Reviews Genetics 7 (2006) 729–735.

the contribution to security of the individual psyche—which in many cases is intractably complex—as well as the possibly even more complex collective psyche of human society, expressed through its economy and institutions.

Efforts have been made to identify a small number of fundamental concepts underlying challenges to security. Two of these were greed and boredom. The question inevitably then arises, is it possible to alter something that is so deep and fundamental? Greed in particular is seemingly so omnipresent that it seems to be impossible to eradicate or even merely attenuate. Yet in fact greed carries with itself its own destruction, as was so eloquently enunciated by Shakespeare:[34]

> ... and appetite, an universal wolf, ...
> Must make perforce an universal prey,
> And last eat up himself.

Is there then any long term harm in indulging greed for the sake of promoting entrepreneurial innovation, since greed cannot persist? As Weir has eloquently stated (see Footnote 28 in Chapter 14), the engineers of the Industrial Revolution "spent their whole energy on devising and superintending the removal of physical obstacles to society's welfare and development." Motivation arising through the gratification of greed is certainly not a prerequisite for entrepreneurial activity. It is, moreover, a paradox that on the one hand the introduction of mass production and the concomitant standardization of produced items has been largely responsible for bringing the immense wealth of our modern era, but on the other hand personal well-being has been greatly impoverished by the strong limitation of the variety offered to the individual consumer that is a corollary of standardized mass production. It seems that the one thing we are incapable of using our wealth for is the individualization of the satisfaction of our needs, both directly material and in other spheres, for example the medical diagnosis of a patient in hospital is typically restricted to a choice from a finite, and fairly small, number of predefined possibilities, neglecting the amazingly complex ramifications of human physical and mental health.

Perhaps the era of the nanorevolution, if it is ever fully consummated, will enable the production of "mass variety",[35] and thereby resolve the paradox.

[34] Ulysses in *Troilus and Cressida.*

[35] R. Kurzweil et al., Nanotechnology Implications—Essays. Basel: Collegium Basilea (2006); N. Bostrom et al., Nanotechnology Implications—More Essays. Basel: Collegium Basilea (2006).

Complexity and Security
J.J. Ramsden and P.J. Kervalishvili (Eds.)
IOS Press, 2008

Chapter 22

Why governments and companies invest in science and technology

J. Thomas Ratchford

National Center for Technology & Law,
George Mason University School of Law,
Arlington, Virginia

22.1 Introduction

Just our technologies and our social structure have become much more complex, so have solutions. Not only are solutions more complex, the approaches to developing and implementing solutions are likewise increasingly so. Nature often refuses to cooperate in a linear manner.

Complex problems and complex solutions require vocabularies that are not parts of ordinary speech. Experts often find it difficult to communicate with decision makers and ordinary citizens.

Yet, publics in most countries are supportive of science and technology and the education they require. They are supportive because science pays economic and other dividends. But implementing technical solutions to today's problems requires not only intellectual accomplishment but also political skills and public acceptance. Solutions not only need to be constructed intellectually, they need to be marketed to technical advisors and political leaders and—perhaps most importantly—to the general public, or at least to the interested public. The population of this interested public is greater than you might think, and they are found in every country and in every political system.

This paper treats two themes. First, the support science and technology receive from the public and companies and, second, the role individual researchers have in communicating its value.

22.2 Science and technology payoff

As we approach complexity and global security it is important to keep in mind the basic reasons society supports research and development (R&D) and science and technology (S&T). The short answer is that such investments pay off.

Quantifying payoff on R&D

Companies and governments support science and technology because it is perceived to pay off. In an economic sense this is often presented as a return on investment (ROI). Macro-economic measures of the ROI on R&D range from zero to 150% per annum. Private rates of return are less (in the 20–30% range) than the social rates of return (on the order of 30–50%), according to educated projections by experts.

Several prominent economists have explored the relationship between S&T (or R&D) and the economy. Three are Robert Solow, Edward Denison and Edwin Mansfield. Another prominent economist who did his work mostly in the U.S. and the UK was Kenneth Boulding. He was not only a good economist but was also a wonderful human being. He told me once how unfortunate it was that economists used computers. He cited econometrics—which depends almost completely on computational capacity—as a case in point. In fact, he defined econometrics as the "celestial mechanics of a non-existent universe." My point here is that the methodology for analyzing S&T and economic growth is crude and you need to retain a healthy questioning (scientific) attitude.

Having said that, let me quote a few lines from the 1987 Nobel Prize Lecture of economist Robert Solow. He alludes to the "growth accounting" work of the late Edward Denison:

> Gross output per hour of work in the U.S. economy doubled between 1909 and 1949; and some seven-eighths of that increase could be attributed to "technical change in the broadest sense" ... [I]n the thirty years since then ... [t]he main refinement has been to unpack "technical progress in the broadest sense" into a number of constituents of which various human-capital variables and "technological change in the narrow sense" are the most important. ... 34% of recorded growth is credited to "the growth of knowledge" or "technological progress in the narrow sense.

Another economist, Edwin Mansfield, has calculated social rates of return on investments in basic research. Mansfield stated: "For the seventeen innovations in our 1977 study, the median social rate of return [on the supporting basic research] was about 50%. For the two follow-on studies, each including about 20 innovations, the median social rates of return were even higher ... [T]he social rate of return ... was, on the average, at least double the private rate of return to the innovator ...."

One should not worry about social and private returns. What they mean is that both the maker and the user of a dishwasher benefit in a quantifiable manner.

Governmental support of military R&D

Governments support research and development for several reasons. First and historically foremost is the desire to gain a military advantage. Being second best in military technology is not good for most nations. This applies to a wide variety of military technologies, both offensive and defensive.

This is not new, of course. Many artists and architects of antiquity financed their professional careers from defense budgets. The catapult, the Damascene blade, and the crossbow represented substantial advances in the state of military technology in their day.

In the twentieth century, science and engineering have benefited immensely from military expenditures. In the United States much basic, academic R&D was supported by the defense agencies, especially in the 1950s. Had it not been for the Cold War and the space race, U.S. government funding for R&D might well have stagnated in the decades following World War II. The same could be said for Great Britain, France, China and, of course, the former Soviet Union. Many other countries, notably in Asia, avoided military R&D and managed to concentrate government support on civilian R&D.

The role of governments has changed in recent years, especially during the last couple of decades. Although government R&D funding has continued to grow, company investments in R&D have grown much faster, with profound implications for military and civilian technologies alike. One result has been a flow of technology from the civilian sector to the military, reversing the earlier balance.

Other government support of S&T

Government agencies in most countries also support S&T because it is essential to the accomplishment of various missions for which they are responsible. We call this "mission agency" support. Health, energy, standards, transportation, and environmental protection are some.

Basic research is a special case. It can be funded by general science agencies, mission agencies, or both. There are three persuasive arguments in favour of government support of basic research. First there is the difficulty of capturing the results of basic research on behalf of the funding organization through the intellectual property system ("someone else may gain the commercial advantage from our discovery"). Second, the uncertainty in basic research may result in failures that lose a lot of money ("we cannot risk technical failure"). Third, the time required for translating basic research into commercial technologies may be too long ("the discount rate will kill us").

Company investments in R&D

Almost as old as the military imperative, perhaps even older, is the profit motive. Businesses need to acquire knowledge for innovation to gain economic advantage. This has resulted in ever-larger R&D expenditures by the private sector. Some of this finds its way to joint ventures, consortia, and partnerships with universities. Today, in the industrialized countries and many of the developing countries, support of R&D by industries is much greater that support from governments.

One result of this investment is found in the stock values of companies. The explanation, widely accepted in the investment community in the past, is that the productiveness of a company's R&D effort is a key determinant of the company's rate of earnings growth, which in turn leads to higher future prices for its stock. This truism has, in recent years, evolved somewhat. You might say it is more complex in its content.

Central research laboratories have essentially disappeared in most countries. The complexities of intellectual property systems and the capital markets have led to successful high technology companies with "buy, not make" policies for acquiring technology. Some buy or license the knowledge they need. Some enter into alliances with other companies or acquire it illegally. Some purchase other companies with the technologies they want. As the R&D/sales ratio increases, companies need to acquire their technologies more efficiently and at lower cost.

Outsourcing also takes place for people doing R&D. "Research temps" are being hired in developed economies to carry out intermittent research activities or to avoid the high overhead. Whole research functions are being located in countries with low wage rates for highly skilled scientists.

22.3 Global trends in S&T

Science and technology are changing: their institutions, support mechanisms and human capital are evolving all over the world. Here are some of the most important characteristics of the current system and the changes that we see under way today.

22.3.1 Global investments in R&D are large

Most nations are spending a lot on science and technology. R&D as a percent of GDP generally runs between two and three percent for the larger, mostly OECD nations that spend the most. I should point out that it is very difficult to make accurate international comparisons of R&D (or almost anything else, for that matter). In particular there can be large discrepancies between OECD PPP (Organization for Economic Cooperation and Development Purchasing Power Parity) and IMF (International Monetary Fund) exchange rate comparisons for certain countries. Still, these comparisons are helpful.

Figure 22.1 shows the trend for R&D investments between 1990 and 2003 for the world. Note two things in particular. First, R&D investments have about doubled in this period. Second, the world total is now approaching one trillion[1] U.S. dollars.

22.3.2 The role of governments in R&D funding is decreasing

The role of governments in funding R&D is decreasing everywhere. Government support of R&D over the last 20 years relative to that of the private sector is down for all industrialized countries. For the United States, for example, between 1960 and 2000 there was a forty-fold increase in company funding of R&D performed in industry. During the same period federal support of R&D in industry increased a little over three-fold.

[1] 10^{12} (U.S. usage).

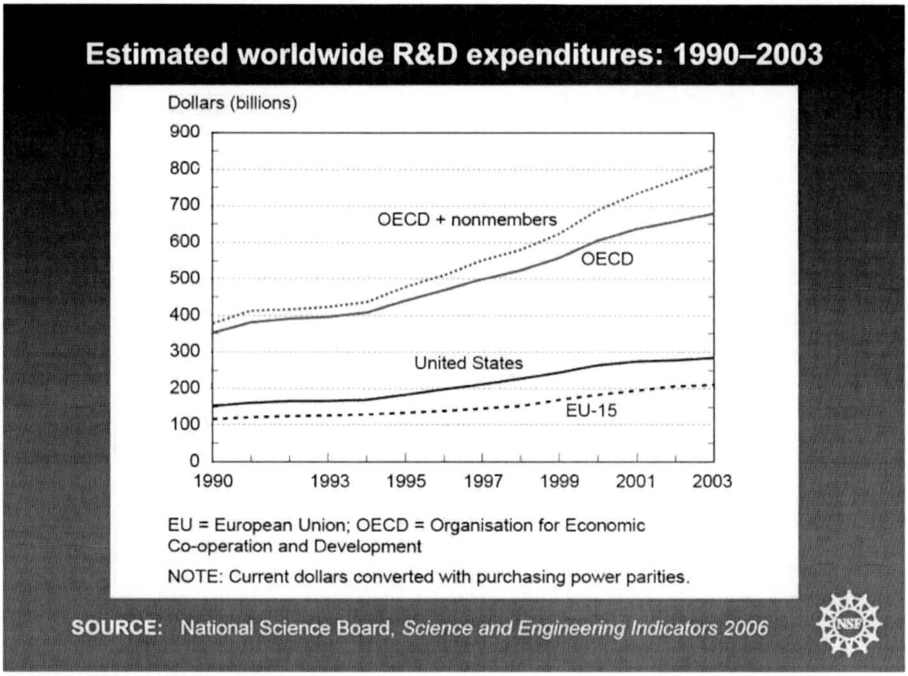

Figure 22.1: R&D investments (in U.S. billions, i.e. multiples of 10^9), between 1990 and 2003 for specified groups of states.

These trends can be demonstrated another way. In the mid-1960s the federal government funded two thirds of all U.S. R&D. Around 1980 the federal and private sector curves crossed, for the first time since before the Second World War. Now the industrial sector's margin is not only large, it continues to increase rapidly. The federal government funds less than one third of total R&D in the USA today.

Figure 22.2 shows government funds as a share of R&D expenditures for the OECD as a whole and for several individual nations since 1990. Although there are difficulties in making accurate comparisons, for the reasons noted above, the trend is clear.

22.3.3 Increased company R&D reflects the technology-intensive global economy

Companies spend more on R&D to keep up with their competitors. This is because the world economy is becoming more technologically intensive. R&D expenditures as a percentage of sales have roughly doubled over the last generation, and in the U.S. this ratio moved from 1.8% in 1975 to 3.7% in 2004 according to the latest NSF figures available.

22.3.4 Technology output is reflected in technology trade

A crucial question is what is the value of R&D? What are the outputs of R&D and how are they valuable to nations and companies?

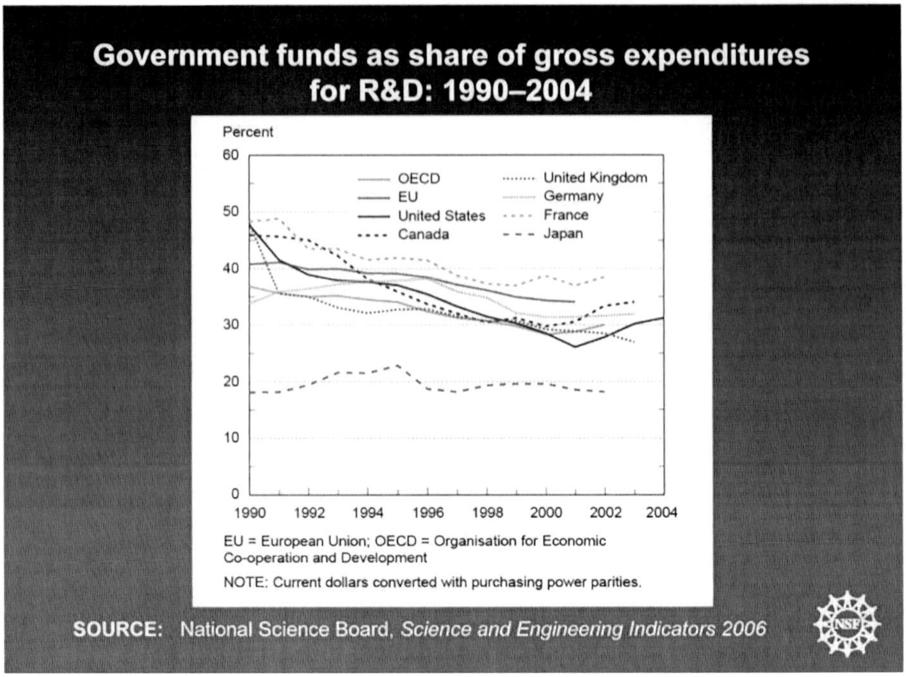

Figure 22.2: Government funds as a share of R&D expenditures for the OECD.

As noted earlier, governments support R&D for a variety of reasons: health, agriculture, energy, and defense among others. Measuring the output of investments by "mission agencies" is difficult. It is the political process that serves this function in most countries. I noted earlier the effects of a good R&D program on stock prices. Having said this, it is true that measuring the economic payoff of R&D at the national level is difficult, and making cross border comparisons is even more so. Trade in high-technology products and services is sometimes used as a measure of technological strength. The number of patents issued to companies and other national institutions is sometimes used. Both have problems.

"Trade in technology" is one interesting measure of comparison. This is not the same as trade in high technology products and services. Rather it measures payments for intellectual property such as royalties and licenses. The U.S. consistently runs a large trade surplus in technology. As a caveat, it should be noted that much of this trade in technology is between related companies, and not all intellectual property is 'technology'. The role of multinationals also complicates national comparisons. But in spite of these caveats this is a trend worth watching. Between 1987 and 2003, according to the latest figures available from NSF (S&E Indicators 2006, Appendix Tables 6–7), the total (positive) U.S. balance of "technology trade" increased from \$8 billion to \$28 billion.

22.3.5 Greater importance of science and engineering education

In a world of global competition that focuses on technology, human resources are the most important resources of all. The percentage of the 24 year old population in 19 mostly OECD countries that have first university degrees in natural sciences and engineering is growing. In the approximately 25 years between 1975 and 1999 the increase in first university degrees in natural sciences and engineering for the U.S. was about 50%. Most countries are growing faster. China approximately tripled. Adequate, high quality scientific and engineering talent is a necessary but not sufficient condition for almost all advances in science and technology and economic growth.

Figure 22.3 shows the growth in first university degrees in regions of the world in recent years.

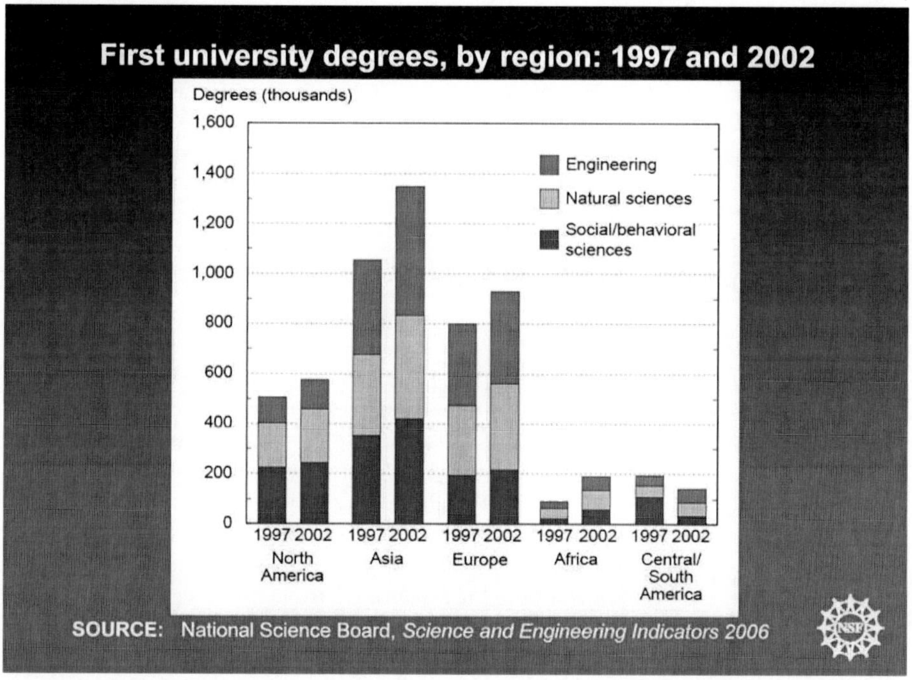

Figure 22.3: Growth in first university degrees in various regions of the world.

22.3.6 Increasing globalization of the R&D enterprise

One indication of the breadth of globalization of industrial R&D is the degree to which domestic firms support R&D overseas and foreign firms support R&D in the home country. In both cases it is not so much cost that is the driving force. Rather it is quality of the research personnel and proximity to markets. This drive to globalization of R&D is measurable, and provides a figure of merit for the scientific and engineering workforce at the national level.

Figure 22.4 indicates the increasing expenditures for R&D since 1990 by foreign-owned firms in the U.S. and vice versa.

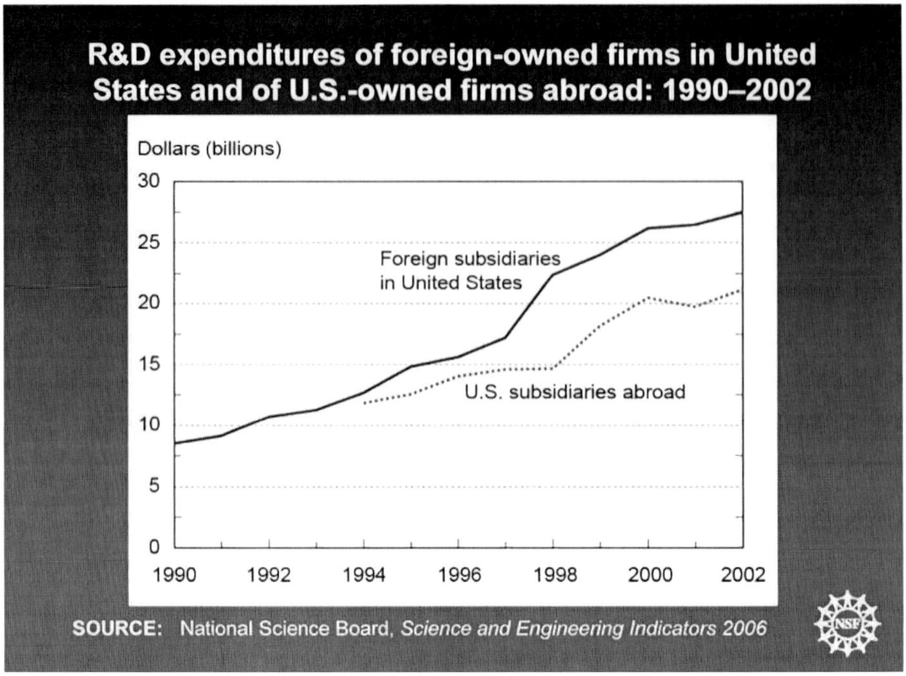

Figure 22.4: increasing expenditures for R&D since 1990 by foreign-owned firms in the U.S. and vice versa.

22.4 Complexity in the global context

Complex systems studies are a wonderful example of what we mean by science. Science is not a collection of facts, figures and theories. It is merely a system for finding truth in the natural universe. It is a method, and therefore is widely applicable. This is what is attractive about the study of complexity. The field targets in its studies such diverse entities as anthills, stock markets and the human nervous system. It is applicable to biological, economic, military and a wide variety of other technological systems.

Scholars, scientists, mathematicians and engineers, among others, study complex systems and the methods used to analyze and describe them for many reasons. Some are attracted by the mathematical elegance. Others are attracted by a desire to obtain tenure. Some want to make money—lots of it—in the financial markets. Others are looking for better weapons systems or a strategy for defending against them, or for more efficient and effective mousetraps (rodent control strategies). The objectives are many and varied.

22.5 Communicating the value of science

Just doing science and research is not enough. Those involved have a duty to make its results and benefits known broadly to decision makers and the general public. In the field of complex systems, implementing technical solutions to problems today requires not only intellectual accomplishment but also sophisticated political skills and public understanding and acceptance. Solutions not only need to be constructed intellectually, they need to be marketed to technical advisors and political leaders and—perhaps most importantly—to the general public, or at least to the interested public.

I would like to end with some comments on the importance of ommunicating the value of science and specifically the field of complex systems.

Many in the scientific and engineering community have bemoaned the lack of understanding of science and technology by the general public. Longitudinal studies of public understanding of science done in Europe, Asia and North America conclude that basic understanding of scientific principles, concepts and associated facts is sorely lacking. Our schools and the parents and peers of our students do not appear to be very interested in science or the scientific method as a means for obtaining a better understanding of the world around us. The sources of information used by most people generally do not include scientists or scientific institutions.

But the fault in this runs both ways, and the scientific and engineering communities of the world bear a lot of the blame for this state of affairs. If you are striving for tenure at a university it is your so-called scholarly publications that impress your peers and ultimately the faculty panel that recommends tenure. Using plain language in your communications is not very helpful professionally.

Yet it is the ability to communicate in a way the public (including almost all decision makers) understands that is more and more important in today's world. This is especially the case in complex systems. A common characteristic of many, if not most, of these systems is that they address issues (and technologies) that are also complex in their decision-making. This is true whether we are speaking of environmental concerns, terrorism threats, weapons systems or mental illness. These are all terribly important issues. They need attention from our best minds. And they will be addressed in a policy sense by our public officials and, at least as far as ultimate acceptance, by the general population.

Scientists and engineers have a special responsibility to communicate their conclusions, methods and insights to policy makers and the public. Scientific work is not complete until the results are disseminated - not just to peers but also to everyone who has an interest in the problem that has been addressed.

There are many parallel (and not so parallel) paths to decision makers and the public. Effective and successful communications strategies are themselves complex. Scientific institutions and sponsors of research should not fall into the trap of just requiring principal investigators to check off a box. Reports by researchers to their funding agencies should address the fact they have advanced the public understanding of their research, but this is just the beginning. What we need to develop is an attitude, a second nature, on the part of researchers, in complex systems and in other areas, that recognizes a responsibility to communicate results beyond peers and administrators.

Carrying out research and development is in one sense just prologue. The complete process includes looking ahead, beyond the conclusion of the research and investigation. It involves questions not only about the utility of the results but also who in the chain of decision makers needs to understand the research results and their implications. And it always involves making the results available in plain language, without jargon and complicated mathematics, for the interested public.

Complexity and Security
J.J. Ramsden and P.J. Kervalishvili (Eds.)
IOS Press, 2008

Chapter 23

Social entropy, synergy and security

Badri Meparishvili,[a] Tamaz Gachechiladze[b] and Gulnara Janelidze[a]

[a] *Georgian Technical University, 75 Kostava St, 0175 Tbilisi*
[b] *Tbilisi State University, 2 Chavchavadze Ave, 0179 Tbilisi, Georgia*

Abstract. In this paper, we sketch out a new and original concept for a formal description of the complexity of society with respect to the viewpoint of modelling and security, conditioned by the existence of a human being as comprising non-linear and fuzzy factors, with a very high degree of freedom of behaviour. The state of human society as a system is described by the degree of dissatisfaction or satisfaction with the current social, political and economical rules. The chief innovation is in the description of society in a form of neural graph, i.e. an axon-dendrite model with synaptic connexions. Every synapse or interaction between any two social clusters forms a new united cluster, which provokes redistribution of the synergy-entropy balance and fitness. Behavioural diversity of society is conditioned by social homeostasis and heterostasis. In this context, the criterion for the security of society is associated with stability, and from a biological viewpoint, with the idea of a homeostasis or fitness-function.

23.1 The actuality of the problem

The modern world is like a machine with wheels revolving at various speeds and in different directions. Such a machine is unstable and does not develop. The analysis of world history shows that the most complicated path followed by mankind at each stage of its development until today is full of antagonism, conflicts of interests and struggle. Much existing antagonism in the form of conflicts (particularly in recent years after reconstruction of the world's political geography), demographic unbalance or other problems are closely connected with the inequality of economical levels creating tensions. Each imbalance or desynchronization is the most important risk factor for international destabilization.[1]

It is our view that the world's problems cannot be solved without a systems approach, such as living systems analysis, autopoiesis, sociocybernetics, synergetics and complexity theory (Bailey, 1993). The role of system sciences is more and more determined by modelling concentrating on the management of society, as the most complex and potentially chaotic part of the system.

If we want to characterize society as a living system, we will first need to define a living organization in terms of autopoiesis (Greek for 'self-production'). An autopoietic system consists of a network of processes that recursively produces its own components, and thus separates itself from its environment. Contemporary system models are more likely to be non-equilibrium models emphasising the concept of entropy. Entropy has a number of advantages over equilibrium as a concept for social systems. It has led to the development of a number of models, including social entropy theory, synergetics, and complexity theory (Parunak and Bruecker, 2001). Synergy (also called synergic or synergistic science, or synergetics) relies on clusters that have properties (functional effects) different than those of their parts. Without synergy there is no complexity, no life and no humanity (Heylighen, 2002). We will see how these disciplines can enhance each other and merge into an evolutionary metascience.

One of the best examples of the development of civilization is the evolution of biological organisms, where perfectly adapted organisms are formed from unicellular micro-organisms. The first cells were antagonistic to each other due to the self-survival instinct, but in the struggle for existence, weak homeostasis failed to save them and as a result unicellular colonies appeared in the process of evolution. They created a population having collective homeostasis to support the coincidence of interests on the basis of the social heterostasis. When the stability of the system cannot be restored, it seeks external help. Only those species that survive and adapt overcome egotistic instincts and create social heterostasis. In this context, the criterion for the security of a society is associated with stability, and in biological terms, with the idea of homeostasis or of a fitness function.

23.2 Society as a system

Building a model of society based upon physical forces between atoms, or living-cellular physical and chemical interactions, would be quite difficult. Even constructing a model based upon social interactions is too difficult. If we consider society as an interactive, multi-agent, heterogeneous chaotic system with a multidimensional, complicated hierarchic structure, then its modelling is a very complicated problem. This is conditioned by the existence of a human being as a non-linear and fuzzy factor of society, with a very high degree of freedom of behaviour (Balch, 2000). Human social entropy is equivalent to the degree of social disorder of a certain social, economic, or political system.

Society, like every system, is characterized by *structure*, *composition* and *state*. The state of human society as a system is described by the different

[1] In spite of this, the development of international trading has tended to promote the formation of modern, open, democratic societies where confrontation is replaced by cooperation.

degrees of dissatisfaction (manifested for example by riots, political meetings, religious behaviour, wars, etc.) or satisfaction with the social, political and economical rules of a country. Structurally, human civilization can be represented by a tree-like structurogenesis of social fractals (or clusters), i.e. the hierarchy of epistemological levels, every level of which corresponds to the degree of system dimension (Figure 23.1). At the same time, at any level, society may be considered in just two aspects: horizontal (epistemological) and vertical (hierarchical). The more complex the system (or the more multilevelled its structure, the more developed it is. The evolution of society (the recursive sociobuilding process) in general represents ascending process in the hierarchy; transition to an upper level occurs only after the formation of the lower level.

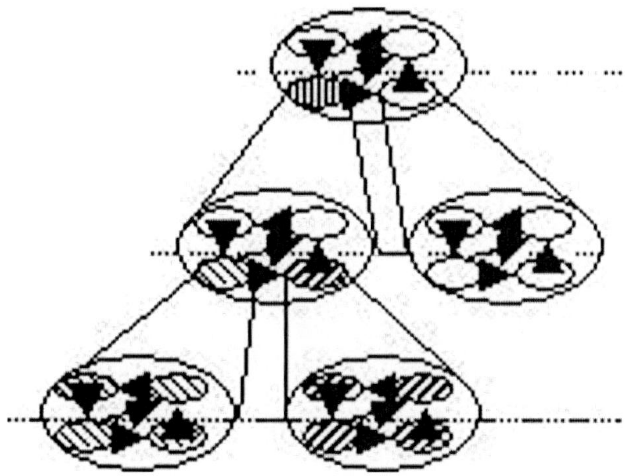

Figure 23.1: Structurogenesis of social fractals.

From the historical viewpoint, the development or building of society is realized in the following sequence: Family → clan → commune → tribe → city-state → ethnos → nation → ... → → empire or superstate → bloc of states → ... → unified civilization → .... The fractal structure of society becomes complex as we move from a human being to human civilization. At the zenith (or nadir?) of the evolution cycle, corresponding to global homeostasis of civilization, the formation of a unified civilization appears to be possible.

23.3 Neural model

Society at any level represents an open system interacting with the environment. Generally, society or its components can be considered as a neural model (Figure 23.2).

Formally, the axon-dendrite model can be represented as a graph:

$$B = \{b_i\} \ , \ i = \overline{1, N} \tag{23.1}$$

with dendrites as the set of society's requirements, needs, also desires and wishes; and axons as the set of possibilities and motivations. Neurons are represented in the following form:

$$B = \{b_{i,k}\} \ , \ k = \overline{1, L} \ . \tag{23.2}$$

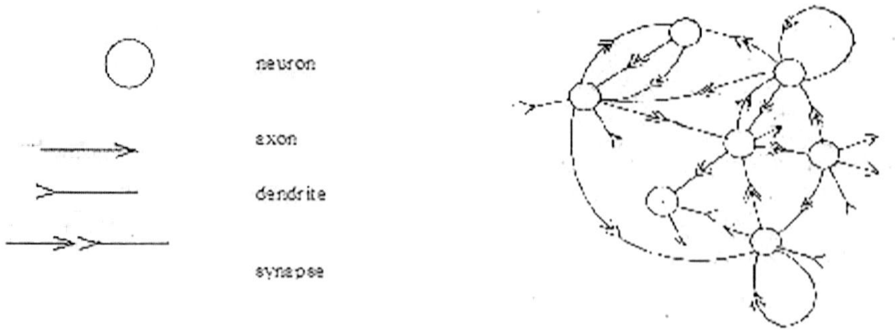

Figure 23.2: (Left) elements (components) of a neural model. (Right) example of a neural model.

Generally, each axon or dendrite is described as a terminal:

$$t_{ik} = \{s_{ik}, d_{ik}, \omega_{ik}\} \tag{23.3}$$

where $s_{ik} \in [-1, +1]$ is the sign of the terminal, $d_{ik} \in D$ is the type of the terminal; $\omega_{ik} \in [-1, +1]$ is the weight coefficient of the terminal. The total number of terminals is given by

$$Q = \sum_{i=1}^{N} \sum_{k=1}^{L} t_{ik} \; . \tag{23.4}$$

The connexion between neurons is realized by synapses

$$C_{ij} = \{t_{ik} \circ t_{kj}\} \tag{23.5}$$

where the symbol $\circ$ represents the synapse or cohesion. Each synapse is established by the conditions:

$$C_{ij} = \{(s_{ik} = -s_{kj}) \wedge ({}^{(l)}d_{ik} =^{(l)} d_{jk}) \wedge |\omega_{ik} - \omega_{jk}| = \min(k)\} \tag{23.6}$$

where $s_{ik} = -s_{kj}$ represents the opposite polarity of terminals; $({}^{(l)}d_{ik} =^{(l)} d_{jk}$ is the identity of types; and $|\omega_{ik} - \omega_{jk}| = \min$ is the minimum difference of weight coefficients, which determines the degree of incompatibility.

Let us consider the environment as a virtual element of the system. The weight coefficients for all its terminals will be $\omega_{or} = 0$, where $r = \overline{1, F}$ and F is the number of free terminals; $F = Q - 2R$, where $R = \text{card}\{C_{ij}\}$ is the number of synapses; every synapse has two values of its terminals (axon-dendrite). We designate ik as $r_{(ik)}$ and the degree of incompatibility as

$$\mu_{r_{(ik)}} = |\omega_{ik} - \omega_{jk}| \; . \tag{23.7}$$

Entropy H is determined by the number of ways a state may be achieved, and is calculated as the following function (Emptoz, 1981):

$$H = -\sum_{r_{(ik)}=1}^{F} \mu_{r_{(ik)}} \log \mu_{r_{(ik)}} - \sum_{r_{(ik)}=1}^{R} P_{r_{(ik)}} (\mu_{r_{(ik)}} \log \mu_{r_{(ik)}} + (1-\mu_{r_{(ik)}}) \log(1-\mu_{r_{(ik)}})) \quad (23.8)$$

where $P_{r_{(ik)}}$ is the probability of the event $r_{(ik)}$.

System behaviour is determined in the areas of external and internal freedom. Compatibility of the synapses is the necessary condition of neuron graph unity. *Generally, a model of society is a multidimensional graph, where the dimension is defined by the number of types of terminals. We can consider a cluster as a subgraph or projection of the graph on any type of terminal. So, the graph is the set of clusters.*

Synergy is the function (Gachechiladze and Criado, 1997):

$$S = \log \sum_{i=1}^{n} \mu_i - \sum_{i=1}^{h} p_i \log p_i \quad (23.9)$$

where h is the number of the orbits of the isomorphic groups, and p is the probability of the orbits of the isomorphic groups (the group of transformations that leaves the connexions between the vertices unchanged). The orbit of the group is determined by the equations $A_j = \{i_\alpha \mid \alpha \in h\}$; the set of A_j are h's orbit.

System stability or social homeostasis at any moment of time is determined as the difference of synergy and entropy:

$$M = S - H \; . \quad (23.10)$$

As a result of the synapses the neurons merge, creating new ensembles that consist of synergic-entropic unions. Every synapse or interaction between any two social clusters recursively form the new entity, the new united cluster, which has mutually modified or provoked redistribution of the synergy-entropy balance and fitness. Creation occurs when entropy converts into synergy and vice versa (breaking up synergy converts into entropy).

23.4 Social behaviour

On the global scale the modern world political processes are characterized by acute confrontational behaviour, often proceeding along the subcritical limit of unbalance. The so called 'strong' social clusters (state or block of states) often try to foster oppression between clusters with "weak homeostasis" striving for world hegemony and generating a new global but unbalanced cluster. For their part, the small clusters try to seek external assistance by social heterostasis, as a means of strengthening their own homeostasis for the survival of the original culture.

Three forms of societal behaviour are observed:

Confrontation caused by antagonism of interests between subjects, when synergy < entropy and $\sum_{i=1}^{n} \mu_i > 0$.

Cooperation or collaboration (low degree of heterostasis) conditioned by coincidence of interests between subjects in the case of internal antagonism, when synergy > entropy and $\sum_{i=1}^{n} \mu_i > 0$.

Consolidation or harmonious coexistence (high degree of heterostasis), which is conditioned by the coincidence of interests between subjects without any internal antagonism (this is an ideal case for the social state), when synergy > entropy and $\sum_{i=1}^{n} \mu_i = 0$.

Social behaviour can be represented as the algorithm of Figure 23.3.

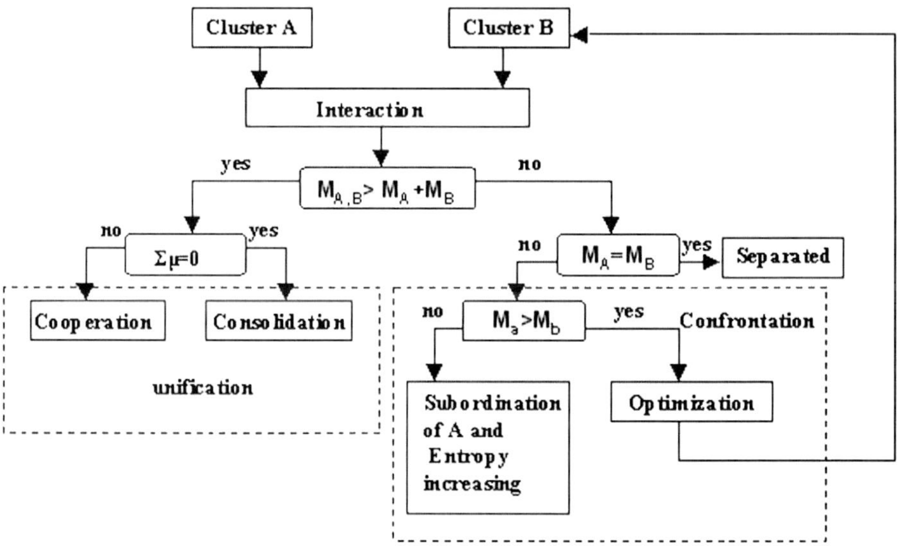

Figure 23.3: Algorithm of social behaviour.

These are the destructive (antagonistic) and beneficial (coöperative) forms of interaction. The very essence of any synergistic behaviour is that the two parts both benefit, and in larger systems all participants should benefit. In each case, the realization of the following versions of optimization is possible by the criterion of stability maximization. There are three modes of optimization:

self-regulation : in the case of constant topological structure and composition; only the weight coefficients of the terminals are varied.

self-tuning : in case of constant composition, the structure and the weight coefficients of the terminals are varied.

self-organization : the topological structure, composition and weight coefficients of the terminals are varied. This is collective heterostasis.

To achieve optimization it may be convenient to use artificial intelligence methods, particularly those of genetic programming (Janelidze and Meparishvili, 2006).

23.5 Hierarchic model

Any level of society can generally be represented in the form of the following scheme (Figure 23.4), in which macrolevel society is considered as a social environment affecting society at the given microlevel.

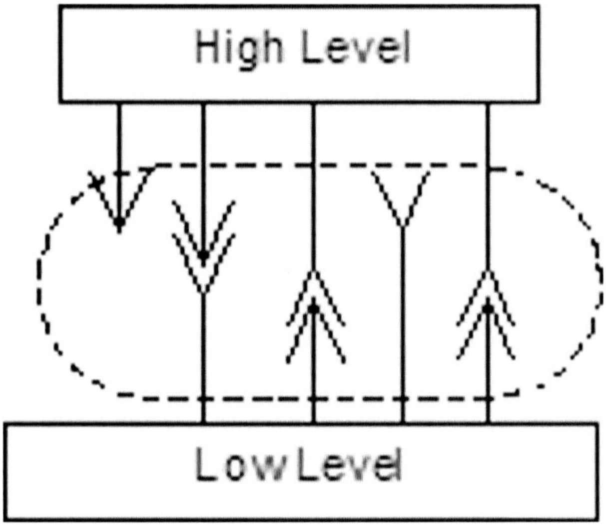

Figure 23.4: Macrolevel society is considered as a social environment affecting society at the given microlevel.

Even in this case, antagonism existing between macro- (superdominant) and microlevel (subdominant) subjects can be described in the form of axon-dendrite synapses or a hierarchic interactive model. Dual-level system stability can be represented in the following form: stability of the macrolevel and of the whole system, as conditioned by microlevel stability as well as by hierarchic interactive stability:

$$M_V = \log \sum_{i=1}^{n} \mu_i - H \tag{23.11}$$

and the stability of the whole system is

$$M_S = M_H M_V M_L \; . \tag{23.12}$$

where M_H is the stability of high-level society, M_V the stability of the vertical interactions, and M_L the stability of low-level society. $M_H > M_L$ corresponds to a dictatorship, $M_H < M_L$ corresponds to anarchy, and $M_H = M_L$, $M_S = \max$ or $\log \sum_{i=1}^{n} \mu_i = 0$ corresponds to democracy.

The analysis of historical processes shows that with hierarchical antagonism there is either dictatorship or anarchy. If more power is concentrated in any group then the resulting imbalance promotes the adverse development of society. But if there is democracy, i.e. synergic balance,[2] then antagonism ceases and

[2] It must be pointed out here that the democracy must be authentic, i.e. some countries or former countries called democratic ("Democratic People's Republic of . . . ") were not in reality. Many so-called representative democracies fall far short of authentic democracy. Switzerland,

social heterostasis appears. The transition to a new stage will not occur without consolidation, because imbalance reaches a crucial limit and the system self-destructs. That is why all empires and all systems united by force sooner or later fall. Here the social entropy accumulates to such an extent that finally a small perturbation suffices to provoke a social cataclysm. Any political system that permanently violates the social, political and economical rules of a nation contributes itself to increase its social entropy, and forces its own demise.

23.6 The complexity of civilization

Since time immemorial, humans have considered that the structure of society has become more and more complex. The use of the term 'complexity' reflects the degree of evolution, structure dimension, and functional diversity. Social synergy exists as much as interests coincide, the necessity of heterostasis exists, and entropy is conditioned by incompatibility. The more the synergy, complexity or diversity the more developed is the society. Homogeneity is unstable because when diversity decreases, the system reverts to the lowest hierarchic level, i.e. it becomes primitive.

Societal development is connected with its structural complexity, inter-contacts,[3] functional symbiosis, holism. The history of civilization can be characterized through the progressive (though non-monotonic) appearance of collective behaviour of larger groups of human beings of greater complexity. Historic changes in the structure of human organizations are self-consistently related to an increasing complexity of their social and economic contexts (Bar-Yam, 2003). The ideal sequence would appear to be "rigid" hierarchy" → hierarchy with lateral interactions → hybrid → network (see Figure 23.5).

Societies with a rigid hierarchy (early civilizations, as far as we know) were unbalanced and presumably characterized by high entropy. Later, the formation of synergetic connexions changed the structure of human organization. The functional complexity increase and imbalance (or entropy) minimization caused the hybrid structure gradually to convert into a balanced network.

23.7 Some reflexions on NATO and Georgia

Among other international institutes, NATO is an effective force balancing world antagonism. That is why small states (Georgia, among them) strive to become members so as to guarantee their own homeostasis by means of collective heterostasis. But, entering the alliance will result in synergic-entropic levelling

with its universal adult suffrage and the subordination of parliament to popular referenda, is probably the only significant example of an authentic democracy in the world as it is in 2007. The clinging of discredited government ministers to office is probably one of the most self-destructive features of many so-called democracies today.

[3]For example, contacts between different groups in society. In many so-called civilized countries today, one notices a significant divide between different social groups, typically separated on the basis of income. There seems to be less social mixing—and hence exchange of ideas and opinions—e.g. of simple workers and company directors in the English 'pub' (public house, or bar). The love of government ministers and other officials for travelling in motorcades, very typical of the Soviet era, is creeping in more and more everywhere, in contrast to those paragons of state service such as Switzerland's Giuseppe Motta, who went on foot or used the tram.

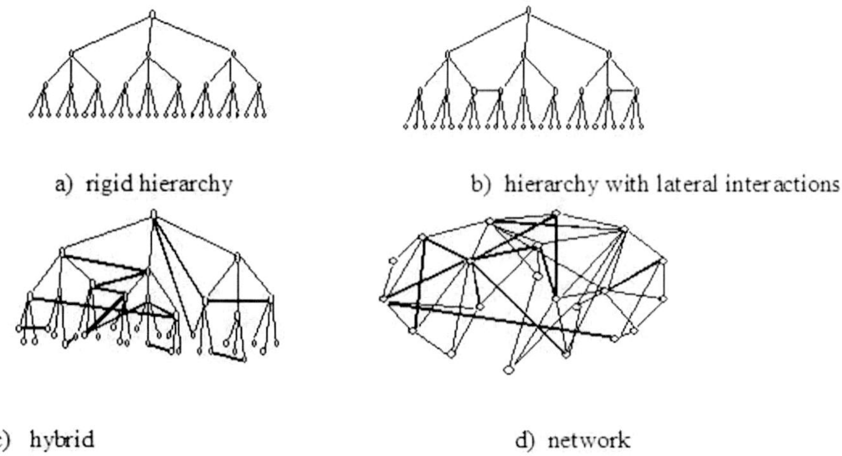

Figure 23.5: Hierarchy.

of the new union, which will cause a certain weakening of collective stability. Hence, any country presenting itself as a candidate for membership should necessarily satisfy minimum synergic conditions and standards, which will probably only be possible by the development of genuine democratic institutions[4] and the growth of societal stability.

23.8 Conclusions

1. This paper has described entropy analysis applying the synaptic graph technique based on the neural or axon-dendrite model. Our analysis shows that every synapse or interaction between any two social clusters recursively forms the new entity, which has mutually modified or provoked redistribution of the synergy-entropy balance and fitness.

2. Dual-level system stability can be represented by the stability of the macrolevel and of the whole system, and is conditioned by microlevel stability as well as hierarchic interactive stability. The analysis of historical processes shows that with hierarchical antagonism there is either dictatorship or anarchy. But if there is democracy, i.e. synergic balance, then antagonism ceases and social heterostasis appears.

3. The formation of synergetic connexions gradually changed the structure of human organization from 'rigid' hierachy through via hybrid structure into eventually a balanced network.

4. It is clear that cultural diversity will be a necessary condition for the creation of a balanced global civilization, because the societal development rate is determined by diversity. Globalization should take place preserving diversity.[5] Globalization also provides the technical and systematic foundation for this

[4]Including a sufficient level of education and general wealth among all those entitled to vote.

[5]Here we note that "digitization", often considered to be a symbiont of globalization, tends to limit diversity. See also Chapter 14.

new thinking, resulting in what would appear to be a self-regulated mechanism, which effectively prevents conflicts, and safeguards the world's long-term peace.

5. Any country being considered as a candidate for membership of NATO should necessarily satisfy minimum synergic conditions and standards, which will only be possible through the development of democratic institutions and the growth of societal stability.

Every society must evaluate its fitness from the viewpoint of the whole and not of the parts. It is unquestionable that the future of civilization should not be sacrificed for the particular interests of a few states. This may seem an impossible dream, especially for the societies of the third world, but sooner or later it will become evident that we, each one of us, are parts of the greater whole and that salvation is only in unity!

23.9 References

1. Bailey, K. D. 1993. Social entropy theory: an application of nonequilibrium thermodynamics in human ecology. Adv. Human Ecol. 2, 133–161.

2. Balch, T. 2000. Hierarchic social entropy: an information theoretic measure of robot group diversity. Autonomous Robots 8, 209–237.

3. Criado, F. and Gachechiladze, T. 1997. Entropy of fuzzy events. Fuzzy Sets and Systems 88, 99–106.

4. Emptoz, H. 1981. Nonprobabilistic entropies and indetermination measures in the setting of fuzzy sets theory. Fuzzy Sets and Systems 5, 307–317.

5. Heylighen, F. 2002. The global superorganism: an evolutionary-cybernetic model of the emerging network society. (preprint).

6. Janelidze, G. and Meparishvili, B. 2006. Evolution algorithm of multi-extreme optimization. Intelekti (Tbilisi) 1, 119–121.

7. Parunak. H.V.D. and Bruecker. S. 2001. Entropy and self-organization in multi-agent systems. Paper presented at the 5th International Conference on Autonomous Agents, Montreal, Canada.

8. Bar-Yam, Y. 2003. Complexity Rising: From Human Beings to Human Civilization. Cambridge, Mass.: New England Complex Systems Institute.

Complexity and Security
J.J. Ramsden and P.J. Kervalishvili (Eds.)
IOS Press, 2008

Chapter 24

An abstract model of political relationships: modelling interstate relations

Irakli Avalishvili

Institute of Cybernetics, 5 Sandro Euli Street, Tbilisi, Georgia

Abstract. In this brief chapter, attempts are made to construct a model of the interstate relationships of human society as nation states based on fundamental sciences. States and state groups act on each other and change each other's political organization. Based on these interactions, a scheme of interstate relationships is constructed in the form of a graph. Whereas basic functions of each state or group of states form a functional scheme in the form of another graph, the interstate relationships graph and the functional graph are isomorphic to each other.

In this chapter we briefly discuss the applications of mathematics, physics, biology, cybernetics etc. in interstate relationships (IR), as well as in sociology. This approach allows historians and politicians to evaluate the past historical events for the foreseeable future, in a new way.

For this purpose, two block-schemes (graphs) are proposed in this chapter, and they are compared with each other. Details of the graphs and the main constructions can be found in Avalishvili (2001).

The first scheme is a functional scheme (FS). It is a self-organization system, which shows the common properties of objects in physics, biology, cybernetics etc. and history, Figure 24.1. (FS) consists of 6 fundamental functions, which are connected with arrows. The arrows express that one function precedes another in time. These functions are interrelated in such a way that they form a self-organization contour.

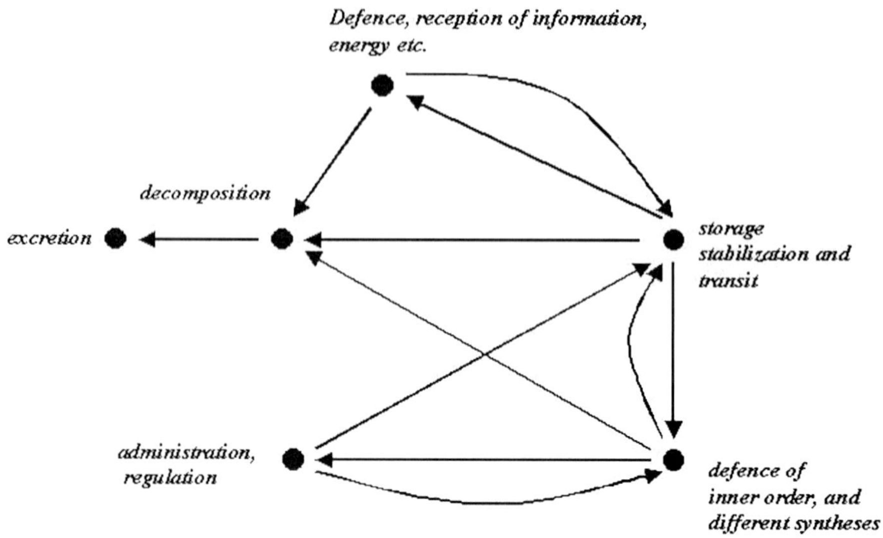

Figure 24.1: The functional scheme (graph) (FS).

In this brief chapter, attempts are made to construct a model of the interstate relationships of human society as nation states based on fundamental sciences. States and state groups act on each other and change each other's political organization. Based on these interactions, a scheme of interstate relationships is constructed in the form of a graph. Whereas the basic functions of each state or group of states form a functional scheme in the form of another graph, the interstate relationships graph and the functional graph are isomorphic to each other. These functions are:

1. Defence, reception of information, energy from outside;
2. Storage, stabilization and transit;
3. Defence of inner order and different synthesis;
4. Administration, regulation, coordination;
5. Decomposition;
6. Excretion (expelling of various products of decomposition).

Although the (FS) is given in a simplified form, it is well founded. The second scheme (graph) is a graph of (IR), Figure 24.2. It is constructed from (FS) for different types of countries and we can correlate these functions according to the actions carried out by the different countries of the world. The arrow expresses the notion that one country takes part in the construction of another country either today, or has done so in the past.

We must note that all the countries possess all these 6 functions, or most of them, but one of these functions is dominant for every country. When we

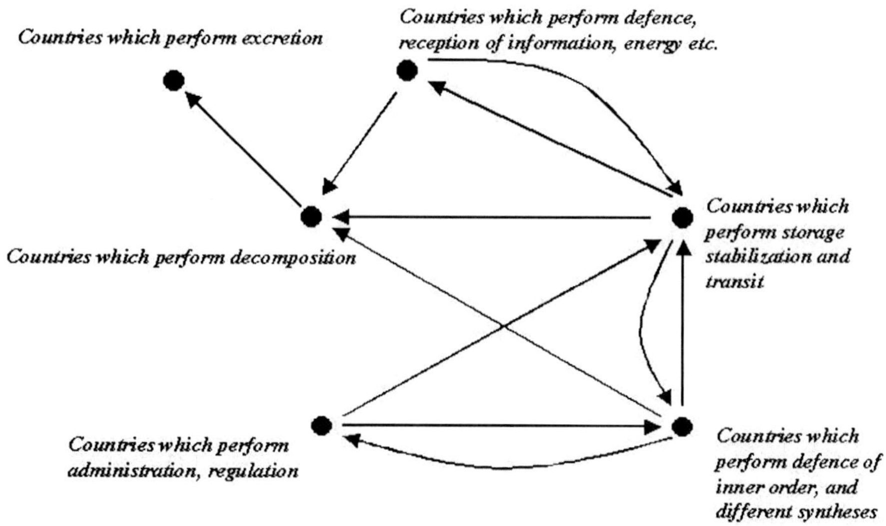

Figure 24.2: Interstate relations (IR).

compare (IR) with real situations and with historical facts, it seems that it captures this phenomenon (cf. Avalishvili, 2001).

Of course, this is only a tentative, provisional first step. (FS) and (IR) should be improved, especially mathematically. From different variants of 6-vertex graphs we must choose those graphs, which possess the minimal chaos, or entropy, i.e. the maximal order (the number of all variants of 6-vertex graphs is 1 540 944—see Harary (1969).

(FS) is one of these graphs that possesses minimal entropy. Another requirement that decreases the number of graphs is the possible existence of cycles in (FS). (Thus for the scheme expressing historical processes it may be necessary to have cycles, etc.)

Changes in (FS) will occur on the basis of ideas from physics, biology etc. And this in turn will influence the structure of (IR) and the best structure of (IR) will correspond to a more peaceful, stable situation in the world.

In future, both (IR) and (FS) will be more complex and refined. So every function in (FS) will be decomposed into subfunctions.

Further, since all these 6 vital functions are realized for every country in the world, it follows that taking into consideration various political, economical traditions we can construct different (FS): $(FS)_1, \ldots, (FS)_6$ and different (IR): $(IR)_1, \ldots, (IR)_6$ for different countries and groups of countries in the world.

Each of these functions in each country is realized by corresponding structures of the country. the totality of the two types of schemes is: (FS) and $(FS)_1, \ldots, (FS)_6$ (this approach will use extensions of graphs—the notion developed by the author in Avalishvili (2001) gives a more complete picture of the situation in the world, cf. Avalishvili and Berishvili, (1977)).

In fact all this is a problem of classification and in problems of classification there is a solution the normal, stable situation, as in other sciences.

References

Avalishvili, I.P. Isomorphism between morphological and functional graphs of the cell and classification of cells by the extensions of graphs. J. biol. Phys. Chem. 1 (2001) 5–9.

Avalishvili, I.P. and Berishvili, G.D. Extensions of automata. Trudi IK AN GSSR (Proc. Inst. Cybernetics Acad. Sci. Georgia) 1977, 183–188. Tbilisi: Mecniereba.

Harary, F. Graph Theory, p. 225. Addison-Wesley (1969).

Complexity and Security
J.J. Ramsden and P.J. Kervalishvili (Eds.)
IOS Press, 2008

Chapter 25

The complexity of the economic transition in Eastern Europe

Fulcieri Maltini

FM Consultants Associates[1]

25.1 Introduction

The fall of the Berlin Wall in November 1989 symbolizes more than any other event the end of communism and the beginning of transition. Much has been achieved since then, even if transition has proved a long and sometimes painful process. The initial effect of market liberalization was a sharp fall in output and an increase in unemployment. Overcoming these problems and creating a new legal framework, a well-functioning financial sector and an effective infrastructure is taking longer than expected. Yet, in most countries, macroeconomic stability has returned and output is now mostly produced by the private sector. The transition has been more complex than initially imagined due to the need to transform the countries and introduce reforms of democracy, economics, legislation, environmental and social integrity, and, above all, introduce good governance and transparency.

A region that was dominated by the Soviet Union and state control until 1989 (Figure 25.1) has entered a new era marked by strong growth, democratic reform and EU expansion.

The entire western world is participating to the reconstruction of the former Soviet empire but the major actor is the European Bank for Reconstruction and Development (EBRD). The EBRD, owned by 61 countries and two intergovernmental institutions, the European Community and the European Investment Bank, aims to foster the transition from centrally planned to market economies

[1]E-mail: fulcierimaltini@gmail.com

in 29 countries from central Europe to central Asia. The EBRD invests in virtually every kind of enterprise and financial institution, mainly in the form of loans and equity. Investments are designed to advance the transition to market economies and to set the highest standards of corporate governance. The Bank does not finance projects that can be funded on equivalent terms by the private sector. In support of its investment activities, the EBRD conducts policy dialogue with national and local authorities to develop the rule of law and democracy.

Figure 25.1: The Soviet Union and Eastern Europe in 1989.

25.2 The transition history

The transition has been marked by a number of significant historical events that took place between 1989 and 2007.

The year 1989 saw the opening of the Austro-Hungarian border by the Hungarian Government to East German refugees, and the subsequent fall of the Berlin Wall; the first democratic election in Poland; and the mooting of the EBRD with the task of helping to develop market economies in Eastern Europe.

In 1991 the Council for Mutual Economic Assistance set up by Moscow was disbanded, the Warsaw Pact was dissolved, and the Soviet Union disintegrated, giving birth to the Commonwealth of Independent States (CIS); the EBRD was inaugurated.

In 1992 Yugoslavia broke up and a new round of conflict erupted in south-eastern Europe; Russia and 11 others countries in the CIS joined the EBRD; civil war broke out in Tajikistan; a ceasefire agreement was reached in Transdniestria[2] in Moldova.

In 1993 the originally artificially created Czechoslovakia split into its natural components, forming the Czech and Slovak Republics.

In 1994 a ceasefire agreement was reached between Armenia and Azerbaijan over Nagorno-Karabakh.

In 1995 civil war ended in Bosnia and Herzegovina; a customs union was established between Russia, Kazakhstan and Belarus.

In 1996 Russia held its first democratic presidential election since the birth of the nation.

In 1997 Russia and Belarus signed a Union Treaty; Georgia, Ukraine, Azerbaijan and Moldova created the GUAM Group to promote regional co-operation in economic, political and security affairs; the Tajik civil war ended.

In 1998 the Russian financial crisis led to a rouble devaluation and partial debt default; EU accession negotiations began with the Czech Republic, Estonia, Hungary, Poland and Slovenia, and the Kyrgyz Republic became the first CIS country to join the World Trade Organization.

In 1999 a conflict began in Kosovo; the EU launched the "Stability Pact for South-eastern Europe" to promote Balkan economic development; EU accession negotiations began with Estonia, Latvia, Lithuania and the Slovak Republic.

In 2000 EU accession negotiations began with Bulgaria and Romania; the Eurasian Economic Community was created by five CIS countries; cumulative EBRD investments in the region exceeded 20 milliard euros.

In 2002 EU accession negotiations were concluded with eight countries in central Europe.

In 2003 the Federal Republic of Yugoslavia was renamed Serbia and Montenegro; the "Single Economic Space Agreement" was signed by Belarus, Kazakhstan, Russia and Ukraine.

In 2004 European Union membership was extended to the Czech Republic, Estonia, Hungary, Latvia, Lithuania, Poland, the Slovak Republic and Slovenia; Estonia, Lithuania and Slovenia joined the European Exchange Rate Mechanism II.

In 2006, Montenegro again became an independent country.

[2] Also known as Transnistria or Trans-Dniester.

In 2007 Bulgaria and Romania joined the European Union; Slovenia joined the euro zone.

The above list of major events points to the complexity of the transformation of a large part of Europe which, following the collapse of communism, was able to transform itself into a number of independent democratic states that have rapidly embraced a western-style economy (Figure 25.2).

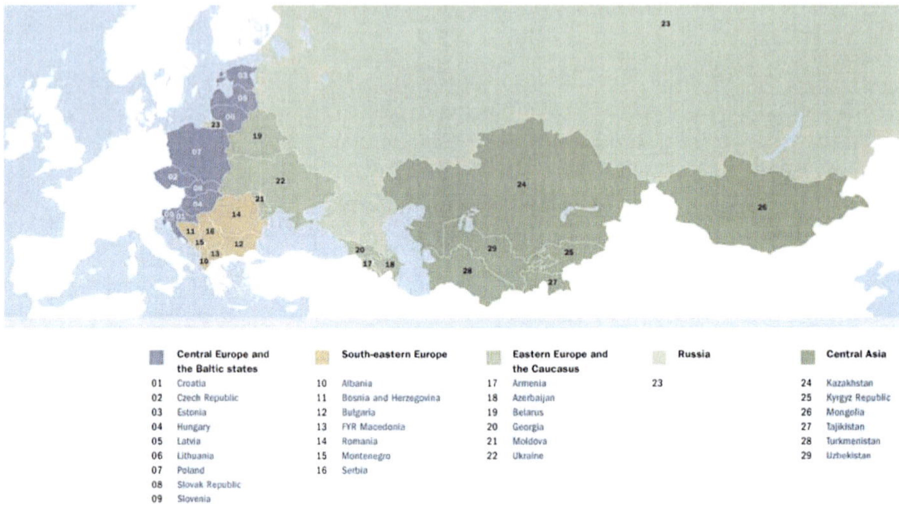

Figure 25.2: Russia and Eastern Europe in 2007.

25.3 Fifteen years to promote economic transition

Alongside the introduction of democracy, legislation and the establishment of private enterprise, the EBRD has been able to carry out investments in the public and private sectors with the participation of western institutions and companies, which have gradually changed the face of the economy of each country (Figure 25.3). This complex transition has become even more complex due to the need to transform and educate the population regarding economic principles such as productivity, competition, marketing, social rights and respect of the environment.[3]

Fifteen years after it was established to promote the transition to democratic market economies, the EBRD undertook a transition of its own, fixing new priorities, adjusting its geographical reach and taking stock of the past decade and a half to project the future of the EBRD region and the Bank (Figures 25.4 and 25.5). Perhaps the most important development in the life of the EBRD in 2006 was the adoption of a strategy to reinforce work in countries where the Bank is most needed and gradually to withdraw from countries where transition is nearing completion. One of the most remarkable successes of the transition is that

[3]The environmental problems, already critical due to poor industrial and heating conditions, were further aggravated by the enormous increase in the use of motor-cars.

€33.3 billion of cumulative commitments, mobilising **€69.6 billion** from other sources, for a total project value of **€102.9 billion**

Annual commitments 2002–06

	2006	2005	2004	2003	2002	Cumulative 1991-2006
Number of projects[1]	**301**	276	265	222	178	2,268
consisting of:						
- standalone projects	***167***	*156*	*141*	*129*	*105*	*1,392*
- investments under frameworks	***134***	*120*	*124*	*93*	*73*	*876*
EBRD commitments (€ million)	**4,936**	4,277	4,133	3,721	3,899	33,348
Resources mobilised (€ million)	**8,915**	6,222	8,799	5,307	4,862	69,571
Total project value (€ million)	**13,851**	10,499	12,932	9,028	8,761	102,919

[1] The calculation of the number of projects signed by the EBRD has been simplified to reflect the actual number of operations to which the Bank made commitments during the year. Previously, the measure was weighted, assigning proportional values to partially signed operations and individual investments under frameworks. An operation that is not linked to a framework and involves only one client is referred to as a standalone project. Operations extended to a number of clients (for example, credit lines to banks) have a framework, which represents the overall amount approved by the Board. Investments under frameworks represent the commitment to individual clients.

Figure 25.3: EBRD Annual commitments 2002-2006.

eight countries of Central Europe (Czech Republic, Estonia, Hungary, Latvia, Lithuania, Poland, Slovakia and Slovenia) joined the European Union in 2004 and will soon 'graduate' from the EBRD. By 2010 all the Bank's investments will be redirected to south-eastern Europe and Ukraine, the Caucasus, Central Asia and across Russia, to continue the transition process.

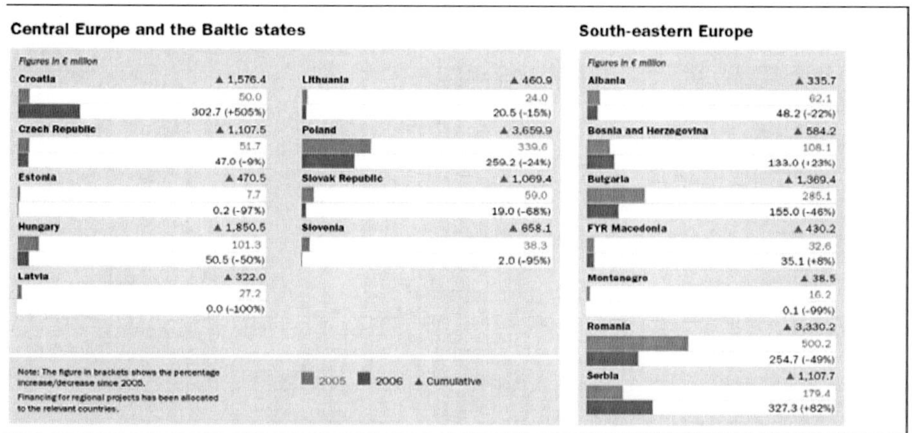

Figure 25.4: EBRD commitments 2005-2006 (1).

The Bank's force is in its knowledge of the region where it operates, its understanding of how countries are evolving and its ability to innovate and adapt. This has underpinned the Bank's success in 2006, with investment of 4.9 milliard euros that reflects the ability to pick projects that the market may be missing or avoiding. The significant profits realized on past transactions offer the ability to take more risk in the future—and the risks will be higher; the challenges are different from those 15 years ago when the Bank was a pioneer in helping to open markets behind the fallen Berlin Wall. However the sense of mission is just as strong today as at its inception and the challenges are probably greater. In the countries east of the present EU the projects are often smaller and more labour-intensive, there is less history of a market culture, and the mission of transition is not driven by the prospect of EU membership in

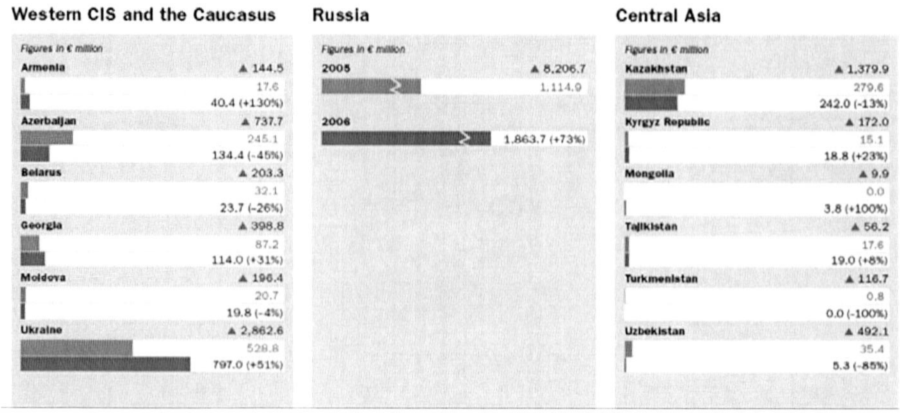

Figure 25.5: EBRD commitments 2005-2006 (2).

most of these countries.

The challenge of focusing work in the countries that were part of the Soviet Union and the Balkans is already well under way. The EBRD undertook 301 projects that promote transition in virtually every sector in 2006, (Figure 25.6), creating jobs in many industries and helping economies to grow and diversify. There was an important multiplier effect on the EBRD's own financing, with co-financing from commercial banks that rose by over 30%.

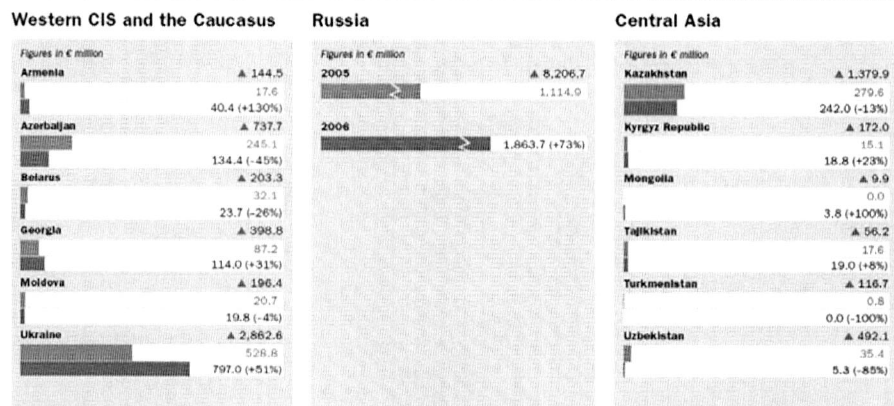

Figure 25.6: EBRD commitments by sector 2005-2006.

Some 45% of the Bank's business was in the financial sector, with one-third of it going to support local banks offering loans for smaller businesses that form the roots of economies and democracies. The Bank also works to develop new financial instruments to meet the growing prosperity and appetites of each country. Mortgage lending, for example, is a reflexion of a burgeoning middle class and the Bank is working with local banks to provide loans and securitize their loan portfolios. Part of the strategy in the Bank's new environment is to engage ever more closely in the projects by becoming active shareholders rather than just providing loans. Twenty per cent of business volume is now in the form of equity stakes, reflecting riskier environments that require more intensive engagement in order to foster transition and to manage the Bank's own risk.

The EBRD has helped to develop capital markets and increasingly, the Bank is able to meet the growing demand for lending in local currencies. Municipalities, for example, are anxious to avoid currency risk and local currency lending means the Bank is more able to support municipal projects that will improve the lives of people, such as upgrading transportation, district heating and treating water. At the municipal level or in industry, investments in energy efficiency are a priority for the EBRD. In a region where there is still enormous waste of resources, there is a compelling case for saving energy whether it is to make enterprises more competitive, boost economic growth or improve national security of energy supply and environmental protection.

The EBRD has led the way in investing in projects that are designed primarily to save energy as well as adding an energy efficiency component to industrial projects, often with added support from donors. The Sustainable Energy Initiative sets targets for efficiency and renewable energy investments and is the focal point for the Bank's efforts to point the private sector to the clear returns from such investments, in business terms as well as environmental rewards.

The initiative to foster sustainable energy reflects the Bank's longstanding approach of working closely to the needs and context of the EBRD region. The EBRD's particular knowledge of the region is well recognised by other international bodies, allowing for fruitful partnerships. Anti-corruption measures are closely coordinated with the other international financial institutions. And an important new agreement with the European Commission and the European Investment Bank provides for mutual financing and management of EIB projects undertaken in Russia, Ukraine, Moldova, the Caucasus and Central Asia.

Knowledge and expertise come from being close to clients and active dialogue with business, governments and civil society in each of the 29 countries where EBRD operates. More and more of the Bank staff is based in the field, and new offices are opened in remote regions of Russia and Ukraine. There has been strong dialogue with Mongolia, which became a new country of operations in 2006, along with Montenegro after its independence from Yugoslavia. The understanding of the region has been enhanced in 2006 on the EBRD's 15th anniversary by some special efforts to look both backwards and forwards. At an anniversary conference people from the operating countries were invited to talk about their impressions of transition—through the eyes of economists, business people, a demographer, investors, and other ordinary people who are living lives shaped by the system of planned economies and the momentous changes that took place in the early 1990s.

Through a survey of households across the region, the EBRD aimed to gain more understanding of how socio-economic status fits with the attitudes that people have to their lives, work and society. The results show support for democracy and market economies, even if a majority do not feel that their lives have yet improved (see below). They do anticipate that things will get better, and young people are the most positive about the future. Corruption seems to be what disturbs people the most and trust in institutions is still not very strong.[4] Follow-on survey work aims to further build the picture of how people

[4]To put this in perspective, trust in institutions in western Europe is currently declining.

see their future. After 15 years, and the adoption of a strategy for the next five years, it is important to probe the longer-term future. A year-long scenario-planning exercise will help the Bank to consider the global trends and regional factors that may have an impact on the way the region evolves. The scenario planning will no doubt highlight factors to feed into the strategic planning of the future.

In its 15th year, the EBRD could be satisfied with both the quality and quantity of its investments in 2006 and, rightly, looked back with pride on its achievements over a decade and a half. But the main message of an anniversary is the future, looking ahead in a region that is still building and a Bank that still has much work ahead. To transform countries who have lived for 40 to 70 years under a communist system is a very complex and difficult challenge. But the experience gained with the 'graduated' countries show that a transition to a free economy is a challenge but not impossible.

25.4 The major achievements

In 2006, eastern Europe continued to be one of the world's fastest growing regions, recording growth of around 6.2%, several percentage points higher than in western Europe. This is being increasingly driven by domestic consumption as countries become more prosperous and people seek a better standard of living. In some countries, growth is also being driven by strong investment and export growth. Countries rich in natural resources continued to benefit from high energy prices during the year (Figures 25.7 and 25.8).

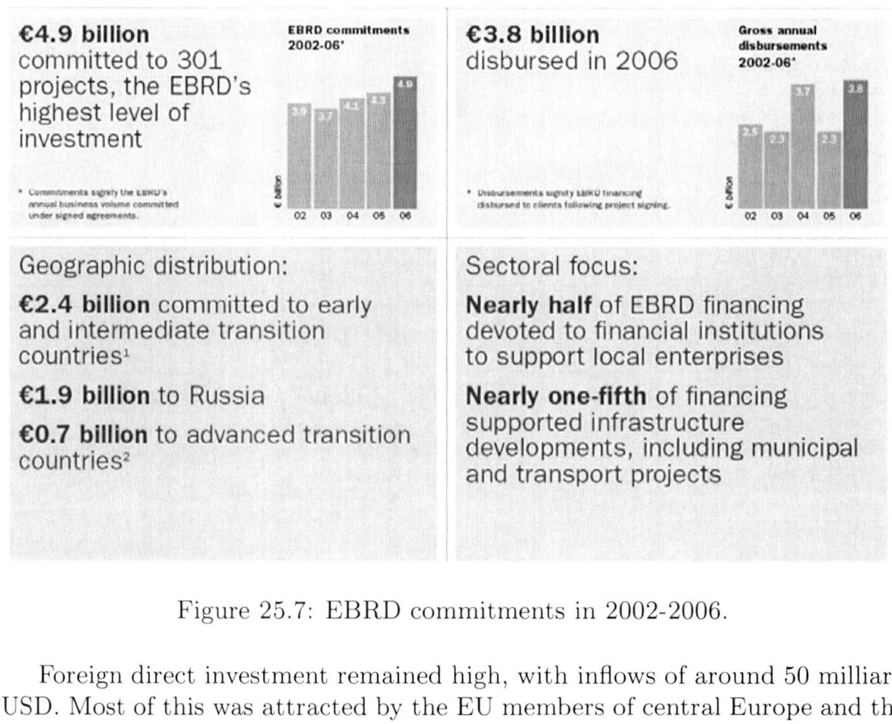

Figure 25.7: EBRD commitments in 2002-2006.

Foreign direct investment remained high, with inflows of around 50 milliard USD. Most of this was attracted by the EU members of central Europe and the two newest members—Bulgaria and Romania—which crossed the EU threshold

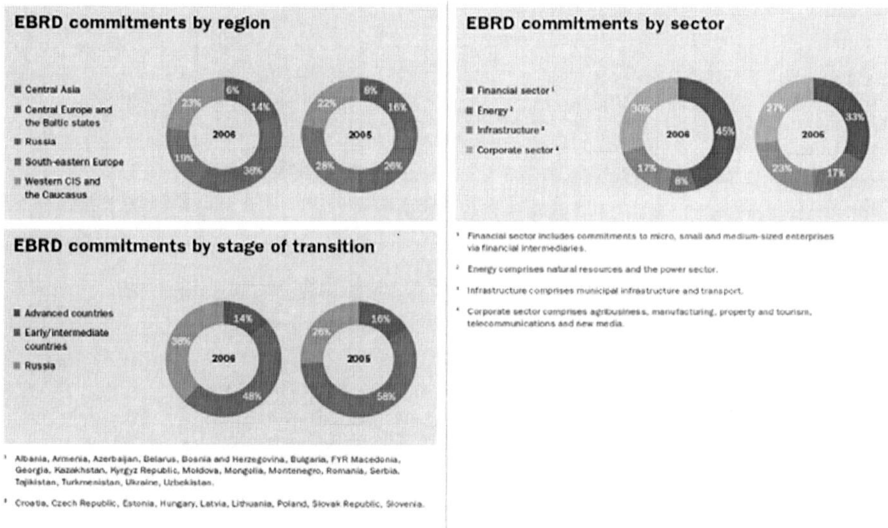

Figure 25.8: EBRD commitments 2005-2006.

in January 2007. Strong economic growth and foreign investment were accompanied by good progress in economic reforms. Much headway was made in south-eastern Europe—not only in Bulgaria and Romania but also in Macedonia (an EU candidate country) and Serbia (from which Montenegro separated in 2006 following a referendum on independence). Another strong reformer was Croatia, which has continued its membership negotiations with the EU.

In some of the countries of central Europe, public support for further restructuring and tight budgetary controls has weakened. As a result, several countries have delayed their timetable for adopting the euro. The exception is Slovenia, which joined the Economic and Monetary Union (EMU) in June 2006 and adopted the euro as the country's official currency in January 2007.

Further east, economic reforms were mainly undertaken in the wealthier countries (Russia, Ukraine and Kazakhstan) while market reforms in other countries have largely been put on hold. Across the region as a whole, progress on reforms was strongest in telecommunications and the financial sector, where lending continues to grow, particularly in the mortgage market.

In Russia the business climate benefitted from strong growth and political stability. However, concerns surfaced about increasing state intervention in various sectors and the security of energy supplies following recent disputes between Russia and some of its neighbours. Both domestic and foreign investment continued to grow dynamically in Russia. Against this backdrop of strong growth, in 2006 the EBRD committed 4.9 milliard euros to projects across its countries of operations from central Europe to central Asia. This represents the Bank's highest level of investment and a 600 million euro increase over 2005. EBRD financing was spread across 301 projects, exceeding the previous year's total of 276. In particular, the Bank increased the number of projects in the very small range of 5 million euros and less. Over one-third of the Bank's signed contracts were within this category, demonstrating the EBRD's commitment to businesses of all sizes.

The largest share of new financing was devoted to the countries at the early and intermediate stages of the transition to market economies. These countries in south-eastern Europe, the Caucasus, Central Asia and the western extremity of the former Soviet Union received 2.4 milliard euros, or 48% of the EBRD's total financing. Russian enterprises received 1.9 milliard euros, representing 38% of the Bank's total business volume and a significant increase on its 26% share in 2005. In the advanced transition countries of central Europe, EBRD commitments reached 701 million euros, or 14% of total funding, a slight decline on 2005. The marked increase in financing committed to Russia is in line with the EBRD's strategy of moving further south and east and reflects a concerted effort to develop new business opportunities through dialogue with senior business executives and with national and local authorities. The share of new projects rated as 'good' or 'excellent' in terms of their potential impact on the transition process totalled 81%. For every euro invested by the EBRD, a further 1.7 were raised from other sources to co-finance the Bank's projects.

The number and volume of equity investments increased significantly in 2006. The number of investments rose to 64, from 61 in 2005, while equity volume increased by 76% to 1.0 milliard euros, up from 572 million euros in 2005. Reflecting a higher average size of individual investment, the equity share of the Bank's annual business volume was 20% in 2006 compared with 13% in 2005. Through these investments, the EBRD is able to use its position on the boards of companies to encourage improvements in corporate governance and sound business standards.

25.5 Does the transition make you happy?

More than 15 years since the collapse of communism in eastern Europe, it is clear that the transition from state control to market economies is delivering benefits. But it is equally clear that, for some people, the transition is not working.

To understand better how the transition process has affected the lives of people in the countries where the EBRD operates, the Bank carried out a survey of people's attitudes to their new way of life. Undertaken in conjunction with the World Bank, the survey involved interviews with approximately 29 000 households across 28 of the EBRD's countries of operations plus Turkey. The aim was to provide a comprehensive assessment of major issues, such as satisfaction with everyday life, living standards, poverty and inequality, trust in state institutions, and satisfaction with public services. It also aimed to assess attitudes to market economies and democracy throughout the region. The survey was undertaken in September 2006 and completed in November 2006. The results show that only a minority of people (approximately 30%) believe that life is better today than it was in 1989. Nevertheless, more people are satisfied with life than dissatisfied, and a majority believes the future will be better for their children. The survey generally shows strong support for democracy and market economies, especially among the better-off, although there is significant variation across countries. Support is particularly strong in central Europe but much more varied further south and east. One in ten still supports a combination of centrally planned economy and authoritarian government.

25.6 Sources consulted

EBRD 1989–2004: 15 Years On.
EBRD Annual Reports 2005, 2006.
EBRD Transition Report 2006.

Complexity and Security
J.J. Ramsden and P.J. Kervalishvili (Eds.)
IOS Press, 2008

Chapter 26

Children and security: "A child has the right to be defended from birth"

Nino Kandelaki and George Chakhunashvili

Zvania Paediatric Clinic, Tbilisi State Medical University

Foreword

Does a child raised in the Caucasus have the best chance in the world of health, education, and safety? Will a child raised in Georgia be able to face the challenges that globalization brings—today and in the future? Securing the well-being of our young people requires greater cooperation and information sharing. Our Chapter shows that we have only a partial picture of how our children are doing—there are significant knowledge gaps, which if better understood could help us make wise and cost-effective decisions in support of children and our youth.

Good health is an essential factor if children are to live to their fullest potential. Children in Georgia share a number of similar experiences when it comes to their health and well-being. While the context of their lives varies, and there are some differences in the health challenges they face, there are surprising similarities across the region. In fact, there are a number of critical health problems that could profitably be addressed through national initiatives.

A more complete understanding of the strength and significance of the relationship between child health and security is necessary to optimally develop policies that improve the quality of life of women and their families The conceptual framework on child health and security is important for at least three reasons: first, there is a need to better understand the relationship between global environmental change and human security; second, there is a need for an international project that facilitates liaisons between researchers, policy makers and non-governmental organizations (NGOs) involved in environment and human security work; and third, work on environment and security thus far suggests that there are significant gaps that must be filled in order to provide useful information to policy makers.

Globalization

Globalization, technology, the "instantaneous" nature of international commerce, the advent of immediate communication, and rapid international travel and transport of people, plants, animals, and goods have all contributed to a world that functions as though it were shrinking in size. Globalization has resulted in increased interdependence, especially in the spheres of information, business, economics, and finance.

Urbanization

Population demographics relating to size, growth, density, geographic distribution, and economic patterns over the past five decades have supported urbanization as a progressive trend in Georgia and the rest of the world. The dominance of developing core cities and their surrounding areas has become the reality. These metropolitan areas, whether large or small, increase in population at a steady pace. The consequences of urbanization include: encroachment of human populations with the geographic spread of cities at their periphery into areas that were natural animal habitats; pollution and adverse environmental outcomes; the overstretching of the available necessary resources (water, clean air, food, sewage systems, healthcare, public safety and emergency services, infrastructure for energy production and distribution/transmission, communications, educational facilities, and social services); and the increasing vulnerabilities of such high population concentrations to disease, economic adversity, and even terrorism partly induced by frustration and helplessness in the face of such pressures.

Trends

Over the past few decades and especially in the last few years, changing patterns relating to human infectious diseases have emerged. SARS, monkeypox, West Nile virus, Dengue fever, plague, AIDS, Hanta viruses, and Lyme disease are prominent examples. Outbreaks of these diseases have occurred in unexpected geographic locations, at unexpected times of the year, and with unexpected severity. The Georgian public in general is unaware of most of these diseases and public health professionals have done little to educate healthcare workers and the public about them. Political events of the last ten years have adversely affected child health in Georgia. Currently the infant mortality rate in the country is very high and varies from 15 to 25 deaths per 1000 live births. In the majority of cases the causative agents are not being identified. As a rule, all newborns with a generalized infection with a wide range of non-specific symptoms, including hypothermia, jaundice, bleeding with associated coagulopathy, respiratory insufficiency, vascular instability, hepatomegaly (enlargement of the liver) and splenomegaly (enlargement of the spleen), are diagnosed as having neonatal sepsis (without identification of etiology) and treated with broad-spectrum antibiotics. Despite this 'empirical' treatment, the mortality rate for neonates with generalized infections of unknown origin is high, estimated to be over 65%.

In addition, neonatal sepsis is frequently associated with infections of the central nervous system—meningitis and encephalitis—which makes prognosis even poorer.

Basic social services

Within the turmoil of the post-Soviet socio-economic transition, the systems and quality of basic social services (health, child/family welfare and education, in particular) have been affected by collapsing infrastructure, overcapacity of certain institutional and human resources versus inefficient management, low budget allotments, and decreased funding. The impact of the economic crisis on the disruption of social services, together with increased poverty, has limited the accessibility of healthcare and education for the general population and overemphasized the institutionalization of vulnerable populations as a viable assistance strategy. The deterioration of the infrastructure and management networks has been further compromised by lack of results-oriented policies and long-term strategies with a vague planning framework incapable of reversing the downturn.

Through analysis of the existing social service systems, including their bottlenecks, quality and accessibility (by the population) of basic social services, have been identified as key challenges.

In terms of health, increasing national capacities for ensuring the fulfilment of the population's rights to basic health services, quality and accessibility of regional health services, enhancement of existing national responses to HIV/AIDS, TB and malaria and the promotion of adequate nutrition have been major programme outcomes prioritized by the UN Country Team (UNCT). The UN contribution towards attainment of the national development goals within the health system will be mainstreamed through advocacy, institutional and human capacity development and programme communication strategies.

Child welfare reform will emphasize the movement from an institution-based system of child welfare to community based services. Policy, standards and programmes promoting a social and family environment for the cognitive, social and emotional development of children will be undertaken within the child welfare system resulting in reduced reliance on residential institutions and increased community-based family and child welfare services. Regarding education, one UNDAF priority will be the inclusion of life skills education into the national curriculum as well as ensuring increased access of marginalized groups to formal and informal education opportunities. This priority will serve as a basis for UN advocacy and policy development assistance.

Our contribution to the welfare of the child

Social work contributions—at clinical and policy levels. Social workers have made an enormous contribution to the child welfare field not only in diagnosis and treatment but in suggesting policy changes such as permanency and concurrent planning, better case management, and training of social workers and nurses. It is both the social work philosophy and methodologies that have allowed the field to develop, to engage in critical thinking and to suggest alternatives to placement as a first option.

The future of child welfare, which will always be a major field of practice for social work, can indeed be a rich one if our focus remains steadily on the child, the family, the community, and a longer range view of needs and services, as opposed to residual approaches that only respond to crises. We have made great strides. We have influenced political leaders and policy makers. We have in large measure delivered the goods. We should be proud of our accomplishments and keep a careful eye on future developments.

Epidemics and pandemics can place sudden and intense demands on health systems. They expose existing weaknesses in these systems and, in addition to their morbidity and mortality, can disrupt economic activity and development. The world requires a global system that can rapidly identify and contain public health emergencies and reduce unneeded panic and disruption of trade, travel and society in general.

The revised International Health Regulations, IHR(2005), provide a global framework to address these needs through a collective approach to the prevention, detection, and timely response to any public health emergency of international concern. An integrated global alert and response system for epidemics and other public health emergencies is based on strong national public health systems and capacities, and an effective international system for coordinated response. National strategies and plans of action are guided by local epidemiology and promote evidence-based, high impact interventions, tailored to national and sub-national levels. The Georgian Social Pediatric Foundation is rapidly expanding its activities to support planning and management at country level, through the development of appropriate tools, and by building capacity for their implementation. Our goals are to:

- Develop standardized approaches for readiness and response to major epidemic-prone diseases (e.g. meningitis, yellow fever, plague);
- Strengthen biosafety, biosecurity and readiness for outbreaks of dangerous and emerging pathogens outbreaks (e.g. SARS, viral haemorrhagic fevers);
- Maintain and further develop a global operational platform to support outbreak response and support regional offices in implementation at regional level.

The catalyst for the project "Role of viral pathogens among infants with systemic infections",[1] which we conduct together with a U.S. partner,[2] was scientific evidence that until now in Georgia, HIV was not seriously considered in paediatric clinical practice. Our activities in this field contribute to the implementation of preventive measures for the HIV epidemic in our country.

Preventing and responding to international public health emergencies

No single country—however capable, wealthy or technologically advanced—can alone prevent, detect and respond to all public health threats. Emerging threats may be unseen from a national perspective, may require a global analysis for

[1] Supported by the Georgian Research and Development Foundation (GRDF).

[2] Department of Microbiology, Virology, Immunology and Molecular Diagnostics at Magee Women's Hospital, University of Pittsburgh.

proper risk assessment, and may necessitate effective coordination at international level. This is the basis for the revised regulations.

As not all countries are able to take up the challenge immediately, the WHO is drawing upon its long experience as a leader in global public health, upon its convening power, and upon its partnerships with governments, United Nations agencies, civil society, academia, the private sector and the media, to maintain its surveillance and global alert and response systems. The building of national capacity will not diminish the need for WHO's global networks, but will result in increased partnerships, knowledge transfer, advancing technologies, strategic planning and management.

Challenges

- Child mortality can be reduced significantly, at low cost, even in very poor countries, but programmes are still vastly underfunded. Governments of developing and donor nations alike must be motivated to make the necessary investment in children who are voiceless and powerless;
- Investments in child health must help build the capacity of health systems to deliver sustainable services over the long term, including improved surveillance, monitoring of progress, and evaluation.

Conclusions

It cannot be overemphasized that a truly effective international preparedness and response coordination mechanism cannot be managed nationally. Global cooperation, collaboration and investment are necessary to ensure a safer future. This means a multisectored approach to managing the problem of global disease, which includes governments, industry, public and private financiers, academia, international organizations and civil society, all of whom have responsibilities for building global child health security. The following points deserve emphasis:

- In achieving the highest level of global child health security it is important that the education sector recognizes its global responsibility. Nonetheless, the building of global child health security must rest on a solid foundation of transparent and benevolent partnerships;
- Global responsibility must be taken for capacity-building within the public health infrastructure of all countries. National systems must be strengthened to anticipate and predict hazards effectively both at the international and national levels and to allow for effective preparedness strategies;
- Cross-sector collaboration within governments is essential for the protection of global child health security, and is dependent on trust and collaboration between sectors such as health, agriculture, trade and tourism. It is for this reason that the capacity to understand and act in the best interests of the intricate relationship between public health security and these sectors must be fostered;
- Increased global and national resources for the training of public health personnel, the advancement of surveillance, the building and enhancing of

laboratory capacity, the support of response networks, and the continuation and progression of prevention campaigns are all necessary to improve future security through our children.

Complexity and Security
J.J. Ramsden and P.J. Kervalishvili (Eds.)
IOS Press, 2008

Subject Index

Note to the user: the words of the chapter and section headings are not indexed.

Complexity and Security
J.J. Ramsden and P.J. Kervalishvili (Eds.)
IOS Press, 2008

Author Index